Late Quaternary Environmental Change
Physical and Human Perspectives

M Bell *Senior Lecturer in Archaeology*   M J C Walker *Professor of Geography*
University of Wales, Lampeter

LONGMAN

Addison Wesley Longman Limited,
Edinburgh Gate, Harlow,
Essex CM20 2JE, England
and Associated Companies throughout the world.

First published 1992
Reprinted 1994, 1995, 1996, 1997, 1998

British Library Cataloguing-in-Publication Data

A catalogue record for this book is available from the British Library

ISBN 0-582-04514-2

Library of Congress Cataloging-in-Publication Data

A catalog entry for this title is available from the Library of Congress.

Coventry University

Set in 8 in 9½ on 12pt Ehrhardt

Produced by Addison Wesley Longman Singapore Pte Ltd
Printed in Singapore

For Jennifer and Gro-Mette

Contents

Preface

The need for this book became apparent during teaching of the Archaeology
and Environmental Studies degree at St David's University College,
University of Wales, Lampeter. This interdisciplinary course, which focuses
on landscape archaeology and related aspects of the natural environment
has, over the years, used a number of texts from archaeology and physical
geography, but no single volume seemed to meet our requirements for a
book that had as its central theme the spatial and temporal interactions
between climate, environment and people. The present book aims to fill this
gap in the literature for its primary concern is with the changes that have
occurred in the physical and human landscape during the closing stages of
the last glacial and over the course of the present interglacial. The book
adopts an ecological approach to archaeology in which the interactive
relationships between people and environment are considered against a
background of climatic change. The material is drawn from a wide range of
sources in the earth and archaeological sciences, and it is hoped, therefore,
that the book will be of interest to undergraduate students in a number of
disciplines including Archaeology, Anthropology, Environmental Science,
Geography, Geology and History.

We are grateful to our colleagues and students in the Departments of
Geography and Archaeology for the opportunity to explore and develop the
themes discussed below, and to St David's University College, Lampeter for
the periods of study leave which made completion of the book possible. At
an early stage we benefited from the advice of Professors I. G. Simmons and
G. Dimbleby, and we are also particularly grateful to Dr K. J. Edwards for
his perceptive and constructive comments on the text as a whole. A number
of colleagues were kind enough to provide suggestions on individual
chapters or sections, including Mrs A. Caseldine, Professor W. Groenman-
van Waateringe, Dr D. Kay, Mr G. Lambrick, Professor L. Louwe
Kooijmans, Dr J. J. Lowe, and Dr R. Macphail. We are of course
responsible for any errors that remain. We would like to thank Mr T. Harris,
who drew more than half of the figures, and Mr I. Clewes who prepared the
photographs. The Pantyfedwen Fund of St David's University College,
Lampeter provided a grant towards the cost of some illustrations. We would
also like to thank the following for providing illustrative material: Dr J.
Andersen, Dr S. Anderson, Dr N. F. Alley, Dr P. Ashbee, Dr R. W.

Battarbee, Dr J-H. Beck, Dr J. Boardman, The Cambridge University Collection (Dr D. Wilson), Dr T. Darvill, Dr D. Drew, Professor B. M. Funnell, Dr H. Heidinga, Dr A. Heyworth, The National Museum of Wales, Mrs E. Proudfoot, Dr I. Renberg, Dr E. Robinson, Dr M. B. Seddon, The Somerset Levels Project, Mr R. Trett, Mr D. Upton. We have benefited greatly from the friendly and helpful advice of our editors at Longman. Finally, M. B. is grateful to Susan Foster and Jack Maple for their hospitality while most of his text was being written on study leave in London.

This book is dedicated to our wives (Jennifer Foster and Gro-Mette Gulbrandsen) and families with thanks for their considerable help and forbearance during the time that this book was in preparation.

Martin Bell
Mike Walker
Lampeter, August 1991

Acknowledgements

We are grateful to the following for permission to reproduce copyright material:

Academic Press and the respective authors for Figs 5.17 from Figs 12.1 & 12.2 (Dumond, 1979), 6.29 from Fig. 96 (Evans, 1972); Academic Press Inc. (London) Ltd. for Fig. 6.17b from Fig. 2 (Burleigh & Kerney, 1982); *Ambio* (The Royal Swedish Academy of Sciences) for Fig. 9.11b from Fig. 5 (Galloway, 1989); American Association for the Advancement of Science and the respective authors for Figs 2.28 from Fig. 4 (Dansgaard *et al.*, 1982), 3.15 from Fig. 12 (Stuiver & Quay, 1980), 9.8 from Fig. 4 (Bradley *et al.*, 1987) Copyright 1980, 1982 & 1987 by the American Association for the Advancement of Science; A.A. Balkema Publishers for Figs 2.18 from Fig. 76, p. 192 (Barber, 1981), 6.23 & 6.26 from Figs 1 & 9 (Kaland, 1986); B.T. Batsford Ltd. and the author, Prof. A. Fleming for Fig. 6.22 from Fig. 34 (Fleming, 1988); Basil Blackwell and the author, Prof. I. Simmons for Fig. 5.2 from Fig. 4.1 (Simmons, 1989); Basil Blackwell for Fig. 6.11 from Fig. 6.12 (Roberts, 1989); the author, Dr. P.C. Buckland for Fig. 6.15 from Fig. 14, p. 76 (Buckland, 1979); Cambridge University Collection for Figs 1.1 & 8.3 copyright reserved; Cambridge University Press and the respective authors for Figs 1.3 from Fig. 2.3, p. 22 (Butzer, 1982), 4.12 & 4.13 from Figs 6.42–6.49 (Huntley & Birks 1983), 5.22 from Fig. 8.1 (Plog *et al.*, 1988); the author Dr. S. Caulfield for Fig. 6.19b from Fig. 2 (Caulfield, 1983); Center for the Study of the First Americas for Figs 6.3a from Fig. 1, p. 163 (Meltzer & Mead, 1985), 6.3b from Fig. 2, p. 7 (Martin *et al.*, 1985); Croom Helm Ltd. for Fig. 4.9 from Fig. 10.6 (Devoy, 1987); Danmarks Geologiske Undersogelse for Figs 6.7 & 6.9 from Figs 29 & 54 (Iversen, 1973); Ecological Society of America and the author, R.E. Moeller for Fig. 6.8b from Fig. 1a & b (Allison *et al.*, 1986); Edinburgh University Press for Figs 6.4b from Fig. 35.1 (Jones, 1989), 7.7 from Fig. 3 (Bell, 1982), 7.11 from Fig. 2 (Joos, 1982); Elsevier Science Publishers (Physical Sciences & Engineering Div.) and the respective authors for Figs 3.1 and Figs 1 & 5 (Williams *et al.*, 1988), 3.11 from Figs 20–23 (Ruddiman & McIntyre, 1981a), 3.18 from Fig. 2 (Hammer *et al.*, 1981); English Heritage for Figs 1.4 (Bell, 1990), 5.16 (Philpott), 6.30 (Proudfoot); the author, Dr. J.G. Evans for Fig. 6.18 from Fig. 37 (Evans, 1975); Evening Argus, Brighton for fig. 1.5; the president, The Foundation for Environmental Conservation for fig. 7.10 from Fig. 7 (Davis

1976b); Geological Society and the author, Prof. G.S. Boulton for Figs 4.1
& 4.3 from Figs 10b & 22 (Boulton *et al.*, 1985; the author, Prof. C.H.
Gimingham for Fig. 8.1 from Fig. 6 (Gimingham, 1972); the editor, Prof. S.
Gregory for Fig. 9.7 from Fig. 3.4 (Jones, 1988); the author, Dr. H.A.
Heidinga for Fig. 7.9; the author, Sir John Houghton for IPCC for Figs 9.1
& 9.6 from Figs 5a & 9 (Houghton *et al.*, 1990), 9.10 from Fig. 9.6 (Warrick
& Oerlemans, 1990); The Institute of British Geographers for Figs 4.10
from Fig. 8 (Shennan, 1983). 4.14 from Figs 6–8 & 11 (Eronen, 1983),
5.27b & c from Figs 6 & 7 (Parry, 1975); International Glaciological Society
and the author, Prof. U. Siegenthaler for Fig. 2.30 from Fig. 1 (Siegenthaler
et al., 1984); the author, Dr. R. Jacobi fro Fig. 5.7b from Fig. 4.31 (Jacobi,
1980); the author, Dr. M.P. Kerney for Figs 6.16 (Kerney, 1976), 6.17a
from Fig. 2 (Kerney, 1968); Kluwer Academic Publishers for Figs 2.1 from
Fig. 2 (Bartholin, 1984), 2.31 from Figs 3 & 4 (Pfister, 1984), 2.33 from Fig.
3 (Comani, 1987), 2.34 from Fig. 1 (Baker *et al.*, 1985), 3.9 from Fig. 9
(Imbrie *et al.*, 1984), 5.6 from Fig. 1 (McAndrews, 1988), 5.27a from Fig. 2
(Parry & Carter, 1985); the author R.M. Lockley for Fig. 5.7a from Fig. 5
(Lockley, 1970); Longman Group UK Ltd. for Figs 3.12 (Lowe & Walker,
1984), 5.1 (Foley, 1987); Leicester University Press for Fig. 2.19 from Fig.
1.2 (Engstrom & Wright, 1984); the author, Dr. F. Lynch for Fig. 5.7c
from Fig. 6.4 (Lynch, 1980); Macmillan Accounts & Administration Ltd. for
Fig. 3.7 from Fig. 26 (Imbrie & Imbrie, 1979); Macmillian Magazines Ltd.
and the respective authors for Fig. 2.11 from Fig. 3 (Atkinson *et al.*, 1987),
2.29 from Fig. 1 (Lorius *et al.*, 1985), 3.10 from Fig. 1 (Lorius *et al.*, 1990),
3.16 from Fig. 1 (Hammer *et al.*, 1981), 3.19 from Fig. 1 (Wollin *et al.*,
1973), 5.18a & b from Figs 1 & 2 (Hammer *et al.*, 1987), 5.18c from Fig. 1
(Baillie & Munroe, 1988), 5.18d from Fig. 1 (Watkins *et al.*, 1978), 5.24a
from Fig. 2 (Dansgaard *et al.*, 1975), 9.5 from Fig. 1 (Jones *et al.*, 1988)
Copyright © 1973, 1975, 1978, 1980, 1985, 1987, 1988, 1990 Macmillan
Magazines Ltd.; Methuen & Co. for Figs 4.4 from Fig. 10.23 (Grove,
1988), 9.11a from Fig. 2.5 (Park, 1987); Museum of New Mexico for Fig.
5.21 (H.D. Walter); The National Museum of Wales for Fig. 6.2; National
Research Council Canada for Fig. 6.8a from Fig. 1 (Anderson, 1974); The
New Phytologist Trust for Fig. 2.9 from Fig. 3 (Kullman, 1988); the editor,
Norwegian Journal of Geography for Fig. 2.3 and Plate 1 (Erikstad & Sollid,
1986); Norwegian University Press for Fig. 6.10 from part Fig. 4 (Berglund,
1985); Oxford University Press for Fig. 5.10 from Figs 139 & 140 (Louwe
Kooijmans, 1987) © Oxford University Press; William Pengelly Cave
Studies Trust for Fig. 2.20 (Sutcliffe, 1970); the editor, *Quaternary Research*
for Figs 2.27 from Fig. 9 (Shackleton & Opdyke, 1973), 3.17 from Fig. 6
(Porter, 1986); the editor *Quaternary Science Reviews* for Figs 2.5 from Fig. 3
(Lundquist, 1986), 3.5 (Behre, 1989), 4.2 (Bowen *et al.*, 1986), 4.15 (Baker
& Bunker, 1985); Routledge for Fig. 5.25 from Fig. 12.2 (Grove, 1988);
The Royal Society and the respective authors for Figs 3.3 from Fig. 2 (de
Jong, 1988), 3.4 from Fig. 7 (West, 1988), 3.14 from Fig. 2 (Stephenson,

1990), 5.14 from Fig. 2a (Allen, 1987); School of American Research Press for Fig. 5.23 from part Fig. 34 (Rose *et al.*, 1981); E. Schweizerbart'sche for Fig. 4.8 from Fig. 6 (Hillaire–Marcel & Ochietti, 1980); Scope for Figs 9.2, 9.3 & 9.4 from Figs 4.2, 4.5 & 4.6 (Bolle *et al.*, 1986), 9.9 from Fig. 1.6 (Bolin *et al.* 1986); Society of Antiquaries for Fig. 5.11 (Louwe Kooijmans, 1980); Somerset Levels Project for Figs 5.13 from Fig. 11 (Orme, 1982), 6.12 (Somerset Levels Project); the author, Prof. L. Starkel for Fig. 7.4 (Starkel, 1981); Universitetsforlaget AS for Fig. 6.20 from Fig. 11 (Solem, 1986); University of Arizona Press for Figs 6.1 from Fig. 17.1 (Martin, 1984), 6.4a from Fig. 29.14 (Horton, 1984) Copyright 1984 by the Arizona Board of Regents; The University of Minnesota Press for Figs 4.6 from Fig. 9.11 (Péwé, 1983a), 4.16 from Fig. 3.9 (Knox, 1984), 6.27 from Fig. 10.5 (Webb *et al.*, 1984) & Table 4.2 from Table 11.2 (Davis, 1984) copyright 1983, 1984 The University of Minnesota Press; Unwin Hyman Ltd. for Fig. 3.6 from Fig. 11.2 (Koerner & Fisher, 1986); Unwin Hyman Ltd. part of Harper Collins Publishers for Fig. 6.13 from Fig. 8.4 (Williams, 1990); John Wiley & Sons Ltd. for Figs 4.7 from Fig. 6 (Mörner, 1980a), 6.5 from Fig. 1 (Bennett, 1989), 7.12 from Fig. 12.8a (Dearing *et al.*, 1990), 7.13 from Fig. 15.6 (Edwards & Rowntree, 1980) & Table 4.3 from Table 12.1 (Rose *et al.*, 1980) Copyright 1980, 1989, 1990 John Wiley & Sons Ltd.; the author, W. van Zeist for Fig. 5.5 (van Zeist & Bottema, 1982); the author, Prof. D. Zohary for Fig. 5.4 from Figs 22.1–22.3 (Zohary, 1989).

Whilst every effort has been made to trace the owners of copyright material, in a few cases this has proved impossible and we take the opportunity to offer our apologies to any copyright holders whose rights we may have unwittingly infringed.

1

Environmental Change and Human Activity

Introduction

The changes that have occurred in the landscape of the northern temperate zone over the course of the past 20 000 years have been nothing short of dramatic. The climatic shift from a regime of arctic severity to one of relative warmth began around 15 ka BP.[1] It led to the virtual disappearance of the continental ice sheets, to the contraction of mountain glaciers and mountain glacier complexes, and to the replacement of barren tundra by mixed woodland over large areas of Europe and North America. Meltwater from the wasting ice sheets raised global sea level by over 120 m, while a combination of climatic and vegetational changes exerted a major influence on a range of other landscape processes including fluvial activity, weathering rates and pedogenesis.

---

[1] Throughout this book, the shorthand form is used for years before present (BP): ka – thousand years; ma – million years. Radiocarbon dates are quoted in uncalibrated form, and the present is taken as 1950 calendar years AD.

This period of natural environmental change is also marked by the evolution of human society, the transition from hunter–gatherer communities to sedentary agriculturalism being accompanied by increasing technological sophistication which culminated in the industrial revolution of the eighteenth century. Economic and social development also had a profound impact on the landscape and within the last five millennia anthropogenic activity in the temperate mid-latitude zones has become almost as important as natural agencies in determining the direction and nature of landscape change. However, there is more to the relationship between people and environment than this, for while on the one hand there can be little doubt about the extent of human influence on the natural environment, on the other hand there is compelling evidence to show that landscape and environment have imposed (and continue to impose) their own constraints on human activity. This is particularly true of climate, and indeed the notion of **climatic determinism** (see below) runs as a thread through much of the literature on historical relationships between people and their environment. Here too, however, there is a degree of reciprocity, for although there is general acceptance that climate is a major independent variable governing human affairs, it is now equally apparent that, with the increased burning of fossil fuels and other forms of atmospheric pollution, human activity is, for the first time in the history of the earth, beginning to dictate the course of future climatic changes.

Landscape, people and climate are three variables which are inextricably linked, and an understanding of the course of recent environmental change requires an analysis not only of the elements themselves, but also of the way in which each influences the other. The purpose of this book is to examine the interactions between people and the natural environment against a background of climatic change. The focus of attention is on the northern temperate zone of Europe and North America where the effects of environmental change have been particularly marked and where the evidence for both natural and anthropogenic historical processes is especially well preserved. The time frame covers the

1

transition from the last cold stage in the Northern Hemisphere (**Late Weichselian** in Europe; **Late Devensian** in Britain; **Late Wisconsinan** in North America), to the present warm episode (interglacial) which began around 10 ka BP (**Holocene** in Europe and North America; **Flandrian/Postglacial** in Britain). The data base is broad-ranging, drawing on material from geology, geomorphology, geography, biology, archaeology, anthropology and history, but the approach to the material is firmly rooted in geography and archaeology in that the emphasis throughout is on landscape as the home of the human race.

Geography and archaeology

The disciplines of geography and archaeology have much in common (Renfrew, 1983; Wagstaff, 1987), being concerned respectively with the spatial and temporal dimensions of the human condition. The prime concern of geography is to understand the processes that operate within the natural environment (physical geography) and to evaluate the ways in which people interact both with their environment and with each other (human geography). Archaeology deals with those aspects of the human past which are mainly elucidated using material remains rather than written sources. Both physical geography and archaeology have been profoundly influenced by the science of ecology, i.e. the relationship of organisms to each other and to their environment. One consequence has been the development of the sub-discipline of environmental archaeology, which is the study of the ecological relationships of past human communities, and the interactions between people and environment through time (Butzer, 1982).

The links between earth science (including physical geography) and archaeology are longstanding. Indeed, it was the cross-fertilization with geology in the nineteenth century which established the antiquity of the human race and laid the foundation for archaeology as a separate academic discipline (Daniel, 1975). This close

working relationship has continued to the present day. Archaeological sites are preserved within sedimentary contexts, a proper appreciation of which can only be obtained by recourse to geological, pedological or geomorphological data. This has led to the development of **geoarchaeology**, archaeological research which draws on the methods and concepts of the earth sciences (Davidson and Shackley, 1976). The relationship is not one-sided, however, for archaeological sites preserve dated contexts that contain information about past environments. These help to provide a past dimension (time-depth) for the studies of environmental processes and of the role of human communities in the development of particular environments. Few truly natural environments remain at the present day, for most have been affected by human activity. These **cultural landscapes** (Birks *et al.*, 1988) comprise landforms, soils, plants and animals which have been modified by people, together with the socially constructed landscape marked by particular arrangements of settlements, fields, tracks, tombs, etc. (Figure 1.1). Reading the cultural landscape demands both an interdisciplinary and a multidisciplinary perspective.

Space, time and scale

Environmental change can be examined across a range of spatial and temporal scales. Many of the changes considered in the following pages operated at a global scale, for example the transition from glacial to interglacial conditions. More restricted in their effects were the macroscale changes which are reflected in the continent-wide development of distinctive climatic/vegetational zones. At an even more restricted spatial scale were mesoscale effects relating to regions and microscale changes affecting individual localities (Dincauze, 1987). Timescales of environmental change also vary from glacial episodes spanning perhaps 100 ka down to earthquakes whose duration can be measured in minutes.

Natural environmental change comes about in

Figure 1.1 A relict landscape on moorland, Horridge Common, Dartmoor, England showing middle Bronze Age fields forming part of a 'reave' system, trackways, and circular huts (photo Cambridge University Collection, copyright reserved)

many ways and on a wide range of timescales. Within this general framework, however, it is possible to distinguish those *processes* which are long term and gradual from those *events* which are sudden and frequently catastrophic. The former category would include such diverse phenomena as mountain building, the movement of the great lithospheric continental plates, climatic change, soil formation and ecological successions in biotic communities. Examples of events include major storms, outbreaks of disease, earthquakes and volcanic eruptions. The distinction between events and processes serves to highlight the importance of

timescale, but it must also be acknowledged that long-term processes may themselves be made up of multiple superimposed events. Hence, long-term changes in sea level may, in reality, take the form of a number of discrete coastal inundation events.

The effects of people on landscale also operate at a range of different scales. Some can be considered as 'events', such as burning, warfare, or the failure of built structures such as dams. More significant, however, has been the impact of gradual processes over time, such as vegetation changes brought about by grazing of domestic stock, or the effects of irrigation, drainage and ploughing. In

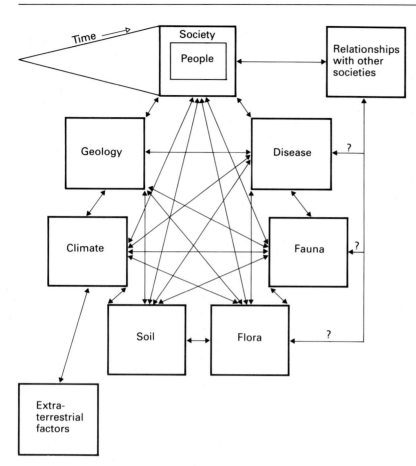

Figure 1.2   A simple model of a human ecosystem

many parts of the northern temperate zone, the most far-reaching impact of human activity during the prehistoric period was the creation of an agricultural landscape which, although a gradual process, was likely to have comprised a series of clearance and burning episodes.

A further factor that needs to be considered in terms of temporal effects of human activity on the landscape is that of **feedback**. Figure 1.2 illustrates the various components of the human ecosystem and their interactions. Each of the various elements (climate, geology, soils, flora, fauna, disease and people) are interlinked such that the impact of one factor can have repercussions throughout the system including a feedback effect on the original factor (Butzer, 1982). Feedback may be negative, i.e. reducing the effect of the originally induced

change thus leading to a resumption of stable **equilibrium** conditions, or it may be positive, reinforcing the consequences of change. One example is the clearance of woodland by prehistoric communities which, through a process of positive feedback, may have serious consequences for geomorphological processes (e.g. water retention in the soil, runoff, erosion) and ultimately on the human communities themselves. Environmental changes may be particularly dramatic in what have been described as 'brittle ecosystems', those so delicately balanced that even relatively minor changes can result in disproportionate repercussions when critical environmental thresholds are crossed (Dimbleby, 1976).

Figure 1.3 shows schematic representations of various types of environmental change and

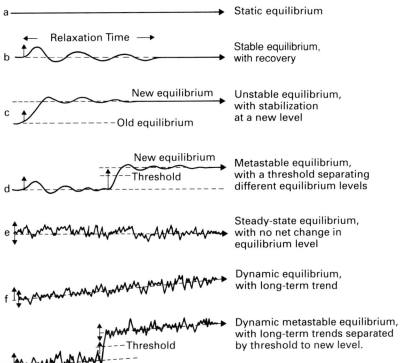

a ————————————————→ Static equilibrium

b ← Relaxation Time →   Stable equilibrium, with recovery

c New equilibrium / Old equilibrium   Unstable equilibrium, with stabilization at a new level

d New equilibrium / Threshold   Metastable equilibrium, with a threshold separating different equilibrium levels

e Steady-state equilibrium, with no net change in equilibrium level

f Dynamic equilibrium, with long-term trend

g Threshold   Dynamic metastable equilibrium, with long-term trends separated by threshold to new level.

Figure 1.3   Diagramatic representation of various forms of environmental change and equilibrium. The vertical arrows indicate change in the controlling variables and the horizontal axis is time (after Butzer, 1982)

equilibrium (Butzer, 1982). Negative feedback often has a self-limiting effect on change and ensures that various forms of equilibrium are maintained (Figure 1.3b). This is the condition known as **homeostasis** in which change occurs within certain defined limits as in Figure 1.3e. Human activity patterns are set within the observed and predicted limits of environmental change. Every so often, however, those limits will be exceeded, either by a chance (**stochastic**) event (Figure 1.3c/d), or because a long-term trend (**dynamic equilibrium**) has crossed a critical threshold (Figure 1.3g), following which a new equilibrium must be established.

A knowledge of the timescale over which environmental changes occur is critical to an understanding of their nature, cause and relationship to human activity. The chronological precision of environmental change, however, is highly variable. In deep ocean cores (p. 39), for example, it is frequently impossible to attain a

resolution of less than 1000 years due partly to slow rates of sediment accumulation on the sea bed and partly to biogenic mixing (Bradley, 1985). Dating based on artefacts (e.g. pots and metalwork) or on the radiocarbon technique (p. 16) can provide an age within 100–300 years and, under ideal conditions, an accuracy of a few tens of years. Even better temporal resolution is now possible as a result of recent advances in dendrochronology (p. 17) and on annually laminated sediments from some of the lakes of the northern temperate zone (p. 19).

The reconstruction of environmental change over the last 20 ka, therefore, involves analysis at a range of spatial and temporal scales within a chronological framework of variable precision. In the following pages natural environmental changes are considered principally at the macro- or mesoscale, whereas human activity is generally evaluated on much smaller scales, sometimes even in the context of short-term events in the lives of particular communities. In these cases it is not possible, nor indeed desirable, to seek law-like generalizations regarding the effects of environmental change on people or of their reciprocal effects on the environment. Much depends on the social context within which particular changes occur (Hodder, 1986). From a geographical perspective, much of the interest in these human impacts lies in their variability in space and time. A climatic change such as that experienced in north-west Europe during the Little Ice Age (from AD 1550 to 1850) may have crippled some communities but acted as a stimulus to others. Hence, those sections of the book dealing with the human perspectives take the form of a series of case studies arranged in a broadly chronological sequence.

Cycles and chance

Many natural changes follow a regular or cyclical pattern and are astronomically determined, including day and night, the phases of the moon, the tidal cycle (Figure 1.4), and seasons and years. These natural environmental clocks form the basis for human timekeeping and the calendar (Coveney and Highfield, 1990). There are also much longer astronomically determined cycles with wavelengths of *c.* 100 ka, 41 ka and 23 ka which are related to the earth's orbital and axial parameters (Chapter 3). Many natural cycles are superimposed upon others, and when trends on different wavelengths happen to coincide mutual reinforcement often leads to pronounced change. The dramatic transition from the last glacial to the present interglacial period in the Northern Hemisphere, for example, may well

have occurred because of the coincidence of certain orbital parameters (p. 60).

Regularities or periodicities are apparent throughout the recent geological record. On the longer timescale glacial and interglacial episodes have occurred at regular or quasi-regular intervals; cycles of vegetational change are detectable within interglacial episodes; and global sea levels have fluctuated in sympathy with repeated expansions and contractions of the great continental ice sheets. Short-term natural changes are also apparent: advances and retreats of mountain glaciers appear to have occurred at intervals of 1–1.5 ka (Grove, 1988), variations in humification in Danish peat bogs are detectable over a 260-year cycle (Aaby, 1976), while fluctuations in solar activity (sunspots) are evident over cycles of 11 and 22 years.

Other stochastic or chance processes are of an unpredictable nature. Examples include the geomorphological and climatic consequences of volcanic eruptions (p. 77), and perturbations within the atmospheric and oceanographic systems which give rise to hurricanes, storms, floods, tidal surges, blizzards, droughts, etc. Such events can sometimes be detected in the palaeoenvironmental record, such as the storm surge or tsunami that occurred along the North Sea coast of eastern Britain some 7 ka years ago (Smith *et al.*, 1985). Extreme climatic events with devastating human consequences are a distinctive feature of tropical and subtropical areas, but are also experienced during particular secular climatic episodes (major periods of changed climate) in the temperate zone. Destructive storms were a feature of the climate of north-west Europe during the Little Ice Age, and this area has also been affected by a succession of intense storms during the late 1980s. One particular example was the hurricane that devastated parts of southern

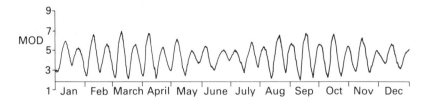

Figure 1.4   Annual changes in high tide levels in the Bristol Channel, England in metres above OD. In times before sea walls such cyclical changes would have been a major influence on the lives of communities in low-lying coastal areas (after Bell, 1990)

Figure 1.5   A wood north of Brighton, England devastated by a storm on 16th October 1987 (photo Evening Argus, Brighton)

Britain on 16 October 1987 (Ogley, 1988) killing 19 people, causing widespread damage to property and blowing down 15 million trees (Figure 1.5). These recent storms and similar extreme weather events that have affected areas such as the interior United States and the Horn of Africa (Pearce, 1989) have fuelled alarm about global climatic change arising from nineteenth- and twentieth-century industrial activity (Chapter 9).

If stochastic phenomena recur sufficiently frequently over a period for which historical records are available, it may be possible to calculate how often events of a specific magnitude are likely to recur (their **recurrence interval**). This approach to hazard prediction is widely used in the field of civil engineering, for example in the design and construction of sea defences or flood prevention works (Handmer, 1987; Shaw, 1988). Perceived

recurrence interval will also have influenced the response of past communities to particular hazards. Those which occurred frequently were most likely to have given rise to major adaptive responses. In the case of infrequent hazards, people would have been more likely to take a chance, as indeed millions do today in the San Andreas Fault earthquake zone of California, or the volcano-dominated yet densely settled Bay of Naples. It must be stressed, however, that estimates of recurrence interval are based entirely on recorded frequency over a particular time period, and that over other periods (both past and future) the frequency of hazards may be very different. Indeed, the new science of **chaos theory** (Chapter 2) carries the implication that reliable prediction of stochastic environmental phenomena is a virtual impossibility!

The effect of environmental change on people

The isolation of cause and effect in the record of past environment/society interactions is one of the most intractable problems confronting the archaeologist or environmental scientist. One complication is establishing whether or not environmental and cultural changes were coeval, and is essentially a problem of chronological resolution. A more difficult proposition, however, is to evaluate the effects of environmental variables on human groups (and indeed on individuals), for this involves a consideration of human environmental perception and an examination of modern analogues, particularly from the work of ecological anthropologists, in order to assess the range of human reactions to specific environmental stimuli such as hazards (Vayda and McCay, 1978; Halstead and O'Shea, 1989). This must ultimately lead to a consideration of how environmental phenomena influenced the thought processes of individual decision makers in prehistory (Mithen, 1990).

During the early years of this century there was a tacit assumption that a particular set of environmental parameters would give rise to only one human response (Wagner, 1977), and this engendered a somewhat simplistic approach to the interpretation of the past which saw many cultural changes as a direct consequence of natural environmental change (Lamb, 1968). This **deterministic** view led, for example, to the suggestion that the collapse of the Mycenaean and other civilizations was a direct result of climatic change (Bryson and Murray, 1977; Bryson, 1988), and similar explanations have been put forward for other cultural events. The deterministic debate in both geography and archaeology has been stimulating, but also controversial (Goudie, 1987). Major doubts have been expressed about this line of thinking, however, for it is apparent that human beings do not simply respond to natural factors, but possess the capacity for free will and for planned, long-term action which frequently has the effect of changing the environment itself (Chapters 6 and 7). People also display a remarkable ability to adapt to

a wide range of environmental changes. That ability varies according to context, depending on the nature of the society, the rigidity of its social organization and economy, its technology and population structure, as well as its communications system for disseminating knowledge about present environmental conditions and past events.

General dissatisfaction with a crude deterministic approach led to the rise of **possibilism**, the view that environments may limit, but not necessarily cause, human behaviour or biology (Hardesty, 1977). Hence it has been argued that the geographical distribution of agricultural types and some cultural traits among Indian communities in North America are *limited and modified* (rather than caused) by environmental factors and that these broadly correspond with climatic zones (Kroeber, 1939). He has commented that '. . . while it is true that cultures are rooted in nature and can therefore never be completely understood except by reference to that piece of nature in which they occur, they are no more produced by that nature than a plant is produced by the soil in which it is rooted. The immediate causes of cultural phenomena are other cultural phenomena'.

Both Determinism and Possibilism originated before the development of ecology (Hardesty, 1977), and are lines of thinking which tend to emphasize the separateness of people from nature. The ecological approach, by contrast, lays stress on the linkages between people and their environment. **Cultural ecology** accepts that human beings have the capacity to adapt in terms of both their biology (i.e. genetically) and culture (the learned pattern of thought and behaviour). These two facets of the human condition, although complementary, operate on very different timescales, cultural change being a more rapid and flexible response than biological evolution (Durham, 1978). Hence culture plays a mediating role between people and environment, for it can represent a form of buffering against the risks and uncertainties resulting both from cyclical environmental change and also the effects of chance events (Butzer, 1982). Factors which can be described as 'cultural' account for the greater adaptability of people by comparison with other elements of the biosphere. Environmental change, therefore, may create the opportunity or necessity for change in human society, but it does not necessarily determine either the precise trajectory or indeed the timescale for that change.

A particular aspect of cultural ecology which has attracted increasing attention in recent years concerns the manner in which human groups perceive their environment, and how different perceptions govern attitudes towards environmental change and resource utilization. This type of approach has been called **cognitive–processual archaeology**, which is an archaeology that is concerned with the processes of cultural change and their interaction with the environment, and which has due regard for the role of human perception and symbolism (Renfrew and Bahn, 1991). It has been argued, for example, that environments are 'not independently given but are constituted in relation to organisms (including human beings) whose environments they are' (Ingold, 1986, 1990). The resources of an area cannot merely be regarded as static elements of the landscape, simply there for the utilization by human communities, but are essentially socially and culturally defined, for some may be favoured while others are eschewed (Abbink, 1986). Hence, in the analysis of the archaeological record, the scientific description of environmental parameters and biotic components needs to be coupled with a consideration of the **environmental perception** of the human group in question. That consideration may be informed by anthropological evidence and by social theory, but conclusions must ultimately rest on interpretation of the group's behavioural response to the environment. **Behaviouralism** is the way in which people perceive and respond to their surroundings, and the 'behavioural environment' may be defined as that part of the perceived environment which influences individual behaviour and decision making (Boal and Livingstone, 1989). In essence, people cannot be regarded simply as responding to changes in the 'actual' environment which can be reconstructed (with greater or lesser accuracy) from the palaeoenvironmental evidence; rather, their response was to their perceived environment which

reflects their own lived world of experience (Butzer, 1982; Brandt and van der Leeuw, 1987). Perception and response, therefore, inevitably embody an historical dimension, and ideas and ethics regarding the environmental past inevitably influence decision making for the future. This constitutes a dialogue between past and present which is part of the *raison d'être* for this book.

Scope of the book

The book has been divided into ten chapters, each of Chapters 2 to 9 dealing with a particular aspect of landscape, climate and society over the past 20 ka. It also falls naturally into three parts. As the landscape forms the stage upon which the drama of human evolution and social development has been (and still is being) played out, the discussion begins with natural environmental changes. In Chapter 2, the different types of evidence that form the bases for environmental reconstruction are introduced and the ways in which these data sources can be obtained and interpreted are assessed. This chapter also considers the methods by which a chronological framework for recent environmental change may be established. Building on that information, Chapter 3 describes the patterns of both long-term and short-term climatic change in north-west Europe and North America, and considers the causes of climatic change over a range of temporal scales. Chapter 4 then examines the impact of the climatic events of the past 20 ka on the biotic and abiotic components of the landscape of the northern temperate zone.

The second part leads into a discussion of people in the landscape, a theme which is developed largely through a series of case studies based on the archaeological and ethno-historical records. Figure 1.6 shows the relationship between climatic stages and selected archaeological periods which are discussed in this book. In Chapter 5, the impacts of climatic and landscape changes on prehistoric and historic communities are explored, while in Chapters 6 and 7 the perspective is reversed to show the ways in which an evolving and

Figure 1.6   Chronological chart for the last 14 ka showing climatic stages and archaeological periods in selected areas

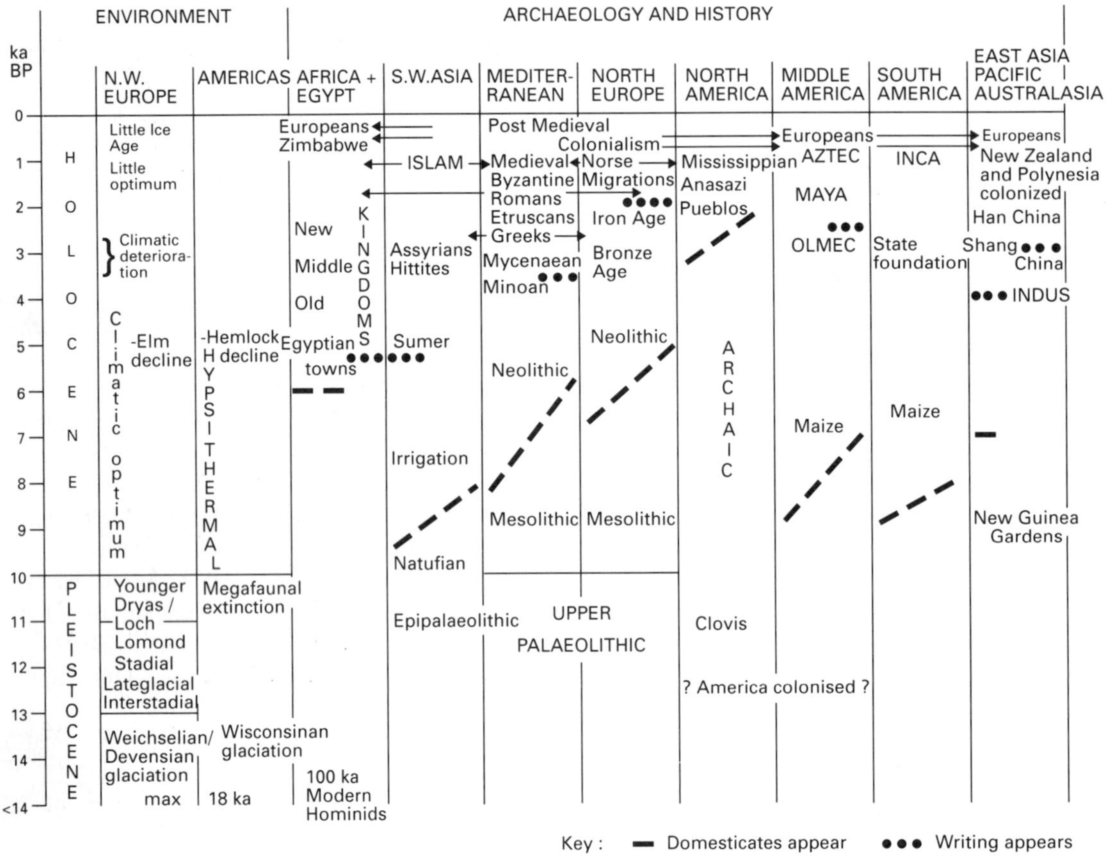

ka BP	ENVIRONMENT			ARCHAEOLOGY AND HISTORY						EAST ASIA PACIFIC
	N.W. EUROPE	AMERICAS	AFRICA + EGYPT	S.W. ASIA	MEDITER- RANEAN	NORTH EUROPE	NORTH AMERICA	MIDDLE AMERICA	SOUTH AMERICA	AUSTRALASIA

ENVIRONMENT — N.W. EUROPE:
- Little Ice Age
- Little optimum
- Climatic deterioration
- Climatic optimum
- -Elm decline
- Younger Dryas / Loch Lomond Stadial
- Lateglacial Interstadial
- Weichselian/ Devensian glaciation max

(left side scale) HOLOCENE ; PLEISTOCENE

AMERICAS:
- Megafaunal extinction
- Hemlock decline
- Wisconsinan glaciation 18 ka
- 100 ka Modern Hominids

AFRICA + EGYPT:
- Europeans, Zimbabwe
- New Middle Old KINGDOMS
- Egyptian towns

S.W. ASIA (with CYPSITHERMAL running vertically):
- ISLAM
- Assyrians, Hittites
- Sumer
- Neolithic
- Irrigation
- Natufian
- Epipalaeolithic

MEDITERRANEAN:
- Post Medieval
- Colonialism
- Medieval, Byzantine, Romans, Etruscans, Greeks
- Mycenaean, Minoan
- Neolithic
- Mesolithic
- UPPER PALAEOLITHIC

NORTH EUROPE:
- Norse Migrations
- Iron Age
- Bronze Age
- Neolithic
- Mesolithic

NORTH AMERICA:
- Mississippian, Anasazi Pueblos
- ARCHAIC
- Maize
- Clovis
- ? America colonised ?

MIDDLE AMERICA:
- Europeans
- AZTEC
- MAYA
- OLMEC

SOUTH AMERICA:
- INCA
- State foundation
- Maize

EAST ASIA PACIFIC AUSTRALASIA:
- Europeans
- New Zealand and Polynesia colonized
- Han China
- Shang China
- INDUS
- New Guinea Gardens

Key: ▬ Domesticates appear   ●●● Writing appears

increasingly sophisticated society has modified the landscape of the mid-latitude regions of the Northern Hemisphere over the last ten millennia.

The third part of the book deals with the present and the future. Preserved within the contemporary cultural landscape is a rich record of the natural and cultural processes that have combined to produce that landscape, and the management of this heritage presents particular problems both for scientists and for politicians. These problems are considered in Chapter 8. Chapter 9 examines the impact of nineteenth- and twentieth-century industrial activity on contemporary climate, and discusses the possible consequences of these anthropogenically induced climatic changes for the global climatic and terrestrial environment. From an initial consideration of the effects of climate on people, therefore, the analysis has moved to an examination of the impact of human society on both present and future climate. To what extent climatic changes arising from human activity will amplify or modulate the natural climatic rhythms discussed in Chapter 3 is one of the most intriguing problems confronting historical scientists, and one to which there are, as yet, no conclusive answers.

2

Evidence for Environmental Change

Introduction

Evidence of past environments and of climatic change comes in a variety of forms. For the relatively recent past, instrumental data on temperature and precipitation are available for limited areas of the world, although the timescale for these records seldom exceeds 300 years. Historical evidence in the form of old diaries, annals, ships' logs, woodcuts, pictures, etc., may yield additional information, although these tend to provide 'snapshots' of former climatic conditions. Some historical records such as harvest dates and crop yields, however, constitute time-series data of climatic change. By far the most widely used bases for environmental reconstructions are **proxy** records. The term proxy is used to refer to any line of evidence that provides an *indirect* measure of former climates or environments (Ingram *et al.*, 1981), and can include materials as diverse as pollen grains, insect remains, glacial sediments and tree rings, as well as data on crop yields, harvest

dates and parish records (see Lamb, 1981). Some of the more widely used proxy records, along with historical evidence and meteorological data, are discussed in this chapter. The difficulties of interpreting these various lines of evidence in the context of environmental and climatic change are also considered. The chapter begins, however, with some methodological considerations, followed by an examination of the dating techniques that are currently employed in the establishment of timescales of recent environmental change.

Methodological considerations

Any attempt to unlock the secrets of the past inevitably involves the palaeoenvironmentalist in **scientific** enquiry. Precisely what constitutes a science has been a matter of debate amongst philosophers for many years, but a general view might be that this is a branch of study that is concerned with the search for truths and the establishment of general laws relating to the natural world, using as a basis the systematic collection of observed facts. Opinion differs, however, over what constitutes the most appropriate **scientific method**, i.e. the procedures employed by scientists to solve problems and hence to search for truths. In this section, two popular views of science, **inductivism** and **falsification**, are examined, for both approaches have underpinned investigations of climatic and landscape change (Haines-Young and Petch, 1986). The related approach of **multiple working hypotheses** is also discussed, and the 'new science' of **chaos theory** is briefly introduced. Finally, because palaeoenvironmental reconstructions using proxy data rely heavily on modern analogues ('the present is the key to the past'), it is necessary to consider the assumptions that underlie what has become known as the **uniformitarian** approach to earth history.

Inductivism

Inductivism is widely regarded as the essence of scientific enquiry, for it appears to provide science with a sound, logical (rational) methodological basis. It still underpins most work in the earth sciences while the quantitative approach, which has characterized a considerable proportion of the research output of geographers, geologists, environmental scientists and archaeologists over the past 25 years, is firmly rooted in inductivism. It rests upon the notion that the accumulation of knowledge relies on experience, and hence scientific statements acquire meaning by virtue of their *empirical* (i.e. experimental or observational as opposed to theoretical) basis. In other words, generalizations that form the basis for scientific laws are derived from observations of reality. Moreover, once laws or theories are established in this way, there is a basis for explanation or prediction by means of *deduction*. The inductive approach, therefore, can be seen as a stepwise ascent in science from observation to theory (O'Hear, 1989) and could, perhaps, be summarized along the following lines:

Collection of data by observation (experience)
→ Ordering of facts
(measurement, classification, definition)
→ Generalizations (Induction)
→ Law/theory construction
→ Explanation (Prediction via Deduction).

This line of reasoning, which argues that real world phenomena can be explained by showing them to be instances of repeated and predictable regularities, underpins much of classical science extending back to the ancient Greeks (Chalmers, 1982), and has frequently been referred to as **classical rationalism**.

Despite its widespread (and perhaps often unconscious) adoption by scientists, inductivism has been the subject of penetrating critical scrutiny (e.g. Russell, 1961; Popper, 1972). A particular problem concerns **verification**, for no number of apparently confirmatory statements can ever show a general proposition to be true. Scientific knowledge can never be more than partial, and hence there must always be the possibility that anomalies exist which remain to be discovered and which may refute a general statement or law derived by inductive reasoning. It is for this reason that many scientists

speak in terms of **probabilities** rather than absolutes. A further difficulty concerns the objective nature of facts. It is implicit in inductive reasoning that observation precedes theory development, that there is a clear distinction between fact and theory and, moreover, that facts obtained by this form of enquiry are objective. However, all observations or measurements are inevitably made in the context of prevailing theory. A considerable body of knowledge has been generated about the way in which the physical or human world is structured and hence that corpus of information will inevitably exercise an influence on new observations, experiments, etc. The objective basis of facts derived from such observations must, therefore, be called into question. A related problem is that observations and measurements will inevitably be influenced by the level of technology that is available at any particular time. In the early years of the present century, for example, it was widely accepted that four separate glacial episodes had occurred in the mid-latitude regions of the Northern Hemisphere during the course of the Quaternary. By the 1950s, technological advances in ocean coring led to the discovery of the oxygen isotope signal in deep ocean sediments (Chapter 3), the interpretation of which eventually increased this number to over 20. The question arises, therefore, as to whether the truth of a fact can ever be satisfactorily demonstrated.

Most contemporary inductivists would acknowledge these difficulties, and would accept that science does not begin with unbiased and unprejudiced statements about reality (Chalmers, 1982). Theories may be conceived of by a variety of routes (accident, inspiration, creative acts, etc.), all of which precede observation but defy logical analysis. This **sophisticated inductivism** lies at the heart of many areas of current scientific enquiry, and is characteristic of much of the research that has been undertaken on environmental change.

Falsification

The most sustained challenge to inductivism is to be found in the writings of Karl Popper (1972,

1974). He accepted the empirical basis of scientific enquiry and also that observations will inevitably be guided by existing theory. He argued that although the truth of a proposition can never be conclusively demonstrated, statements can be rejected. In other words, while hypotheses cannot be *verified* they can, in fact, be *falsified*. Popper's view of scientific investigation, therefore, is a *deductive* one based on conjecture and refutation. Theories are viewed as tentative conjectures which are tested by observation, experiment, measurement, etc.; those that fail are rejected and replaced by further conjectures. In this way, according to Popper, science proceeds by trial and error, with the strongest theories being those that are clear, precise, detailed and broad ranging. Theories that contain more detail, however, are potentially more falsifiable; hence, a theory or hypothesis gains strength the more wide ranging and precise it is, the more falsifiable it is, but most importantly the more it resists falsification and hence constitutes a challenge to science. Because falsification attempts to provide a sound, rational basis for deciding between the relative merits of different theories by testing them critically, the term **critical rationalism** has been applied to this form of scientific reasoning.

Two further aspects of this line of thinking merit consideration. First, failure of a hypothesis or conjecture will not necessarily lead to outright rejection, for theories can be modified to enable them to resist falsification. Secondly, although a theory can never be verified, it can be *confirmed*. The essential difference between verification and confirmation is that the former implies that a theory or hypothesis has been shown to be true, whereas confirmation merely implies that a theory has resisted falsification and hence has been accepted *for the time being*. An example might be the theory of **plate tectonics** which has been supported by a range of geological evidence that has emerged over the last 20–30 years (Frankel, 1988). From the perspective of the critical rationalist, this is a high order theory in so far as it is potentially highly falsifiable, but has so far resisted falsification and can be regarded as being confirmed by the evidence that is currently available.

Critics of falsification have argued that this form of scientific enquiry is too inflexible and does not conform with what scientists actually do. Indeed, it has been suggested that had falsification been rigorously applied, many currently accepted scientific ideas would never have survived simply because they appeared to conflict with prevailing observations (Chalmers, 1982). Moreover, in so far as observations are both theory dependent and fallible, where an observation conflicts with a theory, there is no logical reason why it should not be the former that is in error rather than the latter. Hence it becomes possible to confirm a theory by testing it with a fallible observation. During the nineteenth century, for example, the occurrence in many inland areas of Britain of unconsolidated deposits containing marine shells was used to corroborate the theory of a major marine inundation widely believed to have been the Biblical Flood. Following the adoption of the **glacial theory**, however, it became clear that these shelly deposits were not marine in origin as originally thought, but were glacially transported. The notion of the Great Flood achieved almost universal acceptance but was based on a contemporary interpretation now known to be fallible.

Popper's ideas have found favour with some natural scientists (Moss, 1977; Bishop, 1980), but critical rationalism has not dislodged inductivism as the principal methodological strand in the environmental and historical earth sciences. Times may be changing, however, for Haines-Young and Petch (1980, 1986) have presented compelling arguments for the more widespread adoption of critical rationalism within physical geography and a trend towards the falsificationist position can be detected in biogeography (Ball, 1976) and palaeoecology (Edwards, 1983; Birks, 1986). Whether the 'challenge of critical rationalism' described by Haines-Young and Petch, (1980) is sufficiently robust to displace inductivism from the methodological pinnacle it currently occupies in the earth and environmental sciences remains to be seen.

Multiple working hypotheses

The method of multiple working hypotheses was outlined initially by Chamberlin (1897, reprinted 1965) and involves the formulation of as many hypotheses as possible in an attempt to explain the same phenomenon. Weaker or mistaken theories are progressively eliminated as the hypotheses are tested critically against each other. The aim is to achieve an explanation that is more nearly correct than would have been the case if only a single hypothesis had been considered. Although the method was a precursor to falsification, it has much in common with it (Haines-Young and Petch, 1983, 1986) in so far as decisions have to be made between competing theories, and scientists are encouraged to find evidence that will lead to the elimination of all but one of the working hypotheses. Applications of the approach can be found in geomorphology (Baker and Payne, 1978) and palaeolimnology (Battarbee *et al.*, 1985), while the method of multiple working hypotheses is regarded by Birks and Birks (1980) as one of the fundamental philosophical principles of palaeoecology.

Chaos theory

Much of what has been said so far rests upon the assumption that the natural world operates according to certain physical laws that govern the operation of global systems at a range of spatial and temporal scales. In seeking to formulate and to understand these laws, scientists have traditionally looked for **regularities** (or order) in real world phenomena, and have attempted to use this knowledge to make statements about the future behaviour of natural systems. Classical science, therefore, has both a deterministic and a predictive basis (see Inductivism above). Over the last two decades, however, a conceptual revolution has occurred, particularly in mathematics and physics, which has raised fundamental questions about the philosophical and methodological bases of classical science. This has been the discovery of **chaos**.

In recent years, scientists have begun to realize that small adjustments in the variables of natural

systems can have far-reaching consequences. Tiny differences in input may lead to overwhelming differences in output – a phenomenon which has been referred to as 'sensitive dependence on initial conditions'. The most famous example of this is what has become known as the 'butterfly effect', i.e. in weather patterns, the notion that a butterfly stirring its wings in Beijing can transform storm systems next month in New York (Gleick, 1987). The problem that confronts scientists is that establishing the range of variations in initial conditions is a fundamental impossibility (Reiter, 1984). It is also now apparent that in deterministic systems, random (stochastic) behaviour can exist side by side with order, and hence although systems may obey immutable and precise laws they do not always act in predictable and regular ways. Simple laws may not produce simple behaviour; rather, deterministic scientific laws can produce behaviour that appears random. Hence, order can breed its own kind of chaos (Stewart, 1990).

This is a quite dramatic discovery that is currently transforming the face of contemporary science, for fundamental questions are now being raised about measurement, experimentation and predictability, and about the verification or falsification of theories. The impact of chaos theory on the earth and environmental sciences has so far been small, but the ramifications are clearly far reaching, not only for the way in which research proceeds, but also in the search for explanations of patterns or trends in historical data. The fact that chance may be as significant as physical laws in explaining the operation of earth and atmospheric processes is something with which natural scientists have yet to come fully to terms.

Uniformitarianism

Early attempts to reconstruct patterns and processes of environmental change were rooted in a philosophy that became known as **catastrophism**. Hence landscape change was seen as being brought about by earthquakes, floods, volcanic eruptions and other cataclysmic events (Chorley *et al.*, 1964). The Biblical Flood was considered to have been of widespread significance (the **diluvial** view) and strongly influenced interpretations of the geological record. Underlying this line of reasoning was the almost universally accepted view of Ussher (1650–54) that the Creation had occurred in 4004 BC. This clearly allowed only a limited timescale for geological and geomorphological processes. In addition, the influence of the Church on scientific enquiry was pervasive and divine intervention was frequently invoked to account for otherwise inexplicable phenomena.

The development of the uniformitarian approach to earth history was a radical departure from the catastrophist school of thought. It was first proposed by the geologist James Hutton (1788, 1795) who argued not only for a geological timescale extended beyond that allowed for by Ussher but also for continuity in the operation of geological processes through time. His views are summarized in the well-known aphorism 'the present is the key to the past', in other words, that former changes of the earth's surface may be explained in terms of those processes observed to operate at the present day. Preternatural or catastrophic forces were rejected. This radical reinterpretation of geological history was espoused by Hutton's co-worker John Playfair (1802), and particularly by Charles Lyell (1830–33) who is often regarded as the founder of modern geology. This new approach to earth history came to be known as the **fluvialist** school because of the emphasis on river erosion as a process in the shaping of the earth's surface. Elements of uniformitarian reasoning, notably **actualism** (what exists now also existed in the past) and **gradualism** (geological processes operate at slow rates and in small increments) are implicit in Darwin's work on evolution and natural selection (Stoddart, 1986). In the present century, the sciences of palaeontology (the study of fossil remains) and palaeoecology (the study of the ecological relationships of past organisms) are firmly underpinned by uniformitarian principles (Rymer, 1978).

Although uniformitarian reasoning is now implicit in most studies of earth history, debate about the validity of the approach continues (Hubbert, 1967; Albritton, 1975; Rymer, 1978). One difficulty is that in using the present to interpret the past, environmental scientists are employing **analogy** as their main interpretative argument. Arguably, however, every context is unique, no analogy is exact, and hence no argument from analogy is certain. Moreover, it is becoming clear that many former plant and animal distributions have no modern analogues (see below) in which case uniformitarian reasoning cannot be applied. Moreover, it is no longer appropriate to assume that processes in the geological and biological past have operated at a constant rate, as demanded by the early uniformitarianists. In the same way. it cannot be argued that certain plant species occupied *precisely* the same ecological niche or covered *exactly* the same geographical range in the past as at present. Such strict adherence to uniformitarian principles (**substantive uniformitarianism**) finds little favour with contemporary earth scientists, however, most of whom incline to the view that although currently observable geological and biological processes must have operated throughout history, they would have done so on varying timescales (Chapter 1). Catastrophic events such as floods, earthquakes and volcanic eruptions, which have profound effects on modern landscapes, must have been equally effective in the past. Ironically, while uniformitarianism initially replaced catastrophism as a methodological basis for earth history, cataclysmic events can now be reconciled within a uniformitarian framework. This concept of uniformitarianism has been referred to as actualism (see above) by Hooykaas (1963) and **methodological uniformitarianism** by Gould (1965).

These problems notwithstanding, uniformitarian principles continue to underpin most work in the historical earth sciences and palaeoecology 'which assumes that the world is regular, lawful and therefore intelligible' (Rymer, 1978). Throughout this book methodological uniformitarianism provides the cornerstone for reconstructing the past and is essential for achieving any measure of understanding of the processes and patterns of environmental change.

Dating of proxy records

Four types of method are currently employed in the dating of proxy records. These are: (i) radiometric methods, (ii) incremental methods, (iii) methods that establish age equivalence, and (iv) artefact dating.

Radiometric dating

This is based on the radioactive decay of unstable chemical elements or **isotopes** which emit atomic particles in order to achieve a more stable atomic form. The most widely used of these isotopes is **radiocarbon** (^{14}C), atoms of which are absorbed by all living organisms through the carbon dioxide cycle. Decay of ^{14}C occurs but there is constant replenishment from the atmospheric reservoir. Following death of the plant or animal ^{14}C decay continues, but no replacement can take place. Hence, measurement of the ^{14}C remaining in a fossil will provide an age for the death of that organism. The method is applicable to all organic material and extends back to $c.$ 40 ka BP (Bowman, 1990). Because of technical difficulties in the measurement of radiocarbon activity in the laboratory, dates are always given as a mean age with one standard deviation. Hence a date of 5000 ± 50 indicates that there is a 68 per cent chance that the date lies in the range 4950–5050 years. There is, of course, a one in three chance that the true age might lie outside that range.

A recent advance in radiocarbon dating concerns the application of accelerator mass spectrometry (AMS) to determine the number of ^{14}C atoms in a sample rather than relying on detection of the decay products as in conventional ^{14}C dating (Linick *et al.*, 1989). This approach is especially valuable for establishing the age of very small samples of material such as individual seeds, molluscs, Foraminifera, fragments of hominid remains, or even pollen grains (Brown *et al.*, 1989). Hitherto, dating of such material rested on assumed associations with dated material (e.g. charcoal, organic lake sediments) from the same horizon. Recent applications of the AMS technique have been wide ranging, and include the dating of plant and animal domestication, the first human colonization of America and Australasia (D. R. Harris, 1987), and oceanographic and sea-level changes in the North Atlantic during the last deglaciation (Bard *et al.*, 1987, 1989).

One difficulty with radiocarbon dating is that the amounts of ^{14}C in the atmosphere have varied over the course of the Holocene, levels being significantly higher around 6000 years ago relative to the present day. However, radiocarbon years have been **calibrated** to calendar years by radiocarbon determinations on already dated **dendrochronological** sequences (see below) and these show, for example, that a ^{14}C date of 5000 BP (before present) is equivalent to a calendar age of approximately 5800 years (Pearson *et al.*, 1983). It is now generally accepted practice to calibrate radiocarbon dates in this way so that radiocarbon determinations can be presented as calendar years. Throughout this book, however, radiocarbon dates are quoted in an uncalibrated form unless otherwise specified.

A second and increasingly widely used radiometric technique is that of **uranium series dating** (Ivanovich and Harmon, 1982). ^{230}U and ^{235}U decay to stable lead isotopes through a complex decay chain of intermediate nuclides with widely differing half lives. Uranium and some products of the decay series are soluble, whereas others such as thorium (^{230}Th) and protactinium (^{231}Pa) are not. The latter, therefore, precipitate out and accumulate in sediments on lake floors or on the sea bed. Hence the age of those sediments can be established by measuring the extent to which these daughter nuclides have decayed. By contrast, the ages of carbonate fossils of creatures that absorbed uranium from sea or lake waters during their lifetime can be calculated from the extent to which the decay products of uranium (such as thorium and protactinium) have accumulated (or reappeared) in their carbonate shells. The method has a time range from 1 ka to 350 ka, although with high precision AMS techniques, this may be extended (Schwarcz, 1989). Uranium series dating is applicable to a range of materials including coral, cave speleothem, molluscan shells, marls and lacustrine limestones, bones and teeth, and possibly peats. In addition to providing an independent radiometric dating method, uranium series may also provide a means of calibrating the radiocarbon timescale beyond the time range afforded by dendrochronology (Bard *et al.*, 1990a).

Other radiometric methods for dating the recent past involve the use of **short-lived isotopes** such as ^{210}Pb (range 1–150 years) and ^{137}Cs (1–30 years). The former employs the isotope of lead ^{210}Pb, a decay product of radon, which is the naturally occurring radioactive gas that escapes from the earth. The latter involves the use of ^{137}Cs, an isotope produced artificially as a consequence of nuclear weapons testing since 1945. Both lead and caesium isotopes have been used in the dating of lake sediments (Pennington *et al.*, 1976; Longmore *et al.*, 1983), and ombrogenous peats (Oldfield *et al.*, 1979). More recently, attention has been directed to other isotopes such as ^{32}Si and ^{39}Ar. In the context of dating, these are largely experimental at present, but they may eventually bridge the gap between ^{14}C and ^{210}Pb (Olsson, 1986). The former is a potential dating method for sediments and for glacier ice (Grootes, 1984), while the latter may have applications in dating glacier ice, groundwater and ocean water (Loosli, 1983).

Incremental dating

This group of dating techniques is based on the regular additions of material to organic tissue or to sedimentary sequences, and includes dendrochronology, lichenometry and annually laminated lake sediments.

Dendrochronology (tree-ring dating) involves the measurement of increments of wood that are added annually to the outer perimeter of tree trunks. Annual ring thickness varies with climate (Fritts, 1976; Hughes *et al.*, 1982) and it may be possible to reconstruct former climatic changes from variations in ring width (**dendroclimatology**: Figure 2.1). By using distinctive rings or groups of rings (marker rings), it is possible to correlate wood from different trees and to establish long sequences linking living trees, building timbers, buried timbers from peats or archaeological sites, and other sub-fossil wood (Munaut, 1986). This technique of **cross-dating** enables tree-ring series to be established spanning hundreds and, in some cases, thousands of years (Schweingruber, 1988). The longest continuous tree-ring series is that obtained from the south-west United States where work on the Bristlecone Pine (*Pinus longaeva*; Figure 2.2) has produced a chronology extending back over 8000 years (Brubaker and Cook, 1984). In western Europe, dendrochronological work on oak has produced a continuous tree-ring record extending back beyond 7 ka BP (Pilcher *et al.*, 1984). Dendrochronology provides the basis for calibrating the radiocarbon timescale (see above), and has been used in a wide range of Holocene contexts including the dating of mass balance fluctuations of

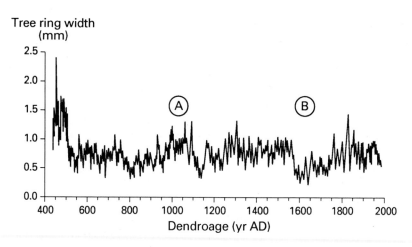

Figure 2.1   *Dendrochronological series AD 400–1970 from the Tornetrask region, northern Sweden. The Little Optimum (A) and Little Ice Age (B) are apparent in the ring-width series (after Bartholin, 1984)*

Figure 2.2   Bristlecone pines (Pinus longaeva) *growing on the semi-arid slopes of the White Mountains, Colorado, USA (photo J J Lowe)*

mountain glaciers in Sweden (Karlén, 1984), lake-level fluctuations in subarctic Quebec (Begin and Payette, 1988), and in the study of soil erosion and stream erosion rates in permafrost areas (Shroder, 1980).

Recent refinements in the technique have meant that dendrochronology has become an increasingly important archaeological method in both Europe and North America. Where suitable timbers survive, events can be dated to the year and sometimes to the season. Prehistoric wooden trackways have been dated in England and Ireland (Hillam *et al.*, 1990; Raftery, 1990), while in the Alpine lake region and in the pueblo area of the American south-west settlement establishment, repair, expansion and abandonment have been dated with great accuracy (Dean, 1986; Coles and Coles, 1989). Such precision is especially valuable in the comparison of archaeological and palaeoenvironmental sequences, and in

Figure 2.3   Lichenometric growth curve for western Norway (after Erikstad & Sollid, 1986). The curve is based on lichen size measurements on newly-deglaciated surfaces of known age (fixed points). The age of unknown surfaces can then be obtained by measuring average lichen size on those surfaces (y axis) and reading off the appropriate age on the x axis

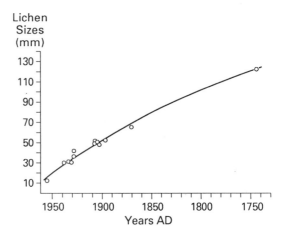

Figure 2.4   Late Holocene varved sediments in a core from Kassjon, Northern Sweden. Each varve, between two thin black layers, is about 0.5 mm. The thin black layers reflect winter sedimentation, the light layers are mainly mineral grains, while the grey layers are organic matter (photo I Renberg)

considerations of the effects of environmental change on human communities (Chapter 5).

Lichenometry is the technique of dating newly exposed surfaces using variations in lichen size. Measurements of lichen size on substrates of known age enable a growth-rate curve to be constructed for a particular locality (Figure 2.3). Reference to this curve enables surfaces of unknown age to be dated on the basis of lichen size. The method has been most widely employed in studies of recent glacier recession (e.g. Mottershead and White, 1972; Karlén and Denton,

1976), but it has also been used in the dating of plant colonization (Matthews, 1978), Holocene raised beaches (Birkenmajer, 1981), proglacial river terraces (Thompson and Jones, 1986), and recent earthquake activity (Smirnova and Nikonov, 1990). An additional application has been in the dating of archaeological features in coastal areas, for example around the Bothnian and Baltic coasts (Broadbent, 1979; Broadbent and Bergqvist, 1986).

Laminations in lake sediments (Figure 2.4) are usually referred to as **rhythmites** or, where the laminations develop because of annual variations in sedimentation they are known as **varves** (O'Sullivan, 1983). Annual laminations in lacustrine sediments are formed as a consequence of seasonal, rhythmical changes in biogenic production, water chemistry and the inflow of mineral matter (Saarnisto, 1986). Seasonal contrasts may be reflected, for example, in the accumulation of diatom remains (Simola, 1977), spring and summer calcareous precipitation (Peglar *et al.*, 1984) or variations in iron precipitation all causing annual laminations (Renberg, 1981). Glaciolacustrine varves (alternating layers of coarse and fine sediment) develop in proglacial lakes due to summer and winter contrasts in sediment input (e.g. Sturm, 1979). Perhaps the best known dating application of varved sediments has been in the development of a chronology for the retreat of the Fennoscandian ice sheet (Lundqvist, 1985; Figure 2.5), although laminated lake sediments have also been used as a chronological basis for studies of vegetational history, landuse history, lake development and climatic change (O'Sullivan, 1983; Saarnisto, 1986).

Age equivalence

This approach to dating the Holocene uses distinctive *marker* horizons within sedimentary sequences, or other characteristics of the stratigraphic record, to establish a chronology of events. Key horizons within a stratigraphic succession enable correlations to be made between different sites and hence form the basis for a *relative* chronology, i.e. whether one deposit is older

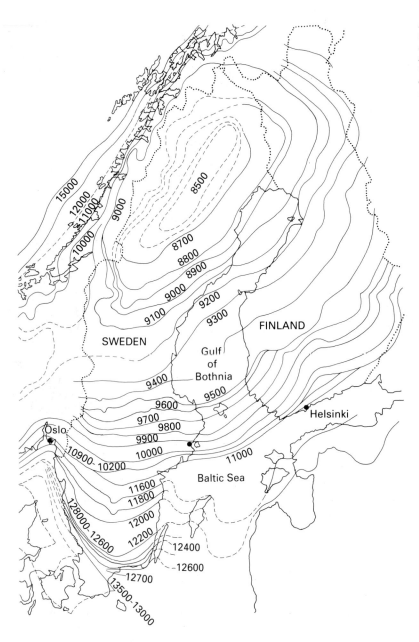

Figure 2.5 Lines of ice recession in Sweden and Finland. The data from north Sweden and southwards along the Baltic Coast are based on clay-varve chronology (after Lundquist, 1986)

or younger than another. Moreover, where distinctive horizons can be dated by one or other of the methods outlined above, the sedimentary sequence may be integrated into an absolute chronological framework. Three examples of this type of approach are tephrochronology, palaeomagnetic dating and pollen analysis.

The term **tephrochronology** refers to the use of volcanic ash (tephra) in the establishment of a chronology of events (Einarsson, 1986). Following a volcanic eruption, volcanic ash is widely dispersed and becomes incorporated into terrestrial and marine sediments (Figure 2.6) and also into glacier ice. Individual tephra units are distinguished on the

Figure 2.6 Mazama ash (c. 6.8 ka BP) exposed in fluvial sediments near Lumby, British Columbia, Canada (photo N F Alley)

basis of their mineral composition and can be dated by radiocarbon determinations on interbedded organic material, or by stratigraphic position in a sequence of annually laminated sediments in lakes or ice cores. The tephra layers effectively constitute time planes in a body of sediment. Hence they enable time–stratigraphic correlation to be made between localities where the ash layers are preserved, as well as forming the basis for a chronology of events. Tephrochronology has been useful, *inter alia*, in studies of vegetational history (Einarsson, 1963), soil formation and erosion, fluvial erosion (Thorarinsson, 1981), archaeological investigations (Keller, 1981), glacier fluctuations (Dugmore, 1989a), and the analysis of ocean-floor sediments (Sejrup *et al.*, 1989).

Palaeomagnetism refers to the record of changes in the earth's magnetic field that may be preserved in a body of sediment. These include **declination** (the angle between magnetic and true north), **inclination** (angle of dip) and **intensity** (strength of the magnetic field). Evidence from lake sediments shows that not only have all three properties of the earth's magnetic field changed during the Holocene, but that these changes are regular and have regional application (Thompson and Oldfield, 1986). The construction of master geomagnetic curves for particular areas, which can be dated by radiometric or incremental means, provides a means by which other lake sediment sequences can be dated and correlated (Figure 2.7). Although the method has been mostly applied to lacustrine sediments, there appears to be considerable potential for dating of other materials in this way, including fluvioglacial sediments, soils, aeolian materials, ombrogenous peats, ice cores (Thompson, 1986) and a range of archaeological features and deposits (Clark *et al.*, 1988).

Pollen records have long been used in chronological studies of the Holocene, working on

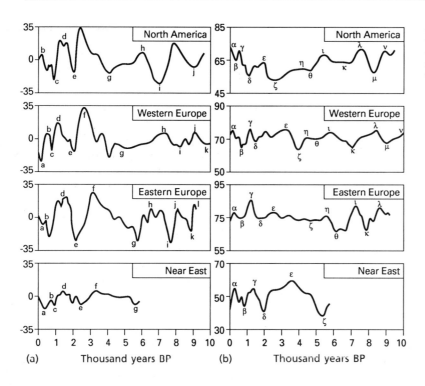

Figure 2.7   Regional declination (a) and inclination (b) master curves. Tree-ring calibrated timescale in calendar years BP (based on Thompson, 1986)

the principle that vegetational changes at the local and sub-regional scale would have been broadly synchronous, and hence assemblage zone boundaries in pollen diagrams represent more or less time parallel horizons. This method of relative dating has been widely used in north-west Europe where the postglacial pollen zones, which represent the successive migrations of forest trees, have been linked to the Scandinavian climatic succession of Blytt and Sernander (Figure 2.17). The approach has long been used to date vegetational and climatic change, and archaeological sites. However, changes in vegetation composition take place relatively slowly so boundaries of pollen zones must also be time transgressive. To a large extent, radiocarbon dating has superseded pollen analysis in the provision of a timescale for the Holocene, although pollen data may still prove a useful means of relative dating within restricted geographical areas (West, 1977a).

Artefact dating

The occurrence of humanly produced objects (artefacts) of pottery, flint, metal, etc., in Quaternary sediment sequences has also provided a basis for dating. Archaeological periods are characterized by specific assemblages of objects. The ages of stone, bronze and iron, first defined in 1836 on the basis of Danish evidence by Christian Thomsen, have subsequently been considerably refined to form the basic framework for prehistory. Types and styles change through time as a result of technological innovation, improved skill levels, and because people wish to convey different meanings, for instance by the way in which they decorate and fashion pots. Where several archaeological horizons are superimposed on one site, they often show marked changes in artefact type. The relative position of types within a sequence establishes a relative chronology. Where artefacts of the same type occur elsewhere in contexts which are historically or scientifically dated, cross-dating is possible. Examples of artefact dating include the use of flint, bone and stone tools to establish an Upper Palaeolithic chronology for parts of France and North America (Champion *et al.*, 1984; Bonnischen *et al.*, 1987), and the well-defined metalwork typologies of Europe which form the basis for continent-wide correlations during the Bronze and Iron Ages (Champion *et al.*, 1984). The discovery of metal artefacts in North European peat bogs (Coles and Coles, 1989) provides a valuable link between human activity and the broader picture of climatic change inferred from the peat bog records (see below).

The accuracy with which an artefact can be dated will be constrained by the overall chronological precision in the period and area concerned, and also by the timescale of innovation. In contexts from the last two millennia in Europe, or since European contact in North America, *some* artefacts may be dated to within a few decades. Coins, for example, can offer particular precision. In prehistory, however, artefact dating is more usually only accurate to within a few centuries. Moreover, there is always the possibility that objects, such as coins or pots, may have been in circulation for some considerable time before deposition. Hence, for much of the archaeological record, artefact dating is far less accurate than dendrochronology, varve chronology and, for older time periods, radiometric dating. Nevertheless, it remains an important technique for the dating of many archaeological sites, and is particularly important in the context of this book in so far as it links environmental sequences and human activity.

Fossil evidence

The term **fossil** is used to describe any organism or part of an organism that is buried by natural processes and subsequently permanently preserved. It includes skeletal material, plant remains, impressions of organisms, trails of organisms, tracks and borings. Human artefacts, however, are not regarded as fossil material (Whitten and Brooks, 1977). The fossilization process involves chemical and/or physical changes to the organic material

which can result in often delicate structures being preserved. Where little or no chemical change occurs subsequent to death (e.g. shell, wood) the term **subfossil** rather than fossil is applied. Fossils are divided into larger **macrofossils** and smaller **microfossils** (Brasier, 1980), this somewhat arbitrary distinction being based on whether or not a microscope is required for study. Detailed examination of macrofossils (e.g. animal bones, wood, molluscs) is often made under a microscope, but they can be seen with the naked eye. By contrast, microfossils can only be detected using microscopy and their study requires the use of a microscope throughout.

Fossil evidence is a central element in environmental reconstruction (Lowe and Walker, 1984). A considerable data base has been assembled over the years on the ecological requirements of present-day plants and animals, although it must be stressed that for a number of groups of biota, there are still major gaps in our knowledge of their contemporary ecology. Nevertheless, sufficient is known about the ecological affinities and associations of many modern species to make inferences about former climatic and environmental conditions by means of uniformitarian reasoning. The following is a sample of some of the more common proxy fossil records currently employed in palaeoenvironmental research.

Macrofossils

Plant remains

Macroscopic plant remains found in Late-Quaternary deposits include fruits, seeds, wood and other parts of plants including leaves, buds, scales and spines (Wasylikowa, 1986). They are best preserved in lake sediments and in peat deposits where anaerobic conditions obtain, but they also occur in riverine sediments, in cave sediments, in buried soils, and in archaeological contexts such as middens and pits from which carbonized remains of fruits and seeds are often recovered (Figure 2.8). Plant macrofossils are frequently deposited close to the original point of growth, and they provide data

Figure 2.8   *Charred cereal grains*

on local vegetation communities (Birks, 1980). They have also been employed in the reconstruction of regional vegetational patterns, providing information on tree migration rates (Davis, 1976a) and variations in the altitudinal limits of the treeline in mountain regions during the Holocene (Karlén, 1983; Kullman, 1987; Figure 2.9). In addition, plant macro-remains have been used to infer former temperature and precipitation

conditions (Iversen, 1954; Davis, 1984), and in the reconstruction of prehistoric economies (van Zeist and Casparie, 1984).

Mollusca

Terrestrial and freshwater Mollusca (Figure 2.10) are preserved in a range of sediments where there are high concentrations of calcium carbonate (Kerney, 1977). These include colluvial, fluvial and lacustrine deposits; cave sediments, aeolian sediments and buried soils. Under acidic conditions, in areas of calcium-deficient bedrock Mollusca are rapidly leached and are usually absent (Ložek, 1986). Land Mollusca provide information about local environments, being particularly responsive to the amount of vegetation cover and shade (Evans, 1972; Preece, 1986). They are a good guide to land-use changes associated with prehistoric sites (Chapter 6) and changes in species distribution may be related to Lateglacial and Holocene climatic change (Kerney, 1968).

Marine Mollusca are preserved in Late Pleistocene and Holocene contexts (Norton, 1977) including boreholes taken from the sea bed; from beach gravels and estuarine sediments, and from localities inland where they have been transported by glacier ice. They provide evidence of former sea-surface temperatures (Peacock, 1989), and they have also proved useful as a medium for dating Late Quaternary events including sea-level changes

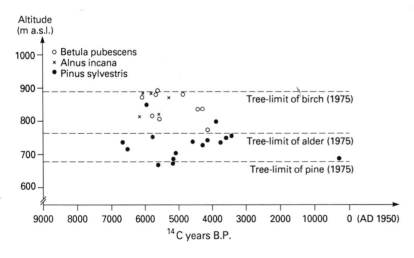

Figure 2.9   *Radiocarbon dates for subfossil* Betula pubescens, Alnus incana *and* Pinus sylvestris *in relationship to altitude in the Scandes mountains, central Sweden. The upper limits for birch, pine and alder in 1975 are also shown (after Kullman, 1988)*

Figure 2.10   Land molluscs (a) Discus rotundatus *(diameter 5.4 mm); (b)* Vallonia costata *(diameter 2.4 mm); (c)* Carychium tridentatum *(length 1.7 mm); (d)* Truncatellina cylindrica *(length 1.7 mm); (e)* Pupilla muscorum *(length 3.5 mm); (f)* Clausilia bidentata *(length 11.6 mm); (g)* Acicula fusca *(length 2.2 mm) (photo I Clewes)*

and glacier advances (Lowe and Walker, 1984). In addition, midden dumps of marine shells around contemporary shorelines provide information on prehistoric coastal exploitation (Bailey and Parkington, 1988).

Fossil insects

Insects are extraordinarily successful animals comprising more than half of the total number of plant and animal species known today (Coope, 1986). The largest order is the Coleoptera (beetles) which have colonized almost every terrestrial and freshwater habitat on earth. Yet many are **stenotypic** tolerating a narrow range of environmental conditions; moreover, climate (particular summer temperature) appears to be a major controlling variable at the regional scale (Coope, 1977a). Their remains are preserved in almost any sediment that contains plant macrofossils, and there is a considerable body of information on their present-day ecological associations and distributions (Buckland and Coope, 1991). Coleopteran evidence has been used in the construction of palaeotemperature curves for the later part of the last cold stage (Atkinson *et al.*, 1987; Morgan, 1987; Figure 2.11) and for the Holocene (Osborne, 1974). In addition, regional and local habitats have been inferred from coleopteran data (Girling, 1982), as have patterns of past human behaviour, most notably living conditions and economic activities in urban contexts (Kenward, 1982).

Mammalian remains

Animal bones occur in many Quaternary deposits including cave sediments (Figure 2.12), fluvial and colluvial sediments; lacustrine and marine deposits; peats and soils; and in burial chambers, middens, hunting sites and other contexts associated with human activity (Stuart, 1982; Davis, 1987). The bones range in size from those of large mammals which, in an archaeological context, have frequently been exploited and deposited by people, to the remains of small mammals, amphibians and birds which are often more valuable as ecological indicators. In some sediments the bones become **permineralized** as salts from circulating groundwaters are deposited in the vacant pore spaces, while in acid peats, where much of the mineral fraction has been lost by decalcification, only the flexible collagen fraction remains. Remarkably preserved human bodies such as Grauballe and Tollund Man from Denmark and Lindow Man from England occur in peat bogs where decay has been inhibited by anaerobic conditions and the chemistry of the peat (Coles and Coles, 1989). Late Quaternary vertebrates reflect former local habitats (grassland, woodland, heathland) and hence regional vegetation patterns. They can be used to infer climatic conditions (Rottländer, 1976; Stuart, 1979) and to provide evidence for the process and diffusion of animal domestication (Clutton-Brock, 1989).

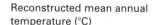

Figure 2.11   Reconstructed values for mean annual temperatures in England, Wales and southern Scotland based on coleopteran evidence. The bold line shows the most probable temperature values (after Atkinson et al., 1987)

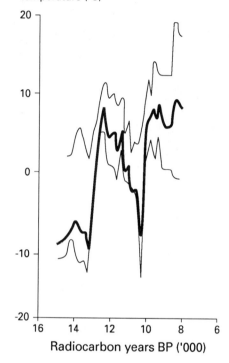

Microfossils

Pollen and spores

Pollen grains derived from the seed-producing plants (angiosperms and gymnosperms) are disseminated over wide areas by wind, water, animals or insects. Spores from the lower plants (cryptogams) are entirely wind dispersed. The grains become incorporated into peats, lake sediments and soils where they are well preserved providing that anaerobic conditions obtain (Figure 2.13). In so far as the composition of the pollen rain is a reflection of regional vegetation cover, fossil pollen and spores obtained from stratified sequences of sediment will provide a record of vegetational (and hence environmental) change

Figure 2.12 Charterhouse Warren Swallet, Mendip Hills, England, showing speleothems and animal bones on the cave floor (photo P Smart)

through time (Moore and Webb, 1978). Pollen and spores constitute one of the most widely used proxy data sources for inferring former environmental conditions and the range of applications of the technique is impressive (Birks and Birks, 1980).

They include the reconstruction of vegetation patterns at a variety of spatial and temporal scales (Godwin, 1975); studies of plant migration and forest history (Huntley and Birks, 1983; Delcourt and Delcourt, 1987); investigations of sea-level

Figure 2.13 Pollen grains. (a) Betula pendula (diameter 35 µm); (b) Picea excelsa (80 × 70 µm); (c) Tilia cordata (35 µm); (d) Erica cinerea (35 µm); Taraxacum officinale (35 µm); Vicia sativa (35 × 30 µm); Typha latifolia (40 µm) (photo I Clewes)

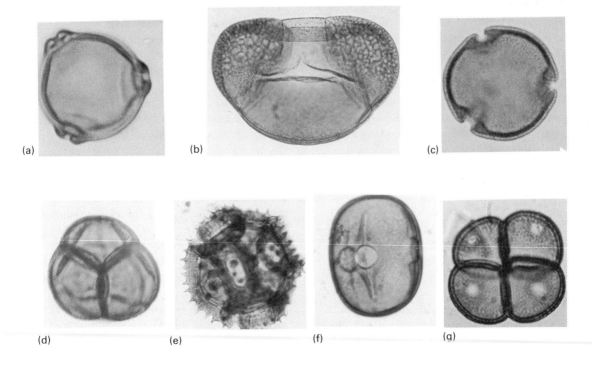

change (Shennan, 1986a); and the relative dating of landslides and other Holocene geomorphological activity (Tallis and Johnson, 1980). Pollen analytical data have also been used as a basis for climatic reconstructions (Birks, 1981; Webb *et al.*, 1987). Finally, much of what is known about the patterns of Holocene woodland clearance and early farming practices has been obtained from pollen analytical evidence (Simmons and Tooley, 1981; Behre, 1988; McAndrews, 1988).

Diatoms

Diatoms are microscopic unicellular algae (Figure 2.14) that live in lakes, ponds, estuaries and the sea, their distribution being controlled by a range of environmental variables including acidity, degree of oxygenation of the water, mineral concentration and especially water temperature and salinity. Fossil diatoms are found in many aqueous sediments and are used to provide evidence, *inter alia*, of lake water levels, vegetation and climatic changes (Brugam, 1980); the developmental history of lakes in relation to regional climatic changes (Rawlence, 1988); and human activity around lake catchments (Bradbury, 1975). Diatom analysis has also been employed in the study of sea-level change to isolate marine and brackish water transgressions (Tooley, 1978), in investigations of recent lake acidification (Battarbee and Charles, 1987; Jones *et al.*, 1989), and in archaeological investigations (Battarbee, 1988). In addition, diatoms have been extracted from deep-ocean cores and have been used, in association with other marine micro-organisms (foraminiferans, radiolarians, coccoliths), as a basis for palaeoceanographical and palaeoclimatic reconstructions (Burckle, 1978).

Cladocera and ostracods

These are microscopic crustaceans that live in both fresh and saline waters (Frey, 1986; Löffler, 1986). Cladocera are the most abundant crustaceans preserved in lake sediments, while ostracods (Figure 2.15) are found in lacustrine, estuarine and marine deposits. Both provide evidence of temperature, salinity and eutrophication changes in

Figure 2.14   Scanning Electron Micrograph of a valve of a diatom Cyclostephanos dubius. This is a freshwater planktonic form found especially in eutrophic lakes (photo R W Battarbee)

lake waters (Crisman, 1978; Carbonel *et al.*, 1988), while ostracods have also been employed as indicators of sea-water temperature and salinity conditions (Robinson, 1980). In addition, ostracods have been used as indicators of past sea level (Penney, 1987) and, in conjunction with other forms of proxy data, in the reconstruction of regional climatic changes (Delorme and Zoltai, 1984).

Foraminifera

Foraminifera (Figure 2.16) are marine protozoans that occupy habitats ranging from salt marshes to the deep oceans of the world (Brasier, 1980). They can be used as indicators of both local and regional environmental conditions, and have also been employed in studies of sea-level change (Kidson *et al.*, 1978) and nearshore marine palaeotemperatures and salinity variations (Nagy and Ofstad, 1980). The open-ocean forms have

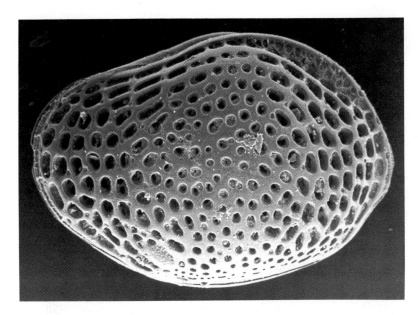

Figure 2.15   Scanning Electron Micrograph of the right valve of Loxoconcha sp, *a marine inshore species of Ostracod. Length* c. *0.6 mm (photo J E Robinson)*

proved particularly valuable in palaeoceanographic research leading to global climatic reconstructions and macroscale patterns of climatic change (Imbrie and Imbrie, 1979).

Charred particles (charcoal)

A characteristic feature of many Late Quaternary sediments (e.g. lake sediments, peats, soils) is the inclusion of microscopic carbon particles resulting from the burning of wood, grass or other vegetation (Patterson *et al.*, 1987). Charcoal can be seen under a microscope, for example in pollen samples, and burning episodes create magnetically enhanced mineral particles which can be detected in sediment cores by magnetic susceptibility measurements (Rummery, 1983). Most charred particles in Holocene sediments reflect burning by people, but some fires may have begun during droughts or may have been caused by lightning (Tolonen, 1986). In certain circumstances, therefore, the occurrence of natural fires can be used as palaeoclimatic indicators (Terasmae and Weeks, 1979). When combined with other forms of proxy data (pollen, plant macrofossils), the analysis of charred particles adds a valuable additional dimension in studies of

land-use history (e.g. Clark, 1988; Macdonald *et al.*, 1991).

Sedimentary evidence

Although interest in Late Quaternary sediments has often been stimulated by their fossil content, valuable information about former climatic and environmental conditions can frequently be derived from the nature of the sediments themselves and from the landforms that they comprise. Physical, chemical and biological properties of sediments can be used to make inferences about the environment of deposition, while stratigraphic relationships and contrasts provide evidence of depositional changes through time. Temporal and spatial changes in sediment accumulation may be governed by climate, and hence may constitute a proxy record of climatic change. However, in some situations (e.g. lakes or valley floors) sediment accumulation has been strongly influenced by human activity, in which case the analysis of the sedimentary sequence provides a proxy record of anthropogenically induced landscape change (Chapter 7).

Figure 2.16   Scanning Electron Micrographs of Foraminifera. (a) Globigerinoides sacculifer: *a planktonic foraminifer characteristic of tropical and subtropical ocean waters. Diameter c. 400 μm (photo B M Funnell)*

(a)

Peats

Peats accumulate in waterlogged localities where the breakdown of vegetal material is reduced by anaerobic conditions. Such areas are known as **mires**, some of which form where drainage is impeded (e.g. enclosed basins or river floodplains), whereas others are initiated and maintained by high atmospheric moisture levels. The latter are termed **ombrogenous mires** and occur as **raised bogs**, domed-shaped accumulations of peat that develop in lowland areas often following the infilling of a lake or pond, and **blanket mires**, namely extensive spreads of peat which cover the landscape in upland areas where rainfall is high. Peat deposits represent one of the most valuable terrestrial 'archives' for palaeoecological research (Godwin, 1981), for not only does the peat constitute an ideal medium for the preservation of fossil evidence but,

in so far as peat development is closely related to climatic conditions (Barber, 1985), the stratigraphy of ombrogenous peat profiles represents a proxy record of climatic change. Examples of the use of peat stratigraphy in palaeoclimatic research include Blytt and Sernander's reconstruction of major climatic episodes for the Holocene (Figure 2.17), the development of surface wetness curves (palaeoprecipitation curves) for upland mires (Barber, 1981; Figure 2.18), and the reconstruction of the pattern of short-term climatic changes during the Holocene (Aaby, 1976). In some peat profiles evidence of human activity is also preserved. For example, lenses of minerogenic sediment provide indications of erosion on hillslopes following woodland clearance (Moore, 1986), while ash and dust particles in ombrogenous peats reflect erosion

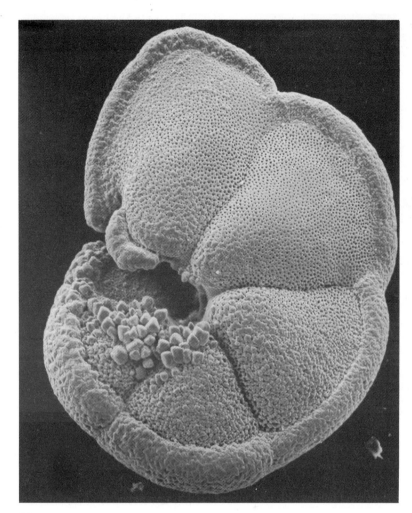

Figure 2.16 Scanning Electron Micrographs of Foraminifera. (b) Globoratalia menardii: *planktonic foraminifer characteristic of tropical ocean waters. Diameter* c. *700 μm (photo B M Funnell)*

(b)

of arable fields by wind and, therefore, constitute evidence for early agricultural activity (Aaby, 1986).

Lake sediments

Lake basins are natural sediment traps and frequently contain a history of deposition spanning thousands of years (Oldfield, 1978). As with peat deposits, lake sediments are ideal media for preserving a range of macroscopic and microscopic fossils, but a considerable amount of palaeoenvironmental information can be derived from the nature of the lake sediments themselves.

For example, in mid-latitude lake sequences, the climatic amelioration at the end of the last cold stage is represented by the transition from minerogenic to organic deposits. This lithostratigraphic change reflects increased organic productivity within the lake ecosystem and also a reduction in mineral inwash as the catchment slopes became stabilized by vegetation (Lowe and Walker, 1984). Reduced inwashing of soils from around the basin catchments is also reflected in the marked decline in concentration of chemical bases (e.g. Ca, Mg, Na and K) in Early Holocene lake sediments. The curves for these bases (Figure 2.19)

Period	Climate	Evidence
Sub-Atlantic	cold and wet	poorly-humified *Sphagnum* peat
Sub-Boreal	warm and dry	pine stumps in humified peat
Atlantic	warm and wet	poorly-humified *Sphagnum* peat
Boreal	warm and dry	pine stumps in humified peat
Pre-Boreal	subarctic	macrofossils of subarctic plants in peat

(a)

Figure 2.17 (a) The Blytt-Sernander sequence of Holocene climatic episodes based on peat bog stratigraphy (b) The Blytt-Sernander climatic episodes and Holocene pollen zones

Years before present	Pollen zone		Blytt-Sernander period	Climate
— 1000	VIII		Sub-Atlantic	Deterioration
— 2000		F1-III		
— 3000			Sub-Boreal	
— 4000	VIIb			
— 5000				Climatic optimum
— 6000	VIIa	F1-II	Atlantic	
— 7000				
— 8000	VI		Boreal	
— 9000		F1-I		Rapid amelioration
	V			
	IV		Pre-Boreal	
— 10000	III		Younger Dryas	Cold
— 11000	II		Alleröd	Rapid amelioration

(b)

provide a proxy record of the occurrence and extent of soil erosion around the lake catchments (Engstrom and Wright, 1984). Reworked minerogenic material in lake sediments and changes in the sediment limit around lake margins provide evidence of fluctuating water levels during the course of the Holocene which have, in turn, been interpreted in terms of temporal variations in precipitation (Digerfeldt, 1988). Evidence of anthropogenic activity is also preserved in many Mid- and Late Holocene lake sediment records, for the accelerated soil erosion that followed woodland clearance by early farmers is reflected in significantly higher rates of sediment yield in lake basins during the Late Holocene (Edwards and Rowntree, 1980).

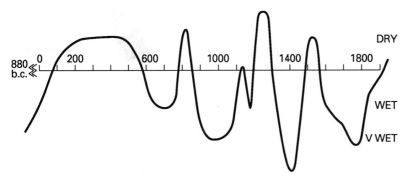

Figure 2.18   Surface wetness curve for Bolton Fell Moss, northern England, based on plant macrofossils and other palaeobotanical evidence (after Barber, 1981)

Cave sediments

Caves also form natural sediment traps and contain materials that originate within the caves (**autochthonous**) as well as sediments that are brought in from outside (**allochthonous**). The former category includes scree, rock rubble and fine-grained materials (cave earth) derived from the weathering of the cave walls and, in limestone areas, secondary mineral deposits of calcium carbonate which are collectively known as **speleothems**, the most common forms of which are stalagmites, stalagtites, flowstones and tufas

(Figure 2.20). Allochtonous materials include fluvial, glacial, colluvial, periglacial and aeolian deposits. In terms of the sedimentary record, deposits of scree material (**thermoclastic scree**) are usually interpreted as the products of frost weathering and are taken as evidence of cold conditions (Laville, 1976), whereas speleothem formation appears to be associated with periods of warmer climate (Atkinson *et al.*, 1986). Caves were favoured by wild animals and early humans, and so are often rich in flint tools and vertebrate remains (Figure 2.12). Detailed physical and chemical

Figure 2.19   Sediment chemistry diagram from Lake Hope Simpson, Labrador, Canada, showing the abrupt decline in concentration of chemical bases during the Early Holocene. Values are expressed in mg/gm dry weight of sediment (after Engstrom & Wright, 1984)

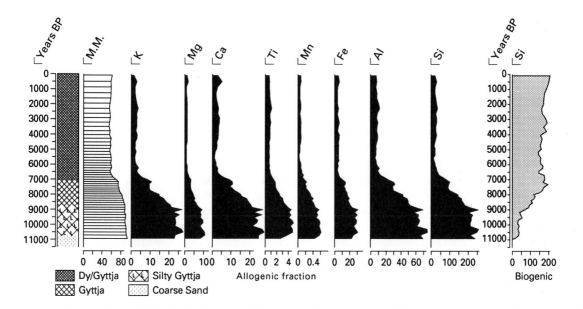

Figure 2.20   Vertical section through an imaginary bone cave illustrating some important types of cave deposit: (1) Water-laid sands and clays; (2a) Deposits of an animal lair; (2b) Hearths of fires made by people; (3) Stalagmite floor; (4) Talus cone with bones of animals which fell down a shaft; (5) Bones and dung of bats; (6) Second talus cone at the cave mouth. The deposits show disturbance at several places (A) Collapse pit (B) Burrow (C) Human burial (D) Washing out and redeposition by a stream (after Sutcliffe, 1970)

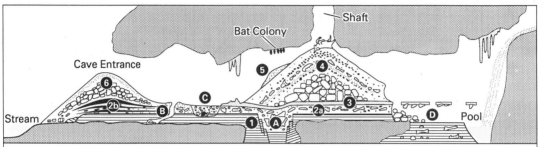

1. Water-laid sands and clays.
2a. Deposits of an animal lair.
2b. Hearths of fires made by man.
3. Stalagmite floor.
4   Talus cone with bones of animals which fell down a shaft.
5. Bones and dung of bats.
6. Second talus cone at cave mouth.

A. Collapse pit.
B. Burrow.
C. Human burial.
D. Washing out and redeposition by a stream.

analyses of cave sediments can provide valuable data not only on climate, but also on the settlement history of cave sites (Farrand, 1979; Butzer, 1981).

Glacial sediments

Glacial sediments cover large areas of the mid-latitude regions of the world, forming an intermittent blanket over one-third of the land area of Europe and around half of the continent of North America (Embleton and King, 1975). The geographical distribution of these glacially derived sediments and their landform assemblages provide evidence of the former extent of the great ice sheets and mountain glacier complexes that developed during the cold stages of the Quaternary. Moreover, patterns of former ice movement can be inferred from the physical and chemical properties of the sediments, and from the orientation or alignment of landforms produced by both glacial erosion and deposition (Lowe and Walker, 1984). Although around two-thirds of the global ice volume present during the Quaternary cold stages disappeared during the Holocene, active glaciers remain in many high-latitude and high-altitude regions of the world, and the distribution of glacigenic sediments and landforms (especially

moraines) in those areas reflects glacier fluctuations during the course of the present interglacial (Davis and Osborn, 1988). Furthermore, an appreciation of the former extent of glacier ice, allied to glaciological principles derived from the study of contemporary ice sheets and glaciers, has enabled increasingly sophisticated modelling of former ice sheets and glaciers (Andrews, 1982; Boulton et al., 1985; Hughes, 1987). Such glaciological reconstructions not only provide evidence of the behaviour of Quaternary ice masses, but they allow inferences about former climatic conditions (Sissons, 1979; Sutherland, 1984a).

Periglacial sediments

The term 'periglacial' is widely used to refer to those high-latitude and high-altitude regions of the world where frost action constitutes the dominant geomorphological process. Cyclic freeze–thaw activity, the growth of ground ice and the presence in many (but not all) periglacial environments of permanently frozen ground (**permafrost**), leads to the development of a suite of highly distinctive sediments, sedimentary structures and landforms (French, 1976; Washburn, 1979). Moreover, the sparse vegetation cover that is characteristic of

Figure 2.21   Fossil ice wedge cast developed in sands and gravel, Stanton Harcourt, Oxfordshire (photo M B Seddon)

much of the periglacial domain means that aeolian and fluvial activity are also highly effective geomorphological processes. Relict periglacial phenomena dating from the cold stages of the Quaternary are found throughout the mid-latitude regions of the Northern Hemisphere, and constitute unequivocal evidence of climatic change. Moreover, by using modern analogues from present-day periglacial environments and data on their controlling climatic parameters, quantitative estimates of former climatic conditions can be derived from relict periglacial phenomena (Karte and Liedtke, 1981). Of particular value in this respect are **ice wedge casts** (sedimentary infillings of thermal contraction cracks in the former permafrost surface: Figure 2.21) and **involutions** (contortions in sediments produced by the action of ground ice: Figure 2.22). These have been used to infer former temperature conditions (Karte, 1987), precipitation levels (Williams, 1975) and permafrost distributions (Worsley, 1987). Also useful are

aeolian materials (**loess** and **coversands**) which provide valuable information on palaeowind directions and which, therefore, offer important data for verification of palaeoclimatic simulations (Koster, 1988).

Slope deposits

A range of sediments occurs on hillslopes and in valley bottom situations as a result of slope processes. These include **head** deposits that develop under periglacial conditions (Harris, 1987), **talus** or **scree** deposits which may also be periglacial in origin; landslide debris; and colluvial and solifluction deposits (Figure 2.23) that are more characteristic of erosion under temperate climatic regimes (Rice, 1988). These geomorphological processes are closely related to vegetation cover, precipitation levels and temperature regimes, and hence a palaeoenvironmental record relating to climatic and land-use change will frequently be preserved in stratified sequences of hillslope sediments (Chapter 7).

Alluvial deposits

The investigation of lake sediments is one aspect of the science of **palaeohydrology**, i.e. the study of water and sediment dynamics in the past (Gregory, 1983; Gregory *et al.*, 1987). The other element of palaeohydrological investigations is concerned with changes in river erosion and deposition, and with temporal changes in river regimes. In certain circumstances inferences can be made about palaeodischarge and palaeovelocity in former river channels (Maizels, 1983a; Clarke *et al.*, 1984). However, in a number of instances, it appears that Late Holocene alluviation is more closely related to patterns of human activity than to climatic change (Chapter 7). Many river valleys were densely settled in prehistoric times and their sediments contain well-preserved archaeological sequences with associated palaeoenvironmental evidence (Limbrey, 1983).

Aeolian deposits

Aeolian activity in the temperate zone is now largely confined to coastal areas and to upland regions

Figure 2.22   Involution structures at Brighton Marina, Sussex, overlain by a Holocene palaeosol and colluvium (photo M Bell)

where wind erosion remains an important agent of geomorphological activity (Ballantyne, 1987). The extensive spreads of windblown sands that are found throughout north-west Europe and North America were largely derived from glacial outwash deposits (Washburn, 1979), and are essentially relict periglacial deposits (see above), despite the fact that in some areas of the United States aeolian activity continued into the Late Holocene (Ruhe, 1984). In north-west Europe, however, aeolian deposition has remained a feature of coastal environments throughout much of the Holocene. Alternating sand deposits and stabilization horizons provide evidence of episodic environmental instability (p. 197) and the sands themselves contain a record of palaeowind directions (Lill and Smalley, 1978). Many coastal sites are calcareous, so molluscs and bones frequently survive in aeolian deposits. Both coastal and inland dunes are rich sources of artefactual material, and well-preserved archaeological sites, sometimes with fields and cultivation marks (Figure 2.24), are found in many parts of Europe.

Palaeosols

Palaeosols may be either **buried soils** below sediments (e.g. alluvium, colluvium, landslip deposits) or below archaeological monuments (Figure 2.25), or they may be currently exposed **relict soils** preserving some characteristics of former pedogenic regimes (Catt, 1979). A considerable amount of useful data has been obtained from buried soils. Pollen grains are preserved in acid soils such as podzols or brown earths (Andersen, 1986), while calcareous soils often contain fossil Mollusca (Evans, 1972). The fossil content of buried palaeosols can provide evidence of former vegetational and climatic environments (Caseldine, 1984). Pedological features of buried soil profiles can also be used to infer environmental conditions (Sorensen, 1977),

Figure 2.23   Colluvial valley sediments of prehistoric date exposed by a coastal landslip at Cow Gap, Eastbourne, England (photo B Westley)

with the study of soil micromorphology based on the analysis of thin sections becoming an increasingly important source of evidence for soil history and landuse change on archaeological sites (Courty *et al.*, 1989).

Marine deposits and landforms

Relict marine deposits and landforms both above and below the contemporary shoreline provide evidence of past changes in sea level. A record of former episodes of low sea level can be found in present offshore areas in the form of submerged coastal landforms (cliffs, caves, reefs, platforms, spits, shingle bars, and river valleys), terrestrial landforms and deposits now covered by the sea (Long and Stoker, 1986; Derbyshire *et al.*, 1985) and the remains of terrestrial fauna and flora recovered from the sea bed and from boreholes in the current sea floor (Jelgersma, 1979; Behre *et al.*, 1985). Evidence for former higher sea levels includes the presence of such erosional features as

clifflines, caves and marine abrasion platforms (Figure 2.26) at altitudes higher than the contemporary shoreline (Trenhaile, 1980), and also the occurrence of marine deposits (beach gravels, sand spits and marine clays) in situations above present-day sea level (Davies and Keen, 1985; Holyoak and Preece, 1983). A history of sea-level change can be reconstructed in those areas where a sedimentary record of marine and terrestrial sedimentation has been preserved. A marine **transgression** is indicated where terrestrial deposits (e.g peats) are overlain by saltwater sediments, whereas a **regression** is implied by a stratigraphic sequence in which marine deposits are succeeded by terrestrial sediments (Lowe and Walker, 1984). The Holocene sequence of sea-level change around the coasts of the British Isles and north-west Europe (Chapter 4) has been reconstructed largely on the basis of this type of stratigraphic evidence (Greensmith and Tooley, 1982).

Figure 2.24   A Bronze Age site within blown sand at Bjerre, North Jutland, Denmark (a) a series of humic stabilization layers with ard (plough) furrows at the base of each, separated by thin layers of blown sand (photo J-H Bech)

Figure 2.24   A Bronze Age site within blown sand at Bjerre, North Jutland, Denmark (b) Bronze Age ard furrows (photo J-H Bech)

(a)

(b)

Isotopic evidence

Isotopes are atoms of an element that are chemically similar *but* have different atomic weights. It now appears that almost all of the known elements have more than one isotope; hydrogen, for example, has two whereas carbon and oxygen have three. Indeed, although there are 92 naturally occurring elements, some 270 naturally occurring isotopes of these elements are known to exist (Gray, 1981). Not all isotopes are chemically stable and from a number there is a spontaneous emission of particles. These are the **radioactive** isotopes,

several of which are used in radiometric dating (see below). However, some stable isotopes, most notably oxygen (^{18}O and ^{16}O) and hydrogen (H and D), are also used in palaeoclimatic research. These isotopes are fundamental constituents of water, but the ratios between them (i.e. $^{18}O/^{16}O$ H/D) will vary over time as changes occur in water state (Bradley, 1985). During evaporation, for example, discrimination against the heavier isotopes (^{18}O and D) due to differences in vapour pressure means that the water vapour will be deficient in the heavier isotopes relative to the original water source. Put another way, the water will be *relatively enriched* in

Figure 2.25   Buried soil below the Neolithic chambered cairn at Hazleton, England which was constructed around 4700 BP, scale 30 cm (photo M Bell)

heavier isotopes by comparison with the atmospheric water vapour. The same effect occurs during condensation with the condensate becoming enriched in the heavier isotopes relative to the water vapour. This process of **isotopic fractionation** is controlled, *inter alia*, by temperature and thus by analysing the isotopic ratios contained within fossils, peats, or glacier ice, it is possible to reconstruct the sequences of isotopic changes that have taken place, and thus derive a proxy record of former climatic conditions. (**Note**: Ratios of oxygen and hydrogen isotopes are measured not in absolute terms but as relative deviations (per millilitre) from the mean ratios of a standard. Hence a $\delta^{18}O$ value of -3 per mille ($^0/_{00}$) indicates that the sample is 0.3 per cent or 3 parts per millilitre deficient in ^{18}O relative to the standard.)

Deep-sea sediments

Micro-organisms that live in the oceans such as the Foraminifera (see above) preserve in their skeletal remains a record of the isotopic composition of the ocean waters at the time they were alive. Hence, by analysing the isotopic content of microfossil remains down a core of deep-sea sediment (Figure 2.27), a record can be obtained of changes in the oxygen isotopic composition of ocean waters over time. Such records may extend back 2–3 million years (e.g. Shackleton *et al.*, 1984). Where the isotopic record has been obtained from organisms that inhabited the upper layers of the water column, the isotopic trace can be interpreted as reflecting changes in sea-water temperature (e.g. Ruddiman *et al.*, 1986a) and, by implication, regional climatic changes.

Figure 2.26   A raised rock platform of interglacial age on the North Devon coast, south-west England (photo M J C Walker)

In deep-ocean waters, however, temperature fluctuations resulting from climatic change are less pronounced than in the upper layers of the water column, and hence the isotopic signal from organisms that formerly inhabited the abyssal depths is taken to reflect variations in the isotopic composition of the ocean waters brought about principally as a result of expansion and contraction of the great ice sheets. Indeed it has been estimated that approximately 70 per cent of the isotopic signal recorded in Late Pleistocene cores from the open oceans is controlled by changes in global ice volumes (Ruddiman and Raymo, 1988). This is because during a glacial stage, large amounts of the lighter isotope ^{16}O would have been preferentially removed from the ocean systems and locked up in the great continental ice sheets. As a consequence, the ocean waters would have been relatively enriched in ^{18}O. The isotopic trace can, therefore, be regarded as representing a **palaeoglaciation** record (Shackleton, 1977).

Ice cores

Stable isotope records have been obtained from cores taken from the Greenland ice sheet and from ice caps in the Canadian Arctic (Figure 2.28). Most extend back some 100 ka (e.g. Dansgaard *et al.*, 1982; Koerner and Fisher, 1986), although continuous isotopic traces have been obtained from Antarctica (Figure 2.29) going back to 150 ka (Lorius *et al.*, 1985; Jouzel *et al.*, 1987). These records reflect changes in the ratios of oxygen and hydrogen isotopes in precipitation falling on the ice sheets which can, in turn, be interpreted in the context of changing temperature regimes in the high latitude regions of the world (Oeschger and Langway, 1989). They provide details of climatic shifts during the Holocene (e.g. Fisher and Koerner, 1980), as well as a proxy record of climatic change throughout the last glacial–interglacial cycle (e.g. Paterson *et al.*, 1977; Paterson and Hammer, 1987). Links have also been established between the isotope traces and recent historical and archaeological record (Dansgaard *et al.*, 1975; Figure 5.24).

Speleothems

Speleothems are mineral deposits found in limestone caves and are composed largely of calcium carbonate that has been precipitated from cave waters. In deep caves speleothems tend to form in isotopic equilibrium with their parent seepage waters, and hence successive layers of

Figure 2.27   Oxygen isotope trace from deep-sea sediment core V28-238. The Brunhes-Matuyama geomagnetic boundary has been dated to c. 735 ka BP. Odd-numbered isotopic stages represent warm (interglacial) episodes, while even-numbered stages indicate cold (glacial) stages (after Shackleton & Opdyke, 1973)

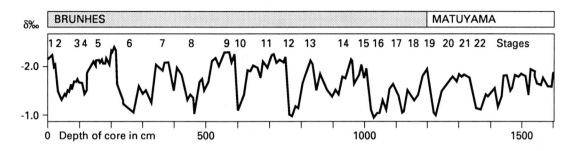

calcium carbonate will contain a record of the isotopic composition of cave waters throughout the period of speleothem accumulation (Hennig et al., 1983). As the $\delta^{18}O$ ratio in cave waters is closely controlled by temperature, the isotopic trace through a section of cave speleothem provides a proxy record of climatic change (Hendy and Wilson, 1968). The method has been used to reconstruct patterns of climatic change over timescales spanning thousands of years (Atkinson et al., 1986; Gordon et al., 1989), but also offers the potential for temperature reconstruction in the relatively recent past on a timescale as small as decades (Wilson et al., 1979).

Tree-rings

Tree-ring sequences reveal variations in the isotopic composition of cellulose. These, in part, reflect variations in the isotopic content of precipitation which are largely determined by former temperature levels. As a consequence, correlations can be made between $\delta^{18}O$ or δD and air temperature or, in some cases, between isotopic variations, temperature and relative humidity (Burk and Stuiver, 1981). Hence, the isotopic trace can be interpreted as a proxy temperature record (Long, 1982), or in terms of changes in atmospheric moisture balance (Friedman et al.,

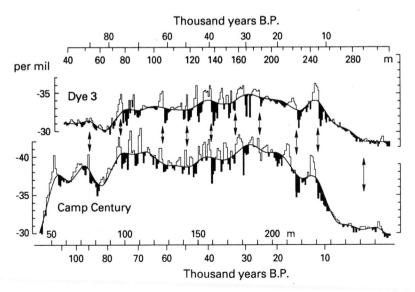

Figure 2.28   Oxygen isotope trace from Dye 3 (above) and Camp Century (below), Greenland (after Dansgaard et al., 1982)

Figure 2.29   The oxygen-18 trace from the 2083 m ice core from Vostok Station, Antarctica (after Lorius et al., 1985)

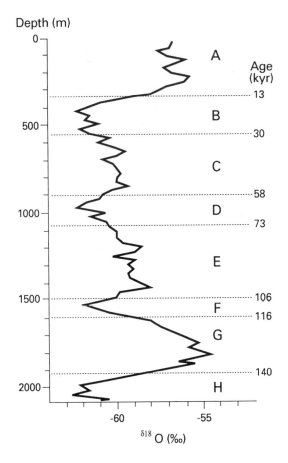

reconstruct variations in temperature and precipitation during the course of the Holocene (Dupont, 1986). However, difficulties in isolating the temperature effect from other variables in the isotopic signal indicate that a number of problems remain to be resolved before this method can be regarded as a wholly reliable technique for palaeoclimatic reconstruction (van Geel and Middeldorp, 1988).

Lake sediments

In lakes where large numbers of submerged aquatic plants and algae use dissolved CO_2 for photosynthesis an insoluble carbonate (marl) is precipitated, the oxygen isotope content of which is related to the isotopic composition of the lake waters at the time of precipitation. In so far as the $\delta^{18}O$ variations of lake waters will be related to former precipitation and temperature levels, analysis of the isotopic record of lake carbonates provides a potential means of reconstructing past climates. The method has been used in the investigation of climatic changes during the Lateglacial and Early Holocene (Eicher and Siegenthaler, 1976; Siegenthaler *et al.*, 1984; Figure 2.30) but, as in the case of peats, difficulties surrounding the isolation of the climatic effect from other environmental variables in the isotopic signal mean that a reliable quantitative relationship between isotopic variation and palaeoclimate remains to be established (Siegenthaler and Eicher, 1986).

1988). The technique has been applied both to living trees (Gray and Thompson, 1976) and to fossil wood (Dubois and Ferguson, 1985). Although some results have been equivocal, the potential of the method appears to be considerable (Bradley, 1985), and stable isotope studies may offer a valuable adjunct to more conventional methods of dendroclimatology (see below).

Peat

Plant cellulose material in ombrotrophic peat bogs contains a record of $\delta^{18}O$ and δD in former precipitation, which can be interpreted as a proxy record of climatic change (Brenninkmeijer *et al.*, 1982), and the method has been used to

Historical evidence

Documentary records constitute a valuable source of information on past climates, particularly for the past 800 years during the period known as the Little Ice Age (Grove, 1988). Three types of data source are usually recognized (Bradley, 1985): observations on weather phenomena; observations relating to natural phenomena closely controlled by weather; and phenological records.

Weather records

These are observations made on weather phenomena such as unusually warm episodes, exceptional snowfall, the duration of frosts, great storms, etc., and can be found in such diverse sources as diaries, annals, chronicles, letters, sagas, personal papers, and administrative and commercial records (Lamb, 1977). Some early Chinese documentary sources date back to the second century BC, intermittent records exist from Greece from about 500 BC, while in northern Europe documentary records begin to appear after about AD 1100 (Ingram et al., 1981). From the sixteenth century onwards, maritime records (ships' logs, etc.) become increasingly abundant and the papers of the great trading and exploration companies (East India Companies, Hudson's Bay Company) also provide material for climatic reconstruction

(Catchpole and Faurer, 1983; Wilson, 1985). Early newspaper reports of unusually severe winters or periods of heavy snowfall (e.g. Pearson, 1973) are also useful.

Weather-dependent natural phenomena

These include droughts, floods, duration of freezing of rivers or estuaries, the timing of ice break-up, and the movement of Alpine glaciers. In so far as all of these are governed by climate, they are sometimes called **parameteorological** phenomena (Bradley, 1985), and constitute another proxy data source. Details of such phenomena are to be found in a variety of records including ancient inscriptions, documents and pictorial records. Examples include the remarkable 400-year record of the opening of the port of Riga which provides evidence of the extent of Baltic Sea ice (Lamb,

Figure 2.30 $\delta^{18}O$ in marl profiles from three lakes in Switzerland. Horizontal axes: $\delta^{18}O$ in per mill.; Vertical axes: depth in cm. Left column in (a) are Lateglacial and Early Holocene pollen zones (after Siegenthaler et al., 1984)

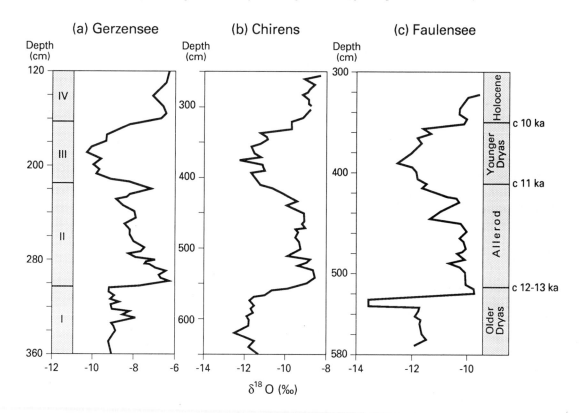

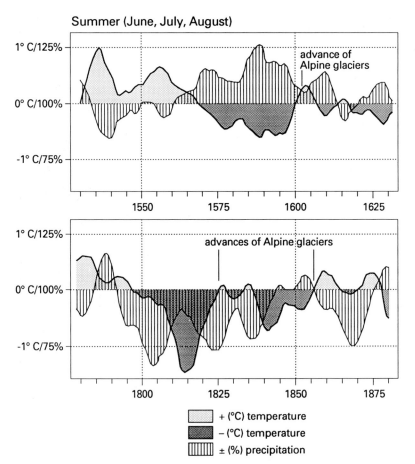

Figure 2.31 Temperature, precipitation and glacier fluctuations in Switzerland 1530–1630 and 1775–1885 based on a range of proxy data including phenological records (after Pfister, 1984)

1982); the reconstruction from various data sources of movements of Swiss glaciers over the past four centuries, and the relationship to Little Ice Age climatic changes (Messerli *et al.*, 1978); and the use of information on the extent of ice cover and barge trip frequency on Dutch canals during the seventeenth and eighteenth centuries (van den Dool *et al.*, 1978).

Phenological records

These relate to the timing of recurrent biological phenomena, and include crop harvest records, flowering and fruiting of plants, and the timing of animal migrations. Data are frequently available over a number of years and such proxy records may constitute time-series data of climatic change. Pfister (1981, 1984), for example, has shown how records of crop yields can be used to reconstruct climatic trends in Switzerland during the course of the Little Ice Age (Figure 2.31), while variations in the date of the grape harvest (Figure 2.32) formed the basis for a reconstruction of climatic change throughout north-west Europe back into the fifteenth century (Le Roy Ladurie and Baulant, 1981). Significantly, a close relationship has been established between the proxy temperature record derived from vine harvest data and the pattern of Alpine glacier movement during the Little Ice Age obtained from other historical sources (Bray, 1982).

Instrumental records

Although measurements of rainfall had been

Figure 2.32   Average annual date for the beginning of the grape harvest in north-east France, French Switzerland and the South Rhineland. At lower right are mean April–September temperatures (°C) in Paris during the period of instrumental records (after Le Roy Ladurie & Baulant, 1981)

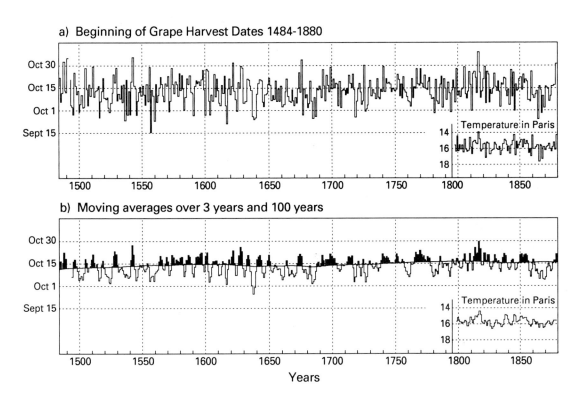

reported from India as early as the fourth century BC, systematic recording of climatic data did not begin until the period of rapid scientific advancement in the seventeenth century AD (Shaw, 1985). The first temperature measurements, using the newly devised thermometer, date from around 1660, while barometric pressure and rainfall records, the latter obtained from carefully designed rain gauges, began some 30–40 years later (Lamb, 1981). Throughout the early and middle years of the eighteenth century, various European weather stations were established and from about 1780 onwards, daily weather maps can be produced for large areas of north-west Europe (Lamb, 1982). Some of the longest climatic records using instrumental observations have been obtained from the British Isles and include the series of mean monthly temperatures for central England extending back to 1659 (Manley, 1974; Probert-

Jones, 1984), and the continuous sequence of rainfall measurements for the East Midlands area from 1726 onwards (Craddock, 1976). These and similar records from France, Switzerland and Italy (Figure 2.33) show unequivocally the climatic amelioration throughout western Europe at the end of the Little Ice Age (Grove, 1988).

Although instrumental series provide ideal information for the reconstruction of former climate, they are restricted to the relatively recent past and the very early records are all from Europe. In Canada, the earliest continuous temperature record is from the Hudson Bay area and begins in 1760 (Ball and Kingsley, 1984), while in the USA the oldest temperature data are from towns on the east coast (New Haven 1780; Baltimore 1817), although a continuous temperature series extending over 160 years is available for Minnesota (Baker *et al.*, 1985; Figure 2.34). With both these and the

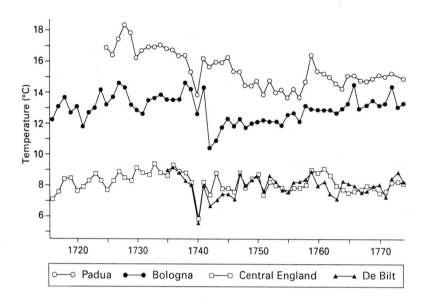

Figure 2.33   Comparison between historical temperature records from Bologna and Padua, and those from De Bilt (Netherlands) and Central England (after Comani, 1987)

European records, however, uncertainties surrounding the accuracy of early instruments, calibration with modern equivalents, and recording practices impose considerable constraints on the use of climatic data from early instrumental records (Ingram *et al.*, 1981).

Assessment of proxy data sources

Although each of the foregoing may be regarded as a tried and trusted means of reconstructing patterns of environmental change during the Late Quaternary, these different lines of evidence are all circumscribed by problems whose ramifications must be appreciated if meaningful inferences are to be made about former landscapes and climatic conditions. Some of these problems are explored in this section. It should be stressed at the outset, however, that the aim is not to undermine the methods themselves, nor to seek to invalidate the evidence upon which palaeoenvironmental reconstructions have been based. Rather, the intention is to encourage a critical and cautious approach when making inferences about past climates and environments based on proxy data sources.

Uniformitarianism

As was noted above (p. 14), uniformitarianism is fundamental to palaeoenvironmental reconstruction, and the idea that former environmental conditions can be interpreted on the basis of what is known about present-day physical, chemical and biological processes is almost universally accepted. Once uniformitarianism became established, it was a relatively straightforward matter to decipher the geological record in terms of contemporary earth-surface processes. However, interpretation of the biological (i.e. fossil) record on the basis of uniformitarian reasoning is more complicated because of the assumptions that need to be made about both the fossil evidence and the present state of knowledge of plant and animal ecology (Lowe and Walker, 1984). If contemporary fauna and flora are to form the basis for the interpretation of the ecology of their fossil counterparts, it is axiomatic that the environmental parameters that influence the distribution of present-day plants and animals are known. Equally, it must be assumed that

Temperature °C

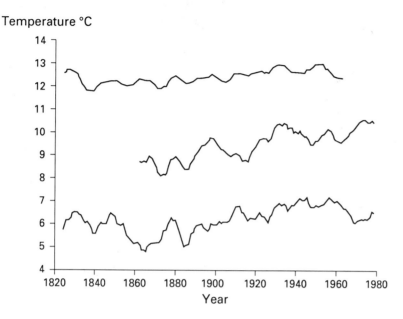

Figure 2.34   Time series of mean annual temperatures from the eastern United States (top), Iowa City (middle) and Minnesota (bottom). Data were smoothed by a normal curve smoothing function, and values are plotted at the midpoint of the smoothing interval (after Baker et al., 1985)

contemporary flora and fauna are in equilibrium with their environmental controls and that the same was true of those elements preserved in fossil assemblages. There is also the presumption that the ecological affinities of plants and animals have not changed through time and, moreover, that former plant and animal communities have present-day analogues. Furthermore, in view of the impact of human activity on natural ecosystems (Chapters 6 and 7), critical appraisal is necessary of the use of contemporary analogues in palaeoenvironmental reconstruction.

Equifinality

The term **equifinality** is employed in the earth sciences to refer to the notion that different processes can give rise to similar end products thus making it difficult to deduce causative process from the geomorphological and/or sedimentological evidence. However, Haines-Young and Petch (1983) have criticized the concept of equifinality as applied in this way, arguing that to describe landforms as being 'equifinal' more frequently reflects either a deficiency in understanding of process, or that landforms *appear* to be similar when, in fact, they are very different. In other

words, the concept of equifinality has been employed as an excuse for methodological and experimental shortcomings. As a consequence, questionable theories are being perpetuated because they are not being subjected to rigorous critical scrutiny and, in particular, competing theories are not being properly evaluated (p. 13). In this respect, they conclude, equifinality is an empty concept. Such a view does not find universal favour (Gerrard, 1984) and even Haines-Young and Petch are prepared to concede (1983: 465) that 'it is logically possible for similar forms to be produced by different processes'. Hence, although there may perhaps have been a tendency to acknowledge too readily equifinality as a problem in the historical earth sciences, awareness of the concept serves as a useful reminder of the difficulties that can arise when inferring process from form on the basis of proxy data sources.

Taphonomy

This concerns the formation processes of the fossil record and, in particular, the way in which living communities have been transformed into the fossil record by, for example, differential burial, preservation, etc. Interpretation of palaeoenvironmental evidence demands careful consideration of how the biota in question came to be incorporated into that context. Sometimes the assemblage taphonomy will be relatively simple, for example molluscs in a buried soil representing species that lived and died more or less on that site. The taphonomy of other assemblages will be more complicated, such as molluscs which have been eroded from soils and subsequently incorporated in valley sediments. Similarly, pollen grains deposited in lake sediments will have followed a variety of depositional pathways (wind, animals, inflowing streams, slopewash, etc.) and will derive from different source areas (local, extra local, regional, long distance). In many cases, it may not be possible to quantify these different components and, as a consequence, the taphonomic complexity of this type of fossil assemblage will impose constraints on subsequent palaeoenvironmental interpretations.

Some of the most sophisticated taphonomic studies have involved work on animal bone assemblages. Distinctive types of assemblage are produced by the activities of carnivores, raptors and bats, and it is possible to distinguish these from the patterning produced by human activity (Binford, 1981). Moreover, taphonomic work on bone assemblages has called into question fundamental assumptions about early hominids, indicating that they may have behaved more like scavengers than hunters (Binford, 1983).

Preservation and contamination

Fossil assemblages vary markedly in degree of preservation. Where anaerobic conditions have obtained since burial and where the sediment matrix is relatively fine, excellent assemblages of fossils which closely resemble the life assemblage may be recovered. Such conditions are rare and it is more common to find fossil assemblages that reflect the post-burial operation of physical and chemical processes. Where the sediment matrix consists of sand-size material, in riverine deposits for example, the more delicate biological remains will frequently be abraded or differentially destroyed. Similarly, if completely anaerobic

conditions are not achieved (e.g. in rapidly accumulating terrestrial peats), oxidation may affect the assemblage and again the more fragile and chemically susceptible elements (pollen grains, diatoms, etc.), will be destroyed. As a consequence the fossil assemblage will be biased towards the stronger, more robust elements, and may bear little relationship to the original living assemblage.

A related problem is contamination by both older and younger material. In peat profiles, for example, percolating groundwaters may carry microfossils down the profile, thereby introducing younger pollen, insects and other fossils into older assemblages. River gravels frequently contain assemblages of both animal bones and Palaeolithic artefacts whose components are of markedly different age reflecting successive episodes of fluvial erosion and deposition and the incorporation of earlier material into later sediments. Similarly, bioturbation (faunal disturbance) on a lake or sea bed may lead to admixtures of contemporaneous and previously deposited fossil assemblages. Problems of contamination are not confined to the interpretation of fossil assemblages, however, for they also affect the movement of archaeological artefacts (Schiffer, 1987). Sediments are also affected; for example, erosion around a lake shore may result in older material being reworked, thereby posing problems for stratigraphic interpretation. Reworking also occurs in hillslope contexts where colluviation, gelifluction (under periglacial conditions) and the processes of sheetwash and rillwash lead to the mixing of older and younger materials.

Climatic inferences from historical data

A number of problems are encountered in the interpretation of historical data as evidence of past climate (Bell and Ogilvie, 1978). With documentary sources, particular problems surround the reliability of subjective and impressionistic records such as diaries and annals. According to Ingram *et al.* (1981) the most serious errors are likely to arise from inaccurate dating; from spurious multiplication of events; from acceptance of distorted or amplified accounts; and from the inclusion of events for which there is no reliable evidence. As for phenological records, the major difficulty is establishing a cause-and-effect relationship between these data and climatic parameters (de Vries, 1981). Variations in crop yield, for example, may be a reflection of climatic change, but they may also have been affected by socio-economic factors including changing agricultural practices, fluctuations in market price, population change, disease or the effects of war. An unambiguous link between historical evidence and climatic change may, therefore, be difficult to establish (Parry, 1978).

Climatic inferences from other proxy data

Reconstructing former climatic conditions on the basis of proxy records from geological or biological sources is a far from straightforward process. Any climatic inference is two stages removed from the original evidence. Take, for example, the interpretation of climate from pollen analytical data. The first stage is to reconstruct former plant communities and vegetational patterns from the pollen evidence. Once that has been achieved, the second stage is to make a climatic inference based upon that reconstruction. In other words, an interpretation (of climate) is being made from an interpretation (of vegetation) and errors can be incurred at both stages in the analysis. In the case of pollen evidence, this problem is compounded by the fact that it is not always possible to identify pollen and spores to individual species. A pollen diagram, therefore, consists of a data bank at a variety of taxonomic levels and this clearly imposes constraints on the reconstructions of former plant communities and hence on climatic inference (Lowe and Walker, 1984). This problem of taxonomic imprecision also arises with other forms of biological evidence.

Because it is often so difficult to reconstruct

plant or animal communities from fossil evidence, climatic inferences tend to be based on particular **indicator species** which can be clearly identified in the fossil record and whose present-day ecological affinities are reasonably well known. A cause-and-effect relationship with climatic parameters is then often inferred. The problem here, however, is that climate is only one of a number of variables governing plant or animal distributions, and such factors as competition, migration rates, threshold levels and local habitat variations may be as important as regional climatic conditions in determining species distribution (Birks, 1981). A straightforward link between fossil occurrence and palaeoclimate may, therefore, be misleading.

A further difficulty in the interpretation of proxy data sources in the context of former climatic conditions concerns the timescale over which geological or biological response to climatic change takes place. Some proxy variables (e.g. Coleoptera) react swiftly to climatic change and the response may be measured over a matter of years. In other cases (e.g. glaciers and treeline fluctuations), there may be a lag in response, the duration of which may not only vary temporally and geographically, but also with the direction of change, i.e. whether the climate is ameliorating or deteriorating. The 'coarseness' of many proxy data records is an additional problem. Slow rates of lacustrine sedimentation, peat accumulation or colluviation, for example, mean that biological records such as pollen, plant macrofossils, diatoms and Mollusca will seldom be interpretable on a timescale of less than 50 years, and hence short-lived climatic variations may go undetected.

Overall, however, the most intractable difficulty in assessing Late Quaternary proxy records is that of distinguishing the climatic signal from the anthropogenic effect, and this problem underpins much of what is to be found on the following pages.

3

Natural Environmental Change

Introduction

The environmental impact of climatic change can be observed at a range of spatial and temporal scales. Of fundamental significance are the long-term global climatic shifts that operate over timescales of 10^4 to 10^6 years, and whose consequences are most spectacularly demonstrated by the repeated expansion and contraction during the course of the last 2 to 3 million years of the great continental ice sheets. Superimposed upon these macroscale climatic changes are short-term climatic fluctuations that occur over timescales of 10^1 to 10^3 years, and which are most notable for their effects on the vegetation cover of the Mid- and Late Holocene, on rates of operation of geomorphological processes, and on prehistoric and historic anthropogenic activity. Although this book is concerned primarily with environmental changes during the relatively recent past, the climatic regimes of the present interglacial reflect the operation of climatic processes over both long and short timescales. A proper understanding of recent climatic and environmental change can only be achieved by examining the way in which climate has fluctuated over a range of temporal scales. Hence, the patterns of both long- and short-term climatic changes, their causal mechanisms, and their impact on the landscape of the mid-latitude regions of the Northern Hemisphere, are examined in this chapter.

Patterns of long-term climatic change

A range of proxy data sources show that throughout the Tertiary period, the earth's climate has gradually deteriorated (Andrews, 1979), with the cooling trend being particularly marked during the Miocene and Pliocene (from *c.* 15 ma onwards). Moreover, pronounced oscillations increasingly became a feature of the global climatic pattern. Hence, although the prevailing climatic mode of the past 2.5 million years or so in the mid- and high-latitude regions of the world has been one of almost unremitting cold, markedly warmer climatic episodes have occurred at quasi-regular intervals. Thus, despite the popular conception of the Quaternary period as the Ice Age, it is now apparent that the **glacial** episodes, marked by expansion of the great continental ice sheets, were interspersed with **interglacial** phases during which global temperatures rose to be as high or even higher than those of the present day. There is also evidence to suggest that relatively short-lived climatic fluctuations occurred within the glacials and interglacials. A short cold episode which resulted in the local expansion of glaciers is usually termed a **stadial**, whereas a period of thermal improvement during a glacial episode when temperatures did not achieve levels comparable with those of the present day is referred to as an **interstadial**. Consequently, the pattern of long-term climatic change over the timescale of the past 2–3 million years is one of major oscillations between glacials and interglacials superimposed upon which are minor climatic fluctuations involving stadial and interstadial episodes.

In the low-latitude regions of the world and areas beyond the influence of glacier ice, cyclic changes in precipitation regimes occurred with phases of higher rainfall (**pluvials**) being interspersed with drier intervals or **interpluvials**. In very general terms, there appears to be a relationship between these climatic oscillations and the glacial/interglacial cycles of the mid- and high latitudes, the glacial episodes usually being equated with interpluvial intervals and the interglacials with pluvial phases (Lowe and Walker, 1984).

Evidence for long-term climatic change

For many years, the most widely used proxy evidence for long-term climatic changes was the sequence of glacial and other cold-climate deposits found throughout Europe and North America. Early investigations of these glacial sediments led to the development of the classical scheme of four glacial stages, Gunz, Mindel, Riss and Wurm in Europe (Penck and Bruckner, 1909), and Nebraskan, Kansan, Illinoian and Wisconsinan in North America (e.g. Chamberlin, 1895), while analysis of organic sediments often found interbedded with the glacial deposits provided evidence of intervening interglacial episodes. The problem, of course, with this type of data is that as the ice sheets covered more or less similar areas during successive cold stages, evidence for earlier glacial episodes has been largely removed by succeeding ice advances. Hence, the terrestrial record from the mid- and high-latitude regions is unlikely to constitute anything other than a partial record of long-term climatic change. Moreover, the fragmentary nature of the evidence presents difficulties both in correlating glacial deposits and in the recognition of separate glacial stages (Bowen, 1978).

The shortcomings of glacial evidence as a proxy record for Quaternary climatic change has been thrown sharply into focus by the evidence that has emerged in the last 20 years from the deep-ocean floors. By contrast with the terrestrial environment, sediments have been accumulating on the ocean floor in a slow but relatively uninterrupted manner throughout the Quaternary. Technological developments in deep-sea coring in the years immediately following the Second World War enabled undisturbed cores of sediment to be raised from the ocean bed in water depths exceeding 3 km

Stage boundary	SPECMAP timescale (Imbrie et al., 1987)	Martinson et al., (1987)	Williams et al., (1988)
	kyr BP	yr BP	kyr BP
1–2	12	1 2050 ± 3140	
2–3	24	2 4110 ± 4930	
3–4	59	5 8960 ± 5560	
4–5	71	7 3910 ± 2590	
5–6	128	12 9840 ± 3050	128
6–7	186	18 9610 ± 2310	194
7–8	245	24 4180 ± 7110	258
8–9	303		313
9–10	339		359
10–11	362		386
11–12	423		430
12–13	478		486
13–14	524		521
14–15	565		544
15–16	620		589
16–17	659		622
17–18	689		658
18–19	726		695
19–20	736		729
20–21	763		743
21–22	790		786
22–23			813

Table 3.1 Age estimates for stage boundaries of the marine oxygen isotope record of the past 800 kyr

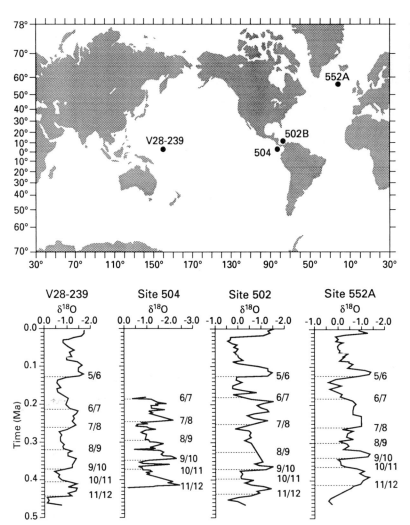

Figure 3.1   Oxygen isotope records for the last 470 ka from the Pacific and Atlantic Oceans. Horizontal lines indicate the positions of stage boundaries (after Williams et al., 1988)

(Imbrie and Imbrie, 1979). The oxygen isotope trace obtained from microfossils taken from successive levels within these cores (Figure 2.27) represents a continuous record of the changing isotopic composition of the ocean waters over the past 2–3 million years (see Chapter 2). As the isotopic signal from organisms that formerly inhabited the open oceans is largely a reflection of fluctuations in volumes of land ice (Shackleton and Opdyke, 1973), the isotopic trace provides a continuous record of changes in global ice volume which, by implication, can also be read as a record of glacial/interglacial fluctuations. Working from the top of the sequence, each isotopic stage has

been given a number, even numbers denoting 'cold' (glacial) episodes while the 'warm' (interglacial) phases are denoted by odd numbers, and attempts have recently been made to establish a chronology for these changes (Table 3.1). Twenty-two isotopic stages can be recognized in the record from approximately the past 800 ka (Imbrie et al., 1984) indicating something in the order of 10 interglacials and 10 glacials or near glacials during that time period. The total number of isotopic stages formally identified in Quaternary deep-ocean cores now numbers 63 (Ruddiman et al., 1986b; Williams et al., 1988), and the record has recently been informally extended to isotopic stage 90 (Kukla,

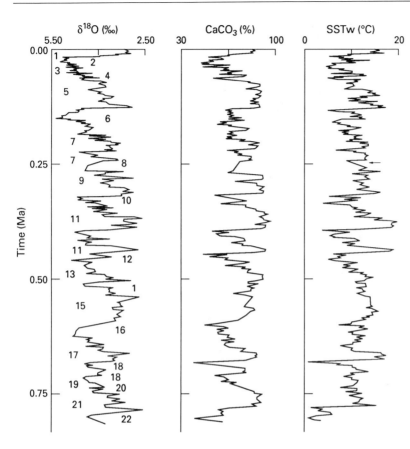

δ¹⁸O (‰)   CaCO₃ (%)   SSTw (°C)

Figure 3.2 Oxygen isotopic, CaCO₃ percent data and winter sea-surface temperature record for the past 800 ka in North Atlantic core V30/97 (41° 00′ N: 32° 58′ W) taken from a water depth of 3427 m. The numbers in the left-hand column represent isotopic stages (after Ruddiman et al., 1986c)

1987). The most impressive feature of the deep-sea oxygen isotope record is that the pattern revealed by the isotopic trace is geographically consistent, and can be replicated in cores taken from different oceanic areas (Figure 3.1). Clearly, therefore, the isotopic record provides a climatic signal of global significance.

In addition to the isotope record, other indicators of long-term climatic change have also been obtained from ocean sediments. These include estimates of former sea-surface temperatures (SSTs) based on fossil assemblages of planktonic Foraminifera and other marine micro-organisms, and downcore variations in CaCO₃ concentrations (Figure 3.2). Relatively low concentrations of CaCO₃ are found in deep-ocean sediments during glacial episodes due to the input into the oceans of large amounts of non-carbonate ice-rafted debris. By contrast, higher concentrations occur during interglacials reflecting

the higher productivity rates of carbonate-secreting plankton (Ruddiman *et al.*, 1986b). The similarity between the SST and CaCO₃ curves on the one hand and the oxygen isotope trace on the other (Ruddiman *et al.*, 1986c; Ruddiman and Raymo, 1988) underlines the significance of these additional parameters as proxy records of long-term climatic change. Although the evidence from the deep-ocean floors provides the touchstone for reconstructions of climatic change over the timescale of the Pleistocene, other forms of proxy climatic data from non-glaciated areas have been correlated with the marine oxygen isotope record. These include the lithostratigraphic and biostratigraphic evidence from deep tectonic basins in which sediment has been accumulating for several million years, and which provide a continuous record of climatic fluctuations extending back into the Tertiary (Cooke, 1981; Hooghiemstra, 1984); long sedimentary sequences spanning the entire Quaternary period from deep lake basins in areas such as Japan, Australia, New Zealand, West Africa and the south-west United States (Horie, 1979; Fuji, 1988) and the successions of loess and interbedded palaeosols that are found in eastern Europe, parts of China and Soviet Central Asia (Catt, 1988) which provide evidence of alternating cold (represented by loess) and warm stages (indicated by soil development) extending back over at least 17 glacial/interglacial cycles (Fink and Kukla, 1977). Palynological data showing oscillations throughout the Quaternary between episodes of woodland development which reflect interglacial conditions, and heath or tundra environments (stadial or glacial conditions) have been synthesized from stratigraphic successions in East Anglia (West, 1980) and The Netherlands (de Jong, 1988) to provide a coherent picture of climatic change extending back some 2.4 million years (Figure 3.3). Moreover, in recent years, painstaking analysis of glacigenic sedimentary sequences coupled with technological advances both in the investigation of sub-surface stratigraphy and in techniques of dating, have led to the development of regional glacial stratigraphies which can be related to the global framework of climatic change provided by the deep-sea oxygen isotope record (Šibrava et al., 1986).

The nature of long-term climatic change

The oxygen isotope, SST and $CaCO_3$ records from deep-ocean sediments provide the most detailed indications of the nature and timing of long-term climatic change. The evidence suggests that over the past 735 ka, the global climate has been fluctuating in a rhythmical manner in a series of cycles, each of around 100 ka duration (Ruddiman and Raymo, 1988). Approximately 90 per cent of each cycle was characterized by glacial conditions in the mid- and high latitudes, with conditions as warm as those of the present interglacial accounting for only around 10 per cent of each glacial/interglacial cycle. Contrasts between full glacial and interglacial conditions were pronounced,

involving annual temperature changes in excess of 15 °C, and marked variations in annual precipitation (Lamb, 1977). Prior to c. 735 ka BP, the cyclical fluctuations in climate appear to have been of lower amplitude, however, with a periodicity of around 41 ka (Ruddiman et al., 1986a). Possible reasons for this shift in climatic pattern are considered below.

A striking feature of the long-term climatic record is the relative rapidity of climatic change (Flohn, 1984). In the isotopic trace from the deep-ocean cores (e.g. Figure 3.1), the glacial stages are typically marked by abrupt **terminations**, i.e. sharp transitions which have carried the earth from maximum glacial to maximum interglacial conditions on a timescale of 7 ka or less (Broecker, 1984). Using faunal and floral evidence from deep-ocean cores, Ruddiman et al. (1977) suggested global warming at the beginning of the last interglacial in the North Atlantic region of around 5.2 °C/1000 years. Even more rapid rates of climatic change have been inferred from terrestrial evidence. Pollen analytical data from the Grande Pile peat bog in northern France, for example, implies that at the end of the last (Eemian) interglacial, temperate forests were replaced by pine, spruce and birch taiga within the space of 150 ± 75 years (Woillard, 1979). In Britain, coleopteran evidence suggests that the rise in mean annual temperature at the beginning of the Lateglacial Interstadial around 13 ka BP (see below) may have been as much as 7.2 °C per century (Atkinson et al., 1987), while similar rapid rates of climatic amelioration at c. 10.2 ka BP are indicated by coleopteran evidence from southern Sweden (Lemdahl, 1991). Most dramatic of all, however, are the heavy isotope and dust concentration profiles from Greenland ice cores. These suggest that at the end of the last cold stage, temperatures in southern Greenland rose 7 °C within the space of 50 years and, moreover, that the climate of the North Atlantic region as a whole changed to a milder and less stormy regime within the space of no more than 20 years (Dansgaard et al., 1989)! Such abrupt climatic changes must have had profound effects on the biosphere and, in particular, on human communities (Chapter 5).

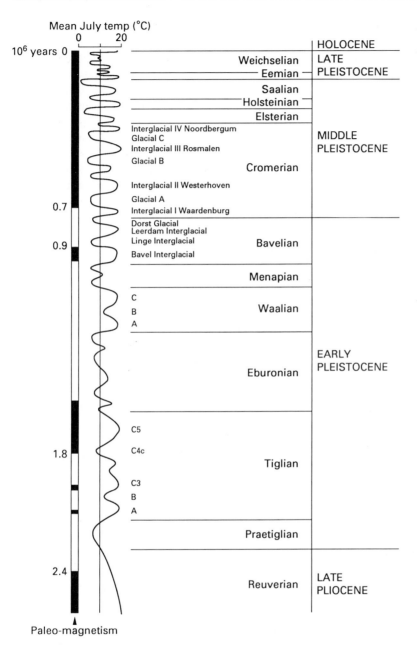

Figure 3.3   Climatic curve and chronostratigraphy of the Netherlands (after de Jong, 1988)

Climatic changes during the last cold stage

The chronology established on the basis of the deep-ocean record suggests that the last cold stage began 115-120 ka BP and ended around 10–11 ka BP (Imbrie *et al.*, 1984). The isotopic trace indicates that within this 100 000 year period, general global cooling, short-lived warm episodes occurred at *c.* 103 ka, 79 ka, 55 ka and 50 ka (Martinson *et al.*, 1987). A number of warm intervals (interstadials) have been recognized in the terrestrial record, although their chronology is less certain as much of the last cold stage lies beyond the range of radiocarbon dating. Nevertheless, similarities are apparent between the marine and terrestrial records, particularly for the middle and later parts of the last cold stage (Woillard and Mook, 1982).

Europe

In north-west Europe, the last cold stage is referred to as the **Weichselian**, while in central and southern Europe it is known as the **Würm**. In Scandinavia, northern Germany and Denmark, sequences of glacial deposits provide evidence of ice advances during the Early and Late Weichselian, the latter being radiocarbon dated to the period between *c.* 21 ka and 13 ka BP (Ehlers, 1983). Similar patterns of glacier activity have been identified in the mountains of central and southern Europe (Šibrava *et al.*, 1986). Beyond the limits of these glaciers and ice sheets, detailed records of the pattern of climatic changes during the last cold stage have been found in The Netherlands, Belgium, and parts of north-west Germany in the form of sedimentary sequences of lacustrine, fluvial and aeolian deposits interrupted by periglacial

structures (West, 1988). In addition, long sequences of organic sediments in deep-lake sites in northern and eastern France have produced virtually continuous pollen records extending back into the last (Eemian) interglacial (Woillard, 1978; de Beaulieu and Reille, 1984). Climatic curves from the last cold stage (Figure 3.4) indicate that, while the prevailing temperatures of the last 100 ka remained well below those of the present day and the environment in northern and western Europe was one essentially of polar desert conditions, significant ameliorations of climate developed at quasi-regular intervals. The warmest episodes occurred during the Early Weichselian/Würm in the Odderade, Brörup and Amersfoort Interstadials of The Netherlands and northern Germany when temperatures appear to have been only marginally below those of the present day. These warm intervals may well be the equivalent of the St Germain I and II Interstadials (the former including two warm episodes separated by a brief deterioration) in the French pollen records from Grande Pile and Les Echets (Guiot *et al.*, 1989). These early interstadials lie beyond the range of radiocarbon dating, but correlation with the deep-

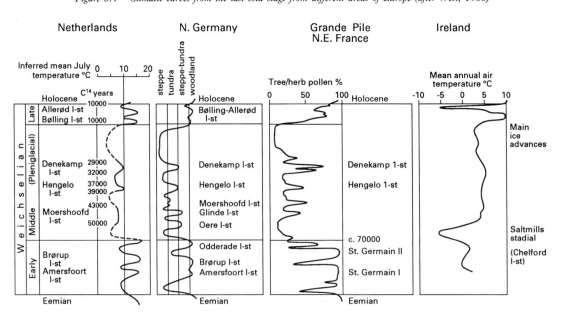

Figure 3.4   Climatic curves from the last cold stage from different areas of Europe (after West, 1988)

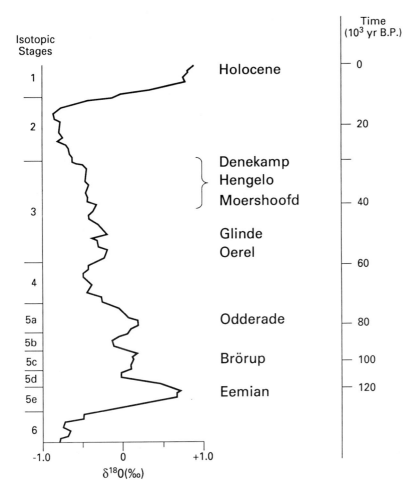

Isotopic
Stages

Holocene

Denekamp
Hengelo
Moershoofd

Glinde
Oerel

Odderade

Brörup

Eemian

$\delta^{18}O(‰)$

Time
$(10^3$ yr B.P.)

Figure 3.5 Correlation between oxygen isotope stages and Weichselian Interstadial episodes in north-west Europe (after Behre, 1989)

sea record suggests ages of *c.* 100 ka BP for the Amersfoort/Brörup and *c.* 80 ka BP for the Odderade warm episodes (Figure 3.5). Well-documented later interstadials include those at Moershoofd, Denekamp and Hengelo, radiocarbon dated in The Netherlands to *c.* 45 ka, 39 ka and 32 ka BP respectively (Behre, 1989), and which may also be present in the Grand Pile pollen record (Woillard and Mook, 1982). By far the most widely recorded episode of climatic warming, however, is the Allerød Interstadial which has been recognized throughout northern and western Europe, and which has been dated to *c.* 13–11 ka BP.

British Isles

The pattern of climatic change during the last **(Devensian)** cold stage in Britain is based on evidence from fossil Coleoptera (Coope, 1977b), supported by a range of sedimentary and other biological data (Lowe and Walker, 1984). There are indications of three warmer episodes, the Chelford, Upton Warren and Lateglacial/Windermere Interstadials, although an additional interstadial that occurred between the Upton Warren and Chelford may be represented in the deposits at Brimpton in Berkshire (Bryant *et al.*, 1983). Radiocarbon dates for the Upton Warren and Lateglacial Interstadials suggest ages of *c.* 43 ka and 13–11 ka BP respectively, while the oldest of a number of radiocarbon dates on the Chelford Interstadial is *c.* 60 ka BP. The view that this represents a minimum date for the interstadial (Worsley, 1980) has recently been confirmed by TL (thermoluminescence) dating which suggests an age in the range 90–100 ka BP for the climatic amelioration (Rendall *et al.*, 1991). Hence, the Chelford Interstadial may well be the equivalent of the Brörup of continental Europe and oxygen isotope stage 5c of the marine sequence (Behre, 1989). The evidence points to summer temperatures of around 15 °C during the Chelford Interstadial (1–2 °C lower than the present day), but reaching 18 °C in the Upton Warren and Lateglacial Interstadials. During the intervening cold periods, summer temperatures remained below 10 °C, with winter temperatures as low as −3 °C and mean annual temperatures typically in the range −8 to −12 °C. For much of the Devensian, large areas of the British Isles resembled arctic tundra, only the northern and western regions being directly affected by glacier ice. Major ice sheets developed in Highland Britain during the early Devensian, but the main phase of glacier expansion occurred during the Late Devensian after *c.* 25 ka BP (Bowen *et al.*, 1986).

Northern United States, Canada and the Arctic

Some of the most detailed proxy records for changing climatic conditions over the past 100 ka or so in the mid- and high-latitude regions of the Northern Hemisphere have been obtained from $\delta^{18}O$ records in ice cores from Camp Century (Johnsen *et al.*, 1972) and Dye (Dansgaard *et al.*, 1982) in Greenland (Figure 2.28), and from the ice cap on Devon Island in the North West Territories of Canada (Paterson *et al.*, 1977; Koerner and Fisher, 1981). The isotopic trends in the cores are broadly similar and show that the main episodes of ice accumulation occurred in the intervals 125–115 ka, 80–60 ka and 40–30 ka BP. These are the intervals during which the deep-sea record indicates substantial build-up of glacier ice (Dansgaard *et al.*, 1982). Close parallels also exist between the isotopic traces in the ice cores and

other proxy data (Figure 3.6).

The terrestrial evidence from Canada and the northern United States finds close parallels with the data from the ice cores. The main phases of expansion of the Laurentide ice sheet, for example, are considered to have occurred around 115 ka, 75 ka and 25 ka BP, with the main phase of ice accumulation in the western cordillera beginning around 75 ka BP (Andrews and Barry, 1978). Important interstadial episodes were the St Pierre Interstade at around 85 ka BP and the long Port Talbot Interstadial dated to between *c.* 70 and 40 ka BP during which the Laurentide ice sheet retreated to a northern core area similar to that which existed around 90 ka BP (Barry, 1983), although a large ice mass may have persisted throughout the Middle Wisconsin over Quebec and Labrador (Andrews, 1987). A range of data suggests that interstadial conditions occurred quite widely in the time interval 22–19 ka BP (Barry, 1984), and there appear to have been several interstadial episodes during the retreat of the Laurentide ice sheet (e.g. the Erie Interstade 16.1–15.5 ka; the Two Creeks Interstade *c.* 11.8 ka) following the glacial maximum around 18 ka BP. Deglaciation may have begun in the Great Lakes region as early as 17 ka BP, but around the northern and eastern margins of the ice sheet glacial retreat from the maximum appears to have been delayed until around 12–8 ka BP (Andrews, 1987). Speleothem data from caves in west Virginia indicate rapid warming around 100 ka BP, with abrupt cooling trends 75–60 ka BP and 30–6 ka BP (Thompson *et al.*, 1974), while similar evidence from caves in Iowa indicates cold conditions from around 70 ka BP with a marked thermal maximum at 60 ka BP and further episodes of climatic warming around 30 and 20 ka BP (Harmon *et al.*, 1978). Temperatures during the early Port Talbot Interstadial in the north-eastern USA may have been comparable with those of the present day, although the vegetational evidence suggests somewhat lower temperatures (3–4 °C) during the later interstadial (Berti, 1975). Estimates of temperature reductions during the last glacial maximum based on proxy data sources range from 3 to 8 °C for the United States as a whole (Barry, 1983), with the largest reductions of 12–15 °C

Figure 3.6   Different forms of proxy climatic data for the period 140 ka to present. The vertical axes are dimensionless, but are drawn to express climate in terms of increasing warmth upwards. Hence, global ice volume is shown in inverse with increasing volume downwards (after Koerner & Fisher, 1986)

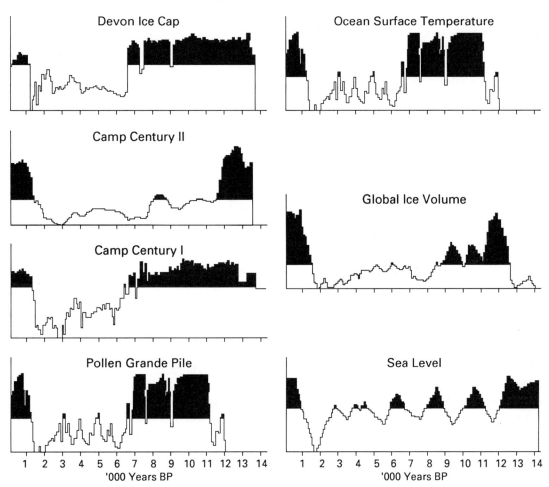

suggested for areas around the margins of the Laurentide ice sheet (Washburn, 1980), and 14–17 °C in parts of the Rocky Mountains (Porter *et al.*, 1983). Slightly lower temperature reductions (5–10 °C) for the area immediately to the south of the Laurentide ice sheet have been inferred on the basis of atmospheric modelling (Kutzbach, 1987).

Causes of long-term climatic change

The cause of global climatic changes over

timescales of 10^4 to 10^6 years has been one of the longest standing and intractable of scientific problems. With the introduction and ultimate acceptance of the glacial theory in the middle years of the nineteenth century, added impetus was given to the search for an explanation of long-term climatic changes of such magnitude as to carry the temperate regions of the world from full glacial to interglacial conditions within what, in geological terms at least, appeared to be a relatively short timespan.

Numerous hypotheses which seek to explain

long-term climatic change have been proposed over the years. Some theories relate to geographical changes on the earth's surface (**terrestrial theories**), including mountain building episodes, changes in the patterns of oceanic circulation and changes in the disposition of the continental land masses (continental drift). Such geographical changes will inevitably affect the distribution of heat across the surface of the earth and hence, it has been argued, could lead to climatic change. Other theories involve possible changes in the earth's atmosphere (**atmospheric theories**), such as variations in the content of carbon dioxide and other gases, changes in the amounts of water vapour, and the occurrence of volcanic dust clouds. These will influence the transmission of solar radiation through the atmosphere, and hence the amount of solar radiation received at the earth's surface. A third group of theories (**solar theories**) involves structural changes that may occur in the sun itself, perhaps leading to variations in the output of radiant heat and hence temperature fluctuations on the earth.

Most of these theories were developed before 1960, and their essentially speculative nature reflects the fact that prior to that date, little firm evidence could be adduced in support of any of them. The chronology of long-term climatic change was uncertain, few reliable long proxy records existed, and the state of knowledge of atmosphere–ocean–terrestrial linkages was relatively rudimentary. Indeed the data base was so limited that it was almost impossible to test the large number of theories that were then proliferating. During the 1960s and 1970s, however, technological improvements and innovations in a number of different areas of Quaternary research prompted a re-evaluation of the nature and causal mechanisms of long-term climatic change. Of particular significance was the development of coring equipment which, for the first time, allowed long sedimentary records to be recovered from terrestrial sites such as deep-lake basins and, more importantly, from the ocean floors where undisturbed sediment sequences spanning the entire Quaternary period were to be found. The discovery of oxygen isotope variations in the

microfossils of deep-ocean sediments (Chapter 2) provided a continuous proxy climatic record, while technical developments in dating techniques, most notably palaeomagnetism, enabled a chronology of long-term climatic change to be established. The deep-sea record suggested a periodicity in long-term climatic change, i.e. cyclical fluctuations in climate over a long time period, that had been suspected but which had hitherto never been clearly demonstrated. This led to a resurgence of interest in the notion that the answer to the problem of the cause of climatic change lies not in the earth nor in its atmosphere, nor indeed in the sun, but rather in the way in which the earth moves around the sun. This is known as the **astronomical theory** of climatic change.

The astronomical theory and its elements

The idea that climatic changes may be triggered by variations in the earth's orbit and axis was first suggested by J. F. Adhemar in 1842 (Imbrie and Imbrie, 1979). The theory was further developed by James Croll in the 1870s, but was really given substance by the Yugoslavian geophysicist Milutin Milankovitch who, in the 1920s, produced a substantial body of data to show that orbital and axial changes (which could be detected by astronomical observations and which had, in fact, been known for some time) would lead to variations in the amounts of radiant energy received at the earth's surface and, moreover, that these data could be depicted graphically to show how insolation at different latitudes had varied over time (Figure 3.7). Milankovitch also suggested that a link might reasonably be postulated between such fluctuations and glacial episodes. Until the early 1970s, however, the astronomical theory was largely disputed, principally because the quasi-periodicities of the earth's orbital elements could not be identified in the relatively fragmentary geological records then available, but also because of widespread uncertainty within the scientific community about the reliability of the long-term variations of the earth's orbital elements and the predicted insolation effects. Doubts were also expressed about whether a correlation could be

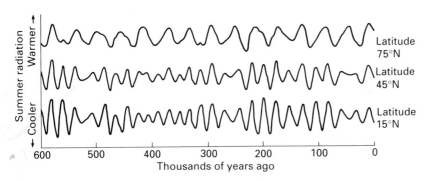

Figure 3.7 Milankovitch radiation curves for different latitudes in the Northern Hemisphere. First published in 1938, these show changes in summertime radiation at 75°N, 45°N and 15°N (after Imbrie & Imbrie, 1979)

detected between the insolation curves and geological data, and also whether the predicted insolation changes could have induced climatic changes of the magnitude of those evident in terrestrial, geological and biological records (Berger, 1980).

In 1976, however, Hays *et al.* convincingly demonstrated that the astronomical frequencies were present in the isotopic record from the deep oceans and it is now apparent that variations in the earth's orbit and axis are also reflected in other geological and biological data (see below). In essence, a scientific revolution has occurred to the extent that all other theories of long-term climatic change have been relegated to a secondary role, with variations in the earth's orbit and axis now widely accepted as the 'Pacemaker of the Ice Ages'. The elements of the astronomical theory are discussed below.

The precession of the equinoxes

The earth revolves around the sun in an elliptical orbit (Figure 3.8A) and each complete revolution takes one year. At times, therefore, during the course of a year the earth passes closer to the sun than at others. The period of closest passage is known as **perihelion** and the time of year when the earth is furthest from the sun is referred to as **aphelion**. Clearly, when the earth is in perihelion (presently *c.* 147 million km (91.5 million miles) from the sun) there will be a greater intensity of solar radiation receipt than in aphelion (152 million km (94.5 million miles)). However, because the earth is tilted on its axis (currently 23.5°), the heat received by the earth will not be uniformly

distributed, but will vary depending on whether the Northern or Southern Hemisphere is tilted towards or away from the sun. At present, the Northern Hemisphere summer occurs in aphelion and the Southern Hemisphere summer in perihelion. Hence, Southern Hemisphere summers will be hotter than those in the Northern Hemisphere. By contrast, Northern Hemisphere winters (perihelion) should be milder than those experienced in the Southern Hemisphere (aphelion).

This situation is not immutable, however, for because of the gravitational pull exerted on the earth's equatorial bulge by the sun and the moon, the earth wobbles on its axis like a top (Figure 3.8A). As a consequence, the direction of tilt varies over time to the extent that after *c.* 10.5 ka it has become completely reversed. The effect of this axial change is that the Northern Hemisphere summer now occurs in perihelion and the Southern Hemisphere summer in aphelion. Seen from above (Figure 3.8A), the seasons seem to move around the sun in a regular fashion, hence the term **precession of the equinoxes** or **precession of the solstices**. As a consequence, there is a regular cyclical change in patterns of heating and cooling across the globe. After 21 ka or so, the cycle is complete and a new one begins. In recent years, however, it has been established that although the precessional cycle averages around 21 ka, there are in fact two separate and interlocked cycles, a major one at around 23 ka and a minor one at *c.* 19 ka (Berger, 1977).

The obliquity of the ecliptic

The second element in the astronomical theory

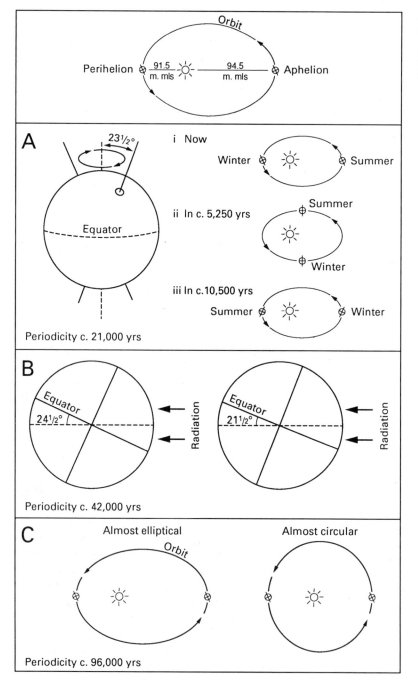

*Figure 3.8   The three
components of the Astronomical
Theory of climate change:
(A) Precession of the equinoxes;
(B) Obliquity of the ecliptic;
(C) Eccentricity of the orbit*

(**obliquity of the ecliptic**) involves changes in the angle of tilt of the earth relative to the plane of the ecliptic (Figure 3.8B). Over time, the tilt will vary from 21° 39′ to 24° 36′ and any change in tilt will heighten or lessen seasonal contrasts. The greater the tilt, for example, the more pronounced will be the differences between summer and winter. Changes in axial tilt occur with a periodicity of around 41 ka. In other words, it takes around 41 ka for the tilt of the earth to change from *c.* 24.5° to 21.5° and back to 24.5° once more.

The eccentricity of the orbit

As already noted, the earth's orbit around the sun is not circular but elliptical, a major consequence of which is that summers and winters in the two hemispheres will be of unequal length. Over time, however, due to planetary gravitational influences, the shape of the orbit will change from being markedly elliptical to less so (Figure 3.8C). The greater the eccentricity, the more pronounced will be the differences in solar radiation receipt and hence summer/winter contrasts. The **eccentricity of the earth's orbit** is, therefore, the third element in the astronomical theory. The periodicity, i.e. the time taken for the orbit to change from one point of greatest eccentricity to the next, is around 100 ka, although again some variation is apparent with cycles of *c.* 95 and 123 ka being identified (Berger, 1978a).

These variables, either singly or in combination, will exert a profound effect, *inter alia*, on spatial variations in solar radiation receipt at the earth's surface, seasonal intensity and duration, and the initiation and development of warm and cold episodes. It is important to stress, however, that only eccentricity will affect the total amount of solar radiation received by the earth; the other astronomical variables contribute simply to a redistribution of that energy at different latitudes (Berger, 1978b). However, because they are interlinked, each will exert an effect on the others, The effects of precessional cycles, for example, on spatial and temporal variations in solar radiation receipts will be modulated by orbital changes, for winter/summer contrasts will be more pronounced at times of maximum eccentricity than during

episodes when the shape of the earth's orbit is less elliptical. Similarly, the greater the angle of axial tilt, the more accentuated will be the seasonal contrasts and the more important will be the seasonal timing of perihelion. Overall, it would appear that in the low and middle latitudes, solar energy receipt and distribution is affected principally by eccentricity variations and by precession, whereas in the higher latitudes the effects of eccentricity are amplified by changes in obliquity.

Evidence in support of the astronomical theory

In the course of the past 15–20 years, these astronomical variables have been detected in a range of proxy records. These include the following.

Deep-sea cores

Because long time-series data can be obtained from deep-ocean cores, the majority of research into linkages between the astronomical variables and terrestrial climatic changes has been focused on these proxy records (Imbrie and Imbrie, 1979). The data suggest that, for the past 730 ka at least, the earth's climatic rhythms have been dominated by the 100 ka eccentricity cycle, modulated or amplified by cycles at 19, 23 ka (precession) and 41 ka (obliquity) intervals (Figure 3.9).

Indeed, following a detailed statistical analysis of the oxygen isotope trace in five ocean cores from the Atlantic, Pacific and Indian Oceans, Imbrie *et al.* (1984) concluded that the Milankovitch variables would explain at least 77 per cent of the amplitude of the $\delta^{18}O$ variations observed in these cores. Other marine parameters such as biological and lithological indicators of surface-ocean variability may also be linked to astronomical influences (Ruddiman and Raymo, 1988). The marine data, therefore, would seem to confirm the hypothesis that the climatic fluctuations of the Late Quaternary are driven by changes in the earth's orbit and axis; in other words, what has become known as **orbital forcing** is the primary

Figure 3.9   Variations in obliquity and precession corresponding to frequency components of the δ¹⁸O record over the past 800 ka. Dashed lines indicate phase-shifted versions of obliquity (a) and precession (b) curves. Solid lines are filtered versions of the δ¹⁸O record (after Imbrie et al., *1984)*

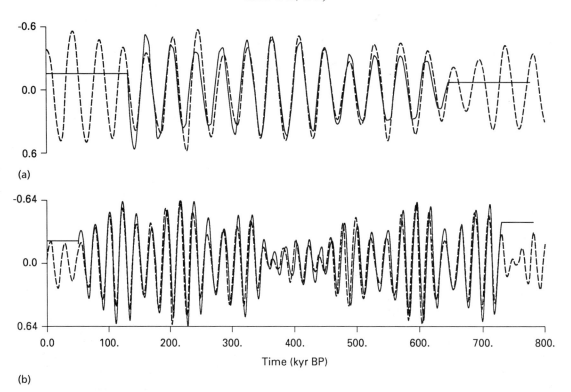

(a)

(b)

Time (kyr BP)

mechanism behind long-term climatic change (Imbrie, 1985).

Coral reef sequences

In Barbados, coral reefs formed during high stands of sea level (global warming?) were dated to around 82 ka, 105 ka and 125 ka, and a possible link was inferred with precessional cycles (Mesolella *et al.*, 1969). Similar dates of 81 ka, 108 ka and 130 ka BP have subsequently been obtained from raised coral terraces in Haiti (Dodge *et al.*, 1983). Data from New Guinea indicate successive reef terrace development at around 130, 107, 85, 60, 45, 40, 29 and 10 ka, the early and later parts of the record corresponding closely with enhanced seasonal insolation (higher summer and lower winter insolation than at present) predicted by the astronomical variables (Aharon, 1984).

Pollen data

Spectral analysis of the 130 ka pollen record from Grande Pile in northern France suggests a close statistical relationship between fluctuations in the herb and pine pollen curves and precessional periodicities (23 and 19 ka). This, in turn, implies a response by continental vegetation to orbital forcing (Molfino *et al.*, 1984). At Lake Biwa in Japan, time-series of palaeotemperature fluctuations over the past 500 ka derived from palynological data show peaks in the spectra at 104 ka, 44 ka, 25 ka and 12.7 ka. It is suggested that the three longer periods correspond with eccentricity, obliquity and precession respectively (Kanari *et al.*, 1984).

Loess sequences

A quasi 100 ka periodicity is apparent in the central Chinese loess record of the past 450 ka (Kukla,

1987), while further east in China spectral analysis of loessic deposits that have accumulated over the past 700 ka shows peaks in the curves at around 41.7 and 25 ka (Lu, 1981), corresponding perhaps to the obliquity of the ecliptic and precession of the equinoxes respectively (Goudie, 1983). Periodicities of *c.* 20 ka and 40 ka have also been detected in loess/palaeosol sequences from Kashmir (Gupta *et al.*, 1991).

Ice cores

The occurrence of *c.* 21 and 41 ka cycles has been detected in long (150–160 ka) oxygen isotopic profiles in ice cores from the Soviet Antarctic station at Vostok in East Antarctica (Genthon *et al.*, 1987; Lorius *et al.*, 1990).

Tropical lake levels

Significant variations in tropical lake levels during the Late Quaternary are reflected in a range of geological evidence, and these have been linked to changes in precipitation regimes (Street and Grove, 1979). An atmospheric general circulation model, which simulates climate over the past 18 ka, suggests that the episodes of increased precipitation apparent in the lake level records are associated with periods of strengthened monsoonal circulation resulting from cyclical changes in the pattern of Northern Hemisphere insolation (Kutzbach and Street-Perrott, 1985). This, in turn, implies a forcing of fluctuations in tropical lake levels by the astronomical variables.

Recent developments associated with the astronomical theory

Although a considerable body of empirical evidence now exists to support the hypothesis that orbital forcing is a fundamental causal component of climatic change during the course of the Quaternary, it is becoming apparent that the astronomical variables are not the only elements in the equation. The ocean core record, for example, suggests that the initiation of moderate-sized ice sheets in the Northern Hemisphere occurred around 2.4 ma (Shackleton *et al.*, 1984) following a progressive, but oscillatory, deterioration in climate

from around 3.15 ma BP. Neither of these is explicable in terms of astronomical forcing alone. Furthermore, recent isotopic, geochemical and biological evidence from the North Atlantic suggests that during the course of the past million years, the climatic cycles of the earth have shifted from a periodicity of around 41 ka to a prevailing rhythm of *c.* 100 ka (Ruddiman *et al.*, 1986b). This change, which became progressively marked between *c.* 900 and 450 ka BP, was accompanied by an apparent intensification of glaciation with the growth of Northern Hemisphere ice sheets to maximum volumes, considerably larger than those attained during the previous 1.6–1.7 ma of the Quaternary. In addition, numerous glacial advances at periods of 41 ka and 23 ka were superimposed on the basic 100 ka cycle (Ruddiman and Raymo, 1988). Again, these changes in periodicity and intensity of climatic cycles cannot be accounted for solely by orbital forcing.

It would appear, therefore, that while long-term climatic changes are largely explicable in terms of the astrononomical variables, other factors serve to modulate or amplify their effects. A widely held view is that global cooling (which can be traced back to the Cretaceous) results from changes in the distribution of the continents accompanied by tectonic uplift, with tectonic changes in particular exerting a profound influence on global climatic developments over the past 3 ma. The closing of the Panama Isthmus, which would have prompted major changes in the oceanic heat and moisture fluxes in the North Atlantic, occurred around 3.5–3.0 ma BP (Keigwen, 1978), close to the beginning of long-term climatic deterioration at 3.15 ma BP. Of greater significance, however, may have been the amount of tectonic uplift that occurred after that time. The Himalayas, for example, have risen by over 3000 m during the past 2 ma (Liu *et al.*, 1986), and significant uplift has also occurred in North America. Climatic modelling suggests that such a marked increase in areas of high land in the middle latitude regions would enhance albedo temperature feedback on a global scale and could lead to significant cooling especially during the autumn and winter seasons (Birchfield and Weertman, 1983). An alternative

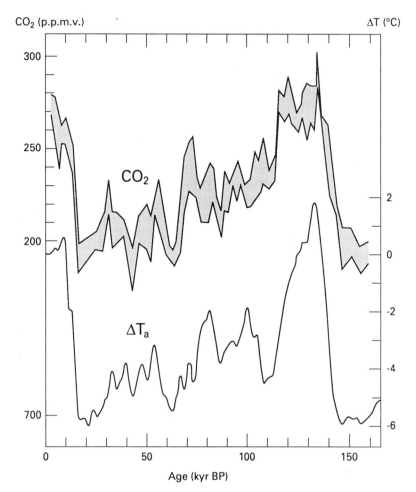

CO$_2$ (p.p.m.v.)

ΔT (°C)

Figure 3.10   Atmospheric variations during the last climatic cycle, as derived from measurements along the 2083 m Vostok ice core, of carbon dioxide concentrations (CO$_2$), and temperature (T) (after Lorius et al., 1990)

Age (kyr BP)

explanation for the initiation and intensification of Northern Hemisphere glaciation is that mountain uplift in Asia and North America has altered the wave structure in the airstreams of the upper atmosphere, the effects of which have been to cool the Eurasian and North American land masses and increase their sensitivity to orbitally driven insolation changes (Ruddiman *et al.*, 1986b; Ruddiman and Raymo, 1988). This hypothesis involves key elevation thresholds which, once attained, could explain both initiation of early Quaternary glaciation and subsequent intensification during the Mid-Quaternary.

Recent work also suggests that the effects of the astronomical variables on global climate could be amplified by changes in atmospheric CO$_2$

(Sundquist, 1987). Evidence from both deep-sea sediments and from polar ice cores indicates that during the last glacial period, atmospheric CO$_2$ levels were significantly lower than during either the present or last interglacials (Shackleton *et al.*, 1983; Barnola *et al.*, 1987). The most detailed record has been obtained from the Vostok ice core in Antarctica (Figure 3.10) which shows dramatic shifts from about 190 to 280 ppmv at the end of the recent and penultimate glacial periods. Moreover, spectral analysis of the CO$_2$ trace reveals a well-defined peak at around 21 ka and a secondary, less well-developed peak, at *c.* 41 ka (Barnola *et al.*, 1987). Similar trends are also apparent in the curve for methane (CH$_4$) (Lorius *et al.*, 1990). The close parallels between the CO$_2$ and temperature profiles

Figure 3.11 Movement of the North Atlantic Polar Front during the Late Devensian/Weichselian Lateglacial period (after Ruddiman & McIntyre, 1981a)

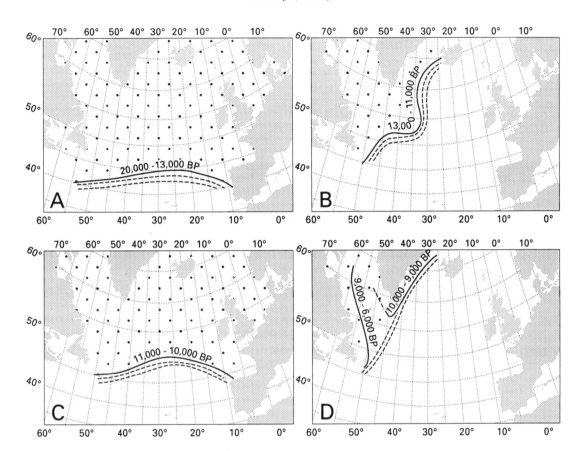

from the Vostok core has prompted the suggestion that CO_2 is a dominant forcing factor in long-term climatic change (Genthon *et al.*, 1987). The results of climatic modelling using data from deep-ocean cores also indicates that CO_2 changes may represent a part of the mechanism by which orbital variations induce changes in climate (Pisias and Shackleton, 1984). The factors underlying these long-term variations in atmospheric CO_2 remain unclear, although they may be related to glacial/interglacial changes in patterns of deep-ocean circulation (Broecker *et al.*, 1985; Broecker and Denton, 1990). Whatever the causes, these new data point very strongly to the importance of CO_2, and perhaps also other trace gases such as CH_4, in the system of climatic feedbacks that amplify the direct effects of insolation changes resulting from orbital forcing.

Patterns of short-term climatic change

Short-term climatic changes over timespans of thousands of years or less are most clearly marked at the end of the last cold stage and during the Holocene. This section examines the pattern and chronology of climatic change over the course of the present interglacial, beginning with the marked oscillation in climate recorded throughout north-west Europe at the close of the

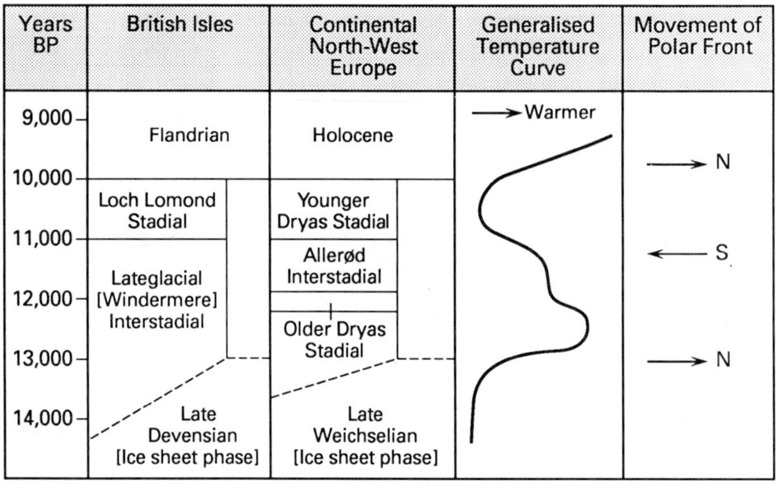

Years BP	British Isles	Continental North-West Europe	Generalised Temperature Curve	Movement of Polar Front
9,000	Flandrian	Holocene	→Warmer	
10,000				→ N
	Loch Lomond Stadial	Younger Dryas Stadial		
11,000		Allerød Interstadial		← S
	Lateglacial [Windermere] Interstadial			
12,000		Older Dryas Stadial		
13,000				→ N
14,000	Late Devensian [Ice sheet phase]	Late Weichselian [Ice sheet phase]		

Figure 3.12   The Lateglacial of Britain and north-west Europe (after Lowe & Walker, 1984)

Weichselian/Devensian cold stage, and extending up to the climatic warming that followed the Little Ice Age of the Medieval period.

The Lateglacial climatic oscillation

Throughout much of the last cold stage, cold surface waters occupied large areas of the North Atlantic, the oceanic Polar Front being situated off northern Portugal around latitude 40 °N (Figure 3.11). By *c.* 13 ka BP, however, the Polar Front and the southern limit of winter sea-ice had receded to a position near Iceland allowing warmer waters to spread northwards around the coastline of western Europe (Ruddiman *et al.*, 1977; Ruddiman and McIntyre, 1981a). The climatic impact of this oceanographic change was dramatic. Prior to 13 ka BP, for example, continuous sea-ice around the British coasts resulted in a climate of extreme continentality, with temperatures of the coldest winter months in the range −20 to −25 °C and summer temperatures below 10 °C. By 12.5 ka BP, however, summer temperatures (as indicated by coleopteran evidence) were typically around 17 °C with winter temperatures in the range 0–1 °C (Atkinson *et al.*, 1987). In other words, the climate of England and Wales at that time was as warm as the present day, although slightly more continental. Evidence for this period of marked thermal improvement has been found throughout north-west Europe, with temperatures similar to those in Britain being experienced along the Atlantic seaboard (e.g. Bohnke *et al.*, 1987), while slightly less equable climatic conditions prevailed in central Europe (Ammann *et al.*, 1984) and in Scandinavia (Berglund *et al.*, 1984). This episode of climatic warming, which began shortly before 13 ka BP and lasted for around 2000 years, is known in Britain as the **Lateglacial** or **Windermere Interstadial**. In continental Europe, however, a short-lived cooler episode (the **Older Dryas Stadial**), which occurred around 12–11.8 ka BP, divides the earlier **Bölling Interstadial** from the succeeding **Allerød Interstadial** (Figure 3.12).

Between approximately 11 and 10 ka BP, polar waters once again spread southwards around the coastline of western Europe (Figure 3.11), with the oceanic Polar Front reaching its maximum southerly position off south-west Ireland (Ruddiman *et al.*, 1977b; Ruddiman and McIntyre, 1981a). The rate of movement of these polar waters appears to have been extremely rapid with velocities in excess of 5 km yr^{-1} being inferred from evidence in cores from the eastern North Atlantic (Bard *et al.*, 1987). The climatic cooling that followed these oceanographic changes is referred to in Europe as the **Younger Dryas Stadial** and in Britain as the **Loch Lomond Stadial**, and indications of this episode of colder climatic conditions are found eastwards into the European USSR (Faustova,

1984; Khotinskiy, 1984a), as far south as Spain and Portugal (Turner and Hannon, 1988), and possibly also in North Africa (Lamb *et al.* 1989). The terrestrial evidence for climatic cooling is impressive. There are indications of renewed glacier activity in mountain regions throughout north-west Europe, while periglacial conditions returned to many lowland areas (see below). The pollen records show that the scattered woodland that had started to develop in response to climatic warming was abruptly replaced by scrub tundra (Rind *et al.*, 1986). Indications of a marked, but relatively short-lived, climatic deterioration between *c.* 11 and 10 ka BP are also apparent in the oxygen isotope traces from Greenland ice cores (Dansgaard *et al.*, 1982) and Swiss lake sediments (Oeschger *et al.*, 1980). Coleopteran evidence suggests that in Britain, summer temperatures fell below 10 °C with winter temperatures in the range −15 to −20 °C (Atkinson *et al.*, 1987), figures that are in broad agreement with temperature reconstructions based on geomorphological evidence (Ballantyne, 1984; Sutherland, 1984b). Elsewhere in Europe, estimates of annual temperature reduction during the Younger Dryas Stadial typically range from 3 to 7 °C (Larsen *et al.*, 1984), the greatest temperature contrasts with the preceding Allerød Interstadial being experienced in areas adjacent to the western European seaboard.

Although the Younger Dryas cooling episode was a climatic event whose effects are apparent in proxy records throughout western Europe, palynological and glacial evidence for a worldwide Younger Dryas cooling event is not convincing (Rind *et al.*, 1986). Indeed, the Younger Dryas Stadial has been described as a 'European climatic anomaly' (Mercer, 1969). However, recent pollen–stratigraphic data from the Maritime Provinces of Canada, and from eastern parts of the United States show clear evidence of abrupt climatic changes around 11–10 ka BP which appear to be the equivalent of the Younger Dryas/Loch Lomond Stadial event of north-west Europe (Mott *et al.*, 1986; Peteet, 1987). This has prompted speculation that the Younger Dryas cooling and its apparent North American equivalent may have resulted from oceanographic changes within the North Atlantic. Data from deep-sea cores point to the chilling of North Atlantic surface waters between 11 and 10 ka BP as a consequence, perhaps, of a major influx of icebergs from disintegrating ice shelves or marine-based ice sheets in the Arctic basin (Rind *et al.*, 1986). A further factor may have been the infusion of huge quantities of meltwater from the wasting Laurentide ice sheet via the Hudson Strait, and following diversion of the Great Lakes drainage from the Mississippi to the St Lawrence (Broecker *et al.*, 1988, 1989). Large volumes of cold freshwater flowing into the Atlantic to the south and east of Newfoundland could have diverted the Gulf Stream southwards so that iceberg influx would have been more effective in chilling North Atlantic surface waters in mid- and high latitudes (Wright, 1989). Although some uncertainties have arisen over the precise timing of meltwater influx into the North Atlantic basin (Fairbanks, 1989), there seems little doubt that a rapid southward expansion of colder waters occurred during the Younger Dryas Stadial, and that this is reflected in the marked climatic deterioration recorded in the terrestrial proxy records from Europe and eastern North America. In those areas of North America not directly affected by North Atlantic oceanographic changes, no climatic oscillation comparable with the European Younger Dryas was experienced during the last glacial/interglacial transition.

The Early Holocene amelioration

The global warming in climate that began at the beginning of the present interglacial around 10 ka BP is recorded in a wide range of marine and terrestrial evidence. Data from the deep ocean floor, for example, suggest that by 9 ka BP, polar waters had retreated to the north-west of Iceland, and that relatively warm surface waters (temperatures above 14 °C) were once more firmly established around the coasts of western Europe. A mean warming rate for this time period of <1 °C per century has been inferred for winter sea-surface temperatures in the mid-latitude areas of the eastern North Atlantic (Bard *et al.*, 1987).

Climatic amelioration is reflected in the marked shift in ^{18}O values in cores from the polar ice caps (Figures 2.28 and 2.29), the isotopic traces suggesting that temperatures comparable with those of the present day had been established by 9 ka BP in the Canadian Arctic (Paterson and Hammer, 1987) and in Antarctica (Jouzel *et al.*, 1987). The rapid wastage of the Laurentide and Fennoscandian ice sheets and the worldwide retreat of mountain glaciers in response to climatic warming is reflected in the abrupt shift in the isotopic trace in deep-ocean sediments at the stage 2/1 boundary (Ruddiman and Duplessy, 1985). In north-west Europe, isopollen maps (Figures 4.12 and 4.13) reflect the rapid northward movement of treeline and the widespread replacement of scrub tundra by birch woodland by 9 ka BP and by mixed woodland within a further 1000 years (Huntley and Birks, 1983). Multivariate analysis of pollen data from sites in France suggests an abrupt increase in temperature after 10 ka BP of around 3–4 °C every 500 years (Guiot, 1987a). More rapid rates of climatic amelioration (1.7 to 2.8 °C every century) have been suggested on the basis of fossil Coleoptera from sites in Britain (Figure 2.11) and France, the evidence indicating that by *c.* 9.5 ka BP, both summer and winter temperatures had reached levels comparable with those of the present day (Atkinson *et al.*, 1987; Ponel and Coope, 1990). In eastern North America, palynological evidence for changing forest composition suggests that the opening of the Holocene at around 10 ka BP was characterized by a marked climatic shift to conditions similar to those of the present, and that by 9 ka BP at least, the climate became warmer than that of today (Davis, 1984). Coleopteran data from sites in east-central North America also show that temperatures comparable with those of today were attained between 10 and 9 ka BP (Morgan, 1987). In the American Midwest, however, the waning Laurentide ice sheet continued to exert an influence on climate so that the Early Holocene (10–8 ka BP) was slightly cooler and more moist than the climatic regime that prevailed in the Late Holocene (Webb *et al.*, 1984).

The Climatic Optimum (the Hypsithermal)

The Climatic Optimum of the present interglacial (the Hypsithermal in North America) occurred between *c.* 9 ka and 4 ka BP. The effects of this climatic warming are reflected in proxy records from many parts of the world, although the dating of the climatic episode is variable and there are geographical differences in the timing of the thermal maximum. In Britain, for example, the period of maximum warmth appears to have occurred between about 8 ka and 4.5 ka BP (Simmons *et al.*, 1981), while in most of northern Eurasia, optimum temperatures were recorded 6 ka to 4.5 ka BP, which is also the climatic optimum in the USSR (Khotinskiy, 1984b). Ice core data from Camp Century in Greenland suggest warmer conditions between 8 ka and 4.1 ka BP, with the thermal maximum around 6 ka to 5 ka BP (Johnsen *et al.*, 1972). In the ice core records from Antarctica, by contrast, the warmest part of the Holocene occurs around 9 ka BP (Jouzel *et al.*, 1987). In the western United States, pollen evidence points to significantly warmer conditions between *c.* 10 ka and 7 ka BP (Barnosky *et al.*, 1987), although in the Midwest, both faunal and floral evidence indicate that the Hypsithermal spans the time interval 8 ka to 5 ka BP (Semken, 1984). Further east, the episode of maximum Holocene warmth appears to have occurred 7 ka to 5 ka BP (Davis, 1984).

A range of proxy data suggests that, at the climatic optimum, temperatures were higher than those of the present day across much of the northern temperate zone. An increase in both summer and winter warmth is reflected, *inter alia*, in the northward migration of mixed woodland (Huntley and Birks, 1983), in the rise of the treeline in mountain regions to elevations well above current levels (Pears, 1968; Kullman, 1988), and in the northward expansion of thermophilous Mollusca (Kerney, 1968). Temperatures at the Climatic Optimum 1–2 °C higher than those of the present day are implied by the postglacial distribution of key indicator species such as holly (*Ilex*), ivy (*Hedera*) and mistletoe (*Viscum*) (Iversen, 1944), and by the occurrence in deposits of Mid-

Holocene age in south-east England, Denmark and southern Sweden of the European pond tortoise (*Emys orbicularis*), whose present-day breeding range is confined to the Mediterranean and eastern Europe (Stuart, 1979). Temperatures at the Climatic Optimum, a degree or so higher than those of the present day, are also indicated in climatic reconstructions based on palynological data from sites in the eastern USSR (Klimanov, 1984), and in isotopic traces from Arctic ice cores (Paterson *et al.*, 1977). In east-central North America, climatic conditions around 2 °C warmer than at present have also been inferred from palaeobotanical and coleopteran evidence (Davis, 1984; Morgan, 1987).

Throughout Europe and North America, there are indications that the early Climatic Optimum (the Boreal period) was markedly drier than the later part of the warm phase. Drier conditions are reflected in lower lake levels (Digerfeldt, 1988) and a general reduction in water-tables (Semken, 1984); in a drying out of mire surfaces (Godwin, 1975); and in palynological evidence of the development of xeric woodland communities. In some areas of the eastern United States abundant charcoal remains in lake sediments older than 7 ka BP indicate extensive burning at a time when precipitation levels may have been more than 100 mm lower than those of the present day (Davis, 1984). In Britain, precipitation values of around 90 per cent of present levels have been suggested for the early part of the Climatic Optimum, compared with estimates of around 110 per cent of present averages in the succeeding Atlantic period (Simmons *et al.*, 1981). Throughout north-west Europe, the Atlantic is characterized by rising water tables and hydroseral developments (Walker, 1970), and the widespread growth of ombrogenous blanket peat (Figure 3.13) from about 6.5 ka BP onwards (Barber, 1981). The relative importance of natural and human factors in peat development is considered in Chapter 6.

The shift to wetter climatic conditions is also reflected in the pollen records, many of which show an expansion of taxa associated with wetland habitats. In eastern North America, a transition from xeric to mesic woodland communities is apparent in the pollen records from *c.* 5 ka BP, while in the western USA a change to cooler and moister conditions is evident from around 6 ka BP (Baker, 1984). In west-central Canada, a similar shift to a wetter climatic regime is indicated by blanket peat growth from 6 ka BP onwards (Zoltai and Vitt, 1990).

The Late Holocene deterioration

From around 5 ka BP, interpretation of the proxy record in north-west Europe becomes more difficult. Changes in forest composition, which had previously been widely employed as an index of climatic change, are now equally likely to reflect the activities of early farming communities. The hydrological changes that accompanied woodland clearance in, for example, runoff, surface water balance and lake sediment budgets, further exacerbate the problems of isolating the climatic signal in proxy records from this and succeeding periods. Despite these difficulties (which are considered further in Chapter 6), it has proved possible to gain some insight into the pattern of climatic change during the Late Holocene.

The evidence indicates that between *c.* 5 ka and 3 ka BP, fluctuating climatic conditions characterized the northern temperate zone, but there was a gradual deterioration in climate which became progressively marked 3 ka to 2.5 ka BP. In North America, for example, there was a southward movement, from about 4 ka BP onwards, of the boreal forest at the expense of mixed conifer–northern hardwood forest (Delcourt and Delcourt, 1984). In the western United States and in the European Alps, treelines have fallen gradually over the past 4500 years (Lamb, 1982), while in Sweden, where the first indications of Holocene temperature decline are recorded around 5.3 ka BP, a marked fall in the upper limit of pine forest (Figure 2.9) indicates abrupt climatic deterioration around 3.3 ka BP (Kullman, 1988). Evidence from peat stratigraphy in western Europe suggests that a series of relatively short-lived climatic changes occurred over the past 5500 years, with a particularly marked deterioration around 2.5 ka BP (Aaby, 1976). This clear shift to cooler

Figure 3.13   Ombrogenous blanket peat on Plynlimon, mid-Wales. Maximum peat thickness 1.5 m (photo M Bell)

and wetter conditions is reflected as a recurrence surface (known as the Grenzhorizont) in peat profiles throughout north-west Europe and has been radiocarbon dated at a number of sites to *c.* 500 BC (Godwin, 1975). In North America, marked cooling at 3–2.5 ka BP in eastern Canada has been inferred from terrestrial evidence (Macpherson, 1985), while in the USA the development of the extensive patterned peatlands of Minnesota is indicative of a cooler and wetter climate after 4 ka BP (Griffin, 1975). Further north, dendroclimatological and palynological data from a network of sites extending from Alaska to Labrador reveal a progressive decline in summer temperatures from around 4 ka BP, the climatic deterioration being especially pronounced after *c.* 2.4 ka BP (Diaz *et al.*, 1989).

Elsewhere in the Atlantic region, climatic deterioration after *c.* 4 ka BP is reflected in the isotopic trace from Greenland ice cores (Dansgaard *et al.*, 1982), in the evidence from the western North Atlantic which shows a marked cooling of surface waters from around 4 ka BP onwards (Balsam, 1981), and in the driftwood evidence from the Arctic basin which indicates a significant increase in sea-ice after 4.2 ka BP (Stewart and England, 1983). In addition, there are indications of renewed glacier activity in Europe and North America after 5.5 ka BP and especially after 2.5 ka BP (Grove, 1979). Overall, a temperature decline of between 1 and 2 °C appears to be inferred from the proxy data sources.

The Little Climatic Optimum

Between *c.* AD 700 and 1300, there is evidence for a short-lived episode of climatic warming which has become known as the Little Optimum. This was a period of glacial retreat in many mountain regions of the Northern Hemisphere (Grove, 1979), and was also a time when a reduction in sea-ice and generally warmer climatic conditions around the Arctic basin allowed the establishment of Viking colonies in Greenland (McGovern, 1981) and northern Newfoundland (Ingstad, 1977) (Chapter 5). The ice core data from central Greenland indicate temperatures 1–2 °C higher than at present between AD 700 and 1000 (Figure 5.24), while in Iceland, significantly warmer conditions obtained from around AD 1050 to 1200 (Dansgaard *et al.*, 1975). In Sweden dendrochronological and lake sediment records point to a warmer climatic regime between AD 950 and 1140 (Karlén, 1984), although not all proxy records from Fennoscandia show such clear evidence of a Medieval warm epoch (Briffa *et al.*, 1990). In southern England, documentary sources for the Early Medieval period (e.g. Domesday Book) provide evidence of extensive viticulture in the twelfth and thirteenth centuries which suggests an increase in both summer and winter warmth. Mean annual temperatures in lowland England between AD 1150 and 1280 have been estimated at 9–10 °C, with maximum summer and winter temperatures of 16.5 °C and 4 °C respectively, these values being a degree or so higher than those of the present day (Lamb, 1977, 1982). Drier conditions in upland Britain are also indicated in surface wetness curves from peat bogs for this time period (Figure 2.18), and there is also evidence for a dry phase with aeolian activity in The Netherlands (Chapters 5 and 7). In the south-west USA, dendrochronological data suggest a significant increase in temperatures around AD 1100–1300 (Brubaker and Cooke, 1984). For reasons which are not currently understood, the thermal maximum of the Little Optimum appears to have occurred some 200 years earlier in Greenland, by comparison with western Europe and the south-west USA.

The Little Ice Age

The episode of cooler climatic conditions that succeeded the Medieval warm phase is an event of global significance (Grove, 1988). It was characterized in the northern temperate regions by markedly cooler summers, colder and wetter winters and a significant increase in storms. In areas of western Europe, for example, temperature reconstructions from a range of proxy data sources suggest a climate 2–3 °C colder than that of the Little Optimum and 1–3 °C cooler than that of the present day (Lamb, 1977). There are also indications of significantly increased summer precipitation. The effects of this climatic deterioration are seen in the re-advance of mountain glaciers, in significant increases in geomorphological activity (landslides, avalanches, etc.) in upland areas, in flooding around the Atlantic coasts, in the expansion of Arctic sea-ice, in vegetational changes in the mountains, and in widespread crop failure and abandonment of settlement. The dating of the Little Ice Age is variable, but it broadly spans the time interval from the fourteenth to the eighteenth century. The Little Ice Age in Europe is considered in more detail in Chapter 5.

Causes of short-term climatic change

The principal causal mechanisms of climatic change over timespans of 10^1 to 10^3 years are generally considered to be variations in solar output, volcanic dust in the atmosphere and, over the last two centuries, changes in atmospheric CO_2 content. Indeed, a climatic modelling exercise in which those parameters were employed as external forcing functions, showed that 93 per cent of the Northern Hemisphere temperature variation between 1881 and 1975 could be explained by the influence of these three variables (Gilliland, 1982). However, in so far as CO_2 is unlikely to have been a major element in pre-industrial high-frequency climatic fluctuations (Porter, 1986), the primary elements involved in the forcing of short-term climatic changes during the course of the Holocene appear to be solar output variations and fluctuations in the quantity of volcanic dust in the atmosphere. In addition, the effects of these two factors on the climatic patterns of the Holocene may, to some degree, be modulated by geomagnetic variations and changes in ocean circulation.

Solar output variations

Changes occur over time in the radiant heat emitted by the sun and involve both fluctuations in solar output as well as variations in the quality of solar energy. The former occurs as a consequence of sunspot activity and solar flares, while the latter reflects changes in the ultraviolet range of the solar spectrum.

Quantitative changes in solar output

It has long been considered that fluctuations in the solar constant (the amount of radiant heat emitted by the sun) are major factors in short-term climatic change, and calculations suggest that a 1 per cent decline in solar output could lower global average temperatures by 1–2 °C (Eddy, 1980). Indeed, many of the climatic variations of the past 10 ka may be explained by changes in solar irradiance of no more than several tenths of a per cent (Wigley and Kelly, 1990). Observations of the surface of the sun (the photosphere) indicate that the sun alternates between active and relatively quiescent phases, the most useful indications of these solar fluctuations being the growth and disappearance of **sunspots**. These are conspicuous dark patches that occur as shallow depressions in the general photospheric level and which may be the central parts of active or disturbed regions reflecting convectional activity within the photosphere. They are markedly cooler than adjacent areas, although more intense radiative outputs have been detected in regions immediately around the sunspots themselves, and they appear to mark the places where intense magnetic fields break through the solar surface (Eddy, 1981). Hot areas (**plages**) have also been detected. In addition, violent eruptions, during which large amounts of energy are released over short timespans, are a regular feature of the outer parts of the sun, and are referrred to as **solar flares**. These short-lived disturbances, which occur from time to time in the sun's atmosphere, are superimposed on the steady output of radiant heat from the sun, and constitute a type of solar weather.

Such short-lived variations in solar activity are not random, but appear to follow a cyclical pattern. In 1843, Heinrich Schwabe suggested a 10–11-year periodicity in the number and activity of sunspots, an observation that has subsequently been confirmed by both astronomical observations and satellite measurements of solar irradiance (Schove, 1983; Smith *et al.*, 1983). Documentary records of sunspots (**sunspot numbers**) reflecting such a cyclical pattern extend back to about AD 1700 (Figure 3.14). Other elements of the solar weather also appear to reach maximum intensity at about 11-year intervals, although shorter term fluctuations in solar activity have also been detected, for example cycles of solar flares with mean periods of 9 years, 2.25 years and 3 months (Landscheidt, 1984). The causes of these apparently cyclical changes in solar activity are unknown, but there has been speculation that they may relate to a change in the thermal structure of the sun under the influence of magnetic fields (Kuhn *et al.*, 1988), or they may in some way be linked to variations in the solar radius (Ribes, 1990).

Over the timescale of the Holocene, longer term fluctuations in solar output appear to have occurred. Sunspot cycles at 22-year (the 'double sunspot' or 'Hale' cycle), 80-year and 200-year intervals have been recognized (Schove, 1984). Proxy data records showing periodicities in climate that may be linked to changes in solar output include tree-ring series, isotopic traces in ice cores and sequences of varved sediments (Schove and Fairbridge, 1984). For example, major oscillatory cycles of 143, 218 and 420 years have been noted in tree-ring records (Stuiver and Braziunas, 1989), periodicities of 88, 440 and 1640 years have been detected in Holocene glacier activity (Bray, 1970), while spectral analyses of the isotopic record in the Camp Century ice core from Greenland showed peaks at around 78, 181, 400 and 2400 years (Johnsen *et al.*, 1970).

Patterns of solar actvity over the 11-year cycle have frequently been linked with terrestrial climatic changes. Connections have been suggested, for example, between sunspot cycles and such diverse phenomena as fluctuations in English temperature

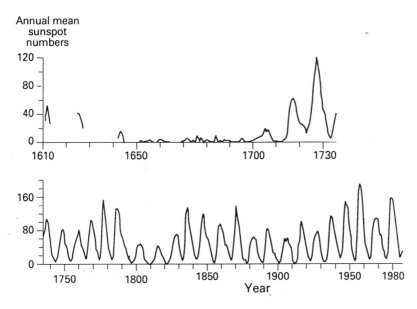

Figure 3.14 Annual mean sunspot numbers: AD 1610–1988 (after Stephenson, 1990)

records, drought in the United States (Lamb, 1982) and floods on the River Nile (Schove, 1983), while solar flares and enhanced eruptional activity on the sun have been related to recent atmospheric circulation changes, rainfall patterns and incidence of thunderstorms (Landscheidt, 1984). A 22-year solar cycle (as well as a 18.6-year lunar nodal tidal cycle) has also been detected in proxy climatic data spanning the time period AD 1700–1979 from central Canada (Guiot, 1987b). Overall, however, attempts to relate terrestrial weather patterns to the 11-year sunspot cycle have met with mixed results, and at the present state of knowledge, the longer cycles appear to be the most informative (Fairbridge, 1984). One way in which longer term solar fluctuations can be established is by using the atmospheric ^{14}C record, for variations in the production rate of atmospheric ^{14}C appear to be related, *inter alia*, to variations in solar output. As a consequence, episodes of reduced solar activity are reflected in higher rates of ^{14}C activity in the upper atmosphere (Harvey, 1980). Using ^{14}C contained in tree rings from sites in California, Sonett and Suess (1984) demonstrated that the solar signal (^{14}C value of wood) was closely correlated with climatic patterns (mean tree-ring width reflecting temperature) over the past five millennia. In a similar study covering the last 860 years, Stuiver

and Quay (1980) found an equally close relationship between sunspot activity and ^{14}C value of tree rings (Figure 3.15). Significantly higher ^{14}C levels correlated with the three sunspot minima of the Medieval periods (Eddy, 1977): the Wolff Minimum (AD 1281–1342), the Spörer Minimum (AD 1450–1534) and the Maunder Minimum (AD 1645–1715). These episodes averaged 79 years in duration, but the succeeding shift from a 'quiet' to an 'active' sun after each relatively quiescent period varied from around 60 to 120 years. Over a longer timescale a 2500-year spectral peak has been detected in reconstructed time-series of ^{14}C production rate deviations and δ^{18}O residuals in Holocene ice core records from Devon Island and Greenland (Fisher, 1982), while a 2300-year cycle is evident in tree-ring records from Ireland and California (Sonett and Finney, 1990). A statistically significant correlation has also been found between global glacier fluctuations over the course of the Holocene, and variations in atmospheric ^{14}C concentrations, which suggests a close relationship between temporal patterns of solar irradiance and glacier activity (Wigley and Kelly, 1990).

Falsifying the hypothesis that solar variability is a major causal mechanism of short-term climatic change is extremely difficult, for while some

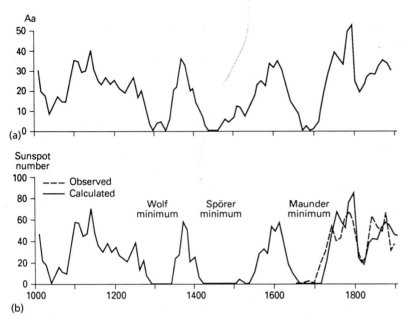

Figure 3.15   Sunspot numbers (b) and Aa indices (a) calculated from the ^{14}C record. The Aa indices reflect geomagnetic changes arising from variations in the solar wind. Changes in the geomagnetic parameters modulate the incoming cosmic ray flux which is responsible for ^{14}C production in the upper atmosphere. The dashed line gives the actual observed averaged sunspot record. Sunspot numbers and Aa indices are averaged over the 11-year solar cycle (after Stuiver & Quay, 1980)

relationships between, for example, sunspot frequency and climate appear to have been established, these are often imperfect (Sofia et al., 1985) and could equally be explained by coincidence. Indeed, the case for no link at all between solar cycles and climatic change has been forcefully stated (Molnar, 1981). Reconstructing the trend of variations in solar output over recent millennia is particularly problematical, for although ^{14}C variations may provide one proxy data source, a statistically significant relationship between regional climatic changes and ^{14}C time-series cannot always be established (Stuiver 1980). Moreover, much remains to be learnt about the operation of the global carbon cycle and of the precise linkages between ^{14}C production and solar emission. On the other hand, in so far as there do appear to have been variations in solar activity over recent millennia, it is difficult to avoid the conclusions that these must, in some way, have exerted an influence on terrestrial climates. Exactly how this has happened remains problematical.

Qualitative changes in solar output

Although the ultraviolet (UV) flux constitutes only a small fraction (about 1.05 per cent) of the total photon energy emitted from the sun, it is a significant factor affecting the heating of the upper atmosphere and it also plays a crucial role in determining atmospheric density and photochemistry. In particular, chemical reactions between UV radiation and oxygen lead to the formation of ozone, the concentration of which is important in controlling stratospheric temperature through UV absorption (Lean, 1984). Measurements of atmospheric ozone content since the late 1960s indicate that concentrations vary with time, and there is evidence of an 11-year fluctuation almost in phase with the sunspot cycle (Angell and Korshover, 1976). This, in turn, has been linked to variations in output of solar UV radiation (Keating, 1978). More recently, satellite observations have confirmed the temporal changes in solar UV irradiance, and indicate that these are associated with the evolution and disappearance of plages, magnetically active regions on the surface of the sun which fluctuate with the 11-year solar activity cycle (Lean, 1984). In addition, there is now evidence of a shorter periodicity in UV irradiance, corresponding with the 27-day solar rotation period (Gérard, 1990).

Once again, however, testing the hypothesis that variations in the quality of solar output can cause climatic change is problematical. On the one hand, there is increasing observational evidence that the upper/middle atmosphere responds both chemically and in terms of temperature to variations in the solar UV flux and, moreover, that changes in solar irradiance follow a cyclical pattern (e.g. Schmidt, 1986). These relationships seem to be supported by numerical simulation models which predict temperature variations of up to 2–5 °C at 50 km, especially in high latitude summer regions (Gérard, 1990). On the other hand, the present data base is limited and only when long-term records over many solar cycles become available will it become possible to evaluate the effects of solar UV variations on the stratosphere, and to estimate with certainty how such changes affect the troposphere and hence climate (Lean, 1984).

Volcanic dust

The infusion into the atmosphere of large amounts of volcanic debris has long been regarded as a possible causal mechanism for producing short-term variations in climate (Lamb, 1977, 1982). For example, the coldest and wettest summers in Britain over the last three centuries appear to have coincided with times of increased volcanic activity, while in the Arctic basin region over the last millennium, there seems to be a correspondence between the extent and duration of sea-ice and the record of volcanic eruptions on Iceland (Lamb, 1972). It was originally believed that a reduction in temperature would result from the screening out of incoming solar radiation by atmospheric dust particles, and that this effect would be amplified by an increase in cloudiness as the particulate matter acted as foci for water droplets. Moreover, it has frequently been suggested that the effects of volcanic activity on climate would be relatively localized. However, it now appears that the amount of dust and ash that remains in the atmosphere for any length of time following a major episode of explosive volcanism is relatively small, and that the primary climatic effects are caused by the release during an eruption of large amounts of volcanic sulphur. Once in the atmosphere, this is converted into sulphuric acid and becomes the dominant aerosol, the effects of which are to cool the lower troposphere by the back-scattering of incoming solar radiation (Pollack et al., 1976). Moreover, there is increasing evidence to suggest that once injected into the atmosphere, these aerosols are disseminated globally (see below). The mean residence time of the sulphuric acid aerosols ranges from one to five years (Schönwiese, 1988).

Considerable empirical evidence exists to support the hypothesis of a link between short-term climatic change and recent volcanic activity. For example, an estimated 1 °C fall in tropical tropospheric temperature followed the eruption in 1963 of Mt Agung on Bali in Indonesia (Newell, 1981), and the effects of this eruption were registered by a temperature decline of 0.8 °C north of latitude 60°N. Other major volcanic events of the last two centuries, such as Tambora (1815), Krakatau (1883) and Santa Maria (1902), may have been associated with global temperature reductions of 0.2 to 0.5 °C for periods of up to five years (Self et al., 1981; Rampino and Self, 1982). Indeed, surface cooling has been detected over land masses in the Northern Hemisphere within months of a major volcanic eruption (Kelly and Sear, 1984; Sear et al., 1987). Overall, the evidence suggests that the amount of surface cooling that can reasonably be attributed to individual large volcanic eruptions probably averages no more than 0.2 to 0.3 °C, although local temperature reductions may be as high as 1.5 °C (Porter, 1986). Other climatic phenomena which show strong associations with increased volcanic activity include tropical storms in the North Atlantic, the Indian Monsoon and possibly also the El Niño/Southern Oscillation (Handler, 1989).

Time-series data on past episodes of volcanic activity have been obtained from polar ice cores. Acidity levels in annual ice layers can be established by electrical conductivity measurements, and these reflect the amount of sulphuric acid washed out of the atmosphere in precipitation over the course of a year. Hence downcore variations in acidity provide an indication of the fluctuations in the amount of atmospheric sulphuric acid and, by implication, of volcanic aerosols over time. These acidity profiles, therefore, constitute a proxy temporal record of volcanic eruptions (Hammer et al., 1980). A long

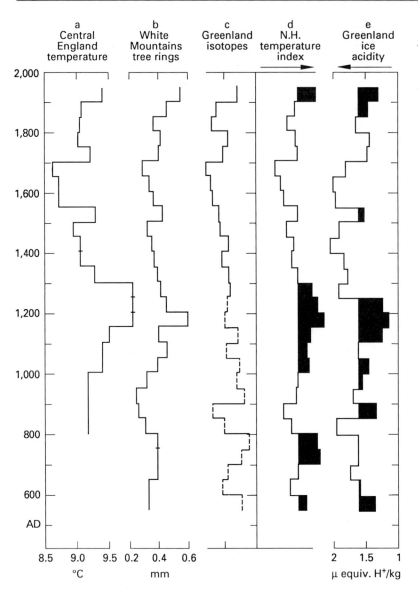

Figure 3.16 Acidity profile from the Crête ice core, Greenland, compared with Northern Hemisphere temperatures for the period AD 550 to present. Episodes of lower than normal volcanic activity correspond to lower acidity levels (shown in black) and these, in turn, can be correlated with warmer climatic phases (also shown in black) (after Hammer et al., 1980)

core from Crête in Greenland spanning the past 1500 years shows how episodes of volcanic activity have varied over that time period, and there appears to be a correlation between the acidity trace and the record of temperature variations in the Northern Hemisphere (Figure 3.16). Overall, episodes of below-average volcanic activity correspond with above-average temperature (Hammer *et al.*, 1980, 1981). The significance of these data lies in the fact that they provide a detailed chronology (accurate to

within <10 years) of volcanic activity against which other proxy records can be assessed. There is, for example, a striking relationship between glacier activity over the last millennium and Greenland ice core acidity (Figure 3.17), which led Porter (1986) to suggest that sulphur-rich aerosols generated by volcanic eruptions are a primary forcing mechanism of recent Northern Hemisphere glacier fluctuations, and hence of climatic change. This conclusion is reinforced by dendrochronological evidence. Tree-

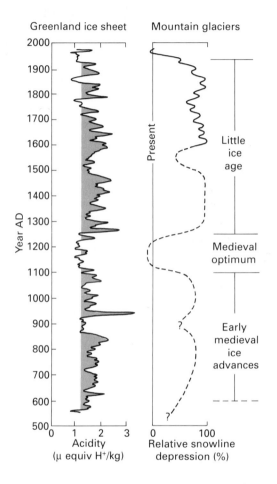

Figure 3.17   Inferred correlation of Greenland Ice Sheet acidity record with evidence of Late Holocene glacier variations in the Northern Hemisphere (after Porter, 1986)

Little Ice Age. The data imply temperature decreases in the order of 1 °C for up to two years after the volcanic eruption (Scuderi, 1990). North American tree-ring records spanning the time interval AD 1602–1900 have also been used to evaluate spatial variations in temperature change following recent volcanic eruptions (Lough and Fritts, 1987).

An additional feature of the proxy volcanic records is the evidence they provide for the global dissemination of atmospheric aerosols. Each of the major volcanic events of the past millennium that occurred north of 20°S, for example, are apparent in the acidity profile from the Crête ice core (Figure 3.18), while the great eruption on the island of Santorini around 3600 BP which destroyed Thera and which may have been a factor in the fall of the Minoan civilization, is registered in such geographically diverse records as Irish and American tree-ring series, and also in the Dye 3 ice core from southern Greenland (Hammer *et al.*, 1987). These data confirm the empirical evidence outlined above that the climatic impact of volcanic activity is likely to be of global significance.

It appears, therefore, that there is now a considerable body of confirmatory evidence to support the hypothesis of a relationship between volcanic activity and climatic change, albeit over very short timescales (decades or less) and for the relatively recent past. However, theories linking climatic change with volcanic activity over the longer timescale of the Pleistocene (e.g. Bray, 1974, 1979) may be less well founded. Moreover, it must be stressed that the evidence for volcanic forcing remains essentially circumstantial and that considerable research is required before the linkages between atmospheric volcanic aerosols, solar radiation balance and climatic change are firmly established.

Geomagnetism

Changes in the earth's magnetic field have frequently been linked to climatic change. Over the period 1925–70, for example, a correlation has been established between terrestrial temperature changes and magnetic intensity, with a weakening

ring series from Northern Ireland contain occasional concentrations of narrow rings (indicating climatic stress) spanning intervals of <20 years, and these appear to correspond with the ice core acidity profiles and other records of volcanic activity (Baillie and Munro, 1988). Similar relationships between tree-ring series and Greenland ice core acidity levels have been found in dendrochronological records from the Sierra Nevada, south-west USA, which extend back to 3000 BP. The ring-width 'events' have also been correlated with historical records from the Mediterranean and with glacier activity during the

Figure 3.18   Major volcanic eruptions over the past millennia recorded in the acidity profile from the Crête ice core, central Greenland (after Hammer et al., 1981)

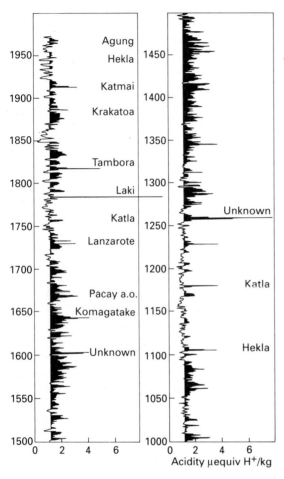

the general circulation (Fairbridge, 1984). However, although there is a certain amount of circumstantial evidence for a connection between changes in geomagnetism and climate, it is difficult to distinguish between cause and effect and it may be that both are reacting to variations in solar activity (Goudie, 1983). Hence, the extent to which terrestrial geomagnetic fluctuations stimulate climatic change remains an open question.

Oceanic circulation

The importance of ocean water circulation in the development of global climatic systems is widely acknowledged, and hence any changes in the nature of ocean water movement are likely to have a climatic effect. Such changes have been cited as a possible causal mechanism of long-term climatic change (e.g. the major oceanographical changes that would have followed the closing of the Gulf of Panama 3.5–3.0 ma BP), but it is also possible that variations in the pattern of oceanic circulation could have occurred during the Holocene. It has been suggested, for example, that in the North Atlantic the Gulf Stream has been pulsating or vibrating over time (Mörner, 1984), and there is evidence in ocean cores from the Denmark Strait of short-term oscillations in ocean water temperature (and possibly, therefore, in climate) throughout the Lateglacial and Holocene (Kellogg, 1984). Moreover, in the north-east Atlantic, there are indications that sea-surface temperatures may fall by as much as 3–5 °C over timescales as short as decades or even years (Lamb, 1979).

Changes in the production of deep water, namely water masses with distinctive temperature and salinity characteristics that develop in the abyssal depths of the world's oceans, may be partly responsible for such observed variations in sea-surface temperatures, for it has been suggested that the formation of such deep-water masses is accompanied by the release of large quantities of heat into the atmosphere. Broecker *et al.* (1985) have discussed the possible climatic effects of deep-water production in the North Atlantic, and have estimated that as a by-product of deep-water formation, the amount of heat currently released

of the geomagnetic field appearing to correspond to increased temperatures (Figure 3.19). A relationship between terrestrial magnetic activity and the 22-year sunspot cycle has been suspected for some time (Chernosky, 1966), and it has been suggested that the solar wind (a more or less continuous flow of protons and electrons emitted by the sun) reacts with the earth's geomagnetic field to trigger geochemical reactions within the gases of the upper atmosphere. These, in turn, control the stratospheric greenhouse effect, the extent of high-altitude cloud cover and other factors that govern

Figure 3.19   Annual means of magnetic intensity (F) at observatories in Greenland, Scotland, Sweden and Egypt correlated with 10-year means of air temperature from nearby weather stations (after Wollin et al., *1973)*

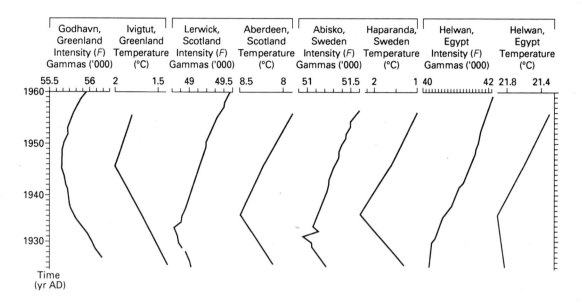

into the atmosphere each year corresponds to approximately 30 per cent of the solar heat reaching the surface of the Atlantic Ocean in the region north of 35°N. Moreover, Watts (1985) has suggested that global temperature changes over timespans of several centuries or less could, in theory, be explained by variations in the rates of bottom-water formation. Indeed, simulation modelling of the Younger Dryas cold episode (11–10 ka BP) suggests that a cessation of North Atlantic deep-water formation would lead to a significant cooling at high latitudes, with winter and summer temperatures over land areas falling by up to 5–6 °C and 2–3 °C respectively (Harvey, 1989). A further element in the equation may be the role

played by the oceans in the global carbon cycle. Surface changes in ocean circulation patterns appear likely to affect the amount of CO_2 released into the atmosphere (Lal and Revelle, 1984), while fluctuations in biological productivity in areas of upwelling may be an additional factor influencing atmospheric CO_2 content (McElroy, 1983; Newell and Hsiung, 1984). In view of the accepted link between atmospheric CO_2 and temperature, it would seem that short-term variations in either of these processes could induce global temperature changes. Recent atmospheric fluctuations in CO_2 resulting from anthropogenic activity are discussed in Chapter 9.

4

Consequences of Climatic Change

Introduction

The global climatic changes discussed in the previous chapter had a profound impact on the landscapes of Europe and North America. The transition from a full glacial to an interglacial climatic regime which occurred between *c.* 13 ka and 10 ka BP was reflected in the rapid wastage of the great ice sheets and mountain glaciers of the Northern Hemisphere, in a contraction of the periglacial domaine, and in the initiation of a vegetation succession which resulted in a change from arctic tundra to closed woodland over much of Europe and North America within the timescale of a few thousand years. Dramatic changes also occurred in fluvial regimes, erosional activity and pedogenesis. Moreover, following the release into the oceans of enormous quantities of meltwater from the wasting ice sheets, global sea level rose by over 100 m, completely changing the configuration of coastal regions in many areas around the Atlantic basin. These geological, geomorphological and

biological responses to climatic change, and their reflection in the landscapes of the mid-latitude regions of the Northern Hemisphere over the past 20 millennia, form the subject matter of this chapter.

The last glaciers in the northern temperate zone

At the height of the last glaciation, around 18 ka BP, there were two major centres of glaciation in mainland Europe, one located in the Baltic region and the other in the Alps, although the two glacier complexes remained as discrete entities throughout the Late Weichselian (Šibrava *et al.*, 1986). In addition, smaller glaciers formed in the mountainous areas of central and southern Europe including the Pyrenees, the French Massif Central, the Apennines, the Dinaric Alps and on the island of Corsica (Andersen, 1981). By far the largest ice sheet was that which developed over Scandinavia which, at its maximum, covered the whole of Norway, Sweden and Finland, and extended southwards into Denmark and Germany, and eastwards into Poland and the north-west USSR. Glaciological modelling (Figure 4.1) indicates that the main ice dome lay over the Gulf of Bothnia where the ice was over 2 km in thickness (Boulton *et al.*, 1985), and that the total volume of the ice sheet was of the order of $7–7.5 \times 10^6$ km^3 (Hughes *et al.*, 1981).

In the British Isles, the major ice accumulation and dispersal centres lay along an axis extending from the mountains of western Scotland to the uplands of Wales. At the glacial maximum the ice sheet covered almost all of Scotland, Ireland, Wales and northern England (Figure 4.2), with only the Midlands and southern England remaining ice-free (Bowen *et al.*, 1986). The modelled volume of the last British ice sheet is around 0.8×10^6 km^3 (Hughes *et al.*, 1981), with a maximum ice thickness in excess of 1.5 km (Boulton *et al.*, 1985). Recent evidence from the bed of the North Sea indicates that, contrary to previously held opinion, the British and Fennoscandian ice sheets did not coalesce during the Late Weichselian/Late

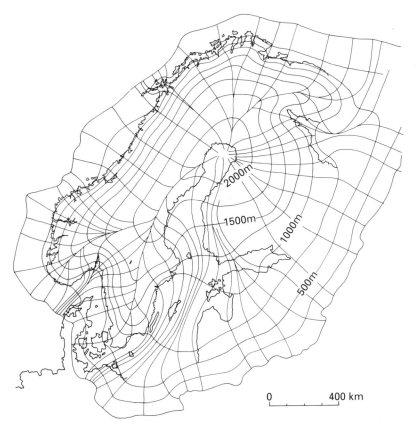

Figure 4.1 Computer model of the Fennoscandian ice sheet at its maximum extent (after Boulton, et al., 1985; reproduced by permission of the Geological Society)

0 _____ 400 km

Devensian (Sutherland, 1984b), but that an ice-free corridor over 100 km wide separated the two great ice masses (Sejrup *et al.*, 1987).

The largest glacier complex in the Northern Hemisphere, however, was the Laurentide ice sheet of North America. Centred on Hudson Bay, the ice sheet extended southwards across the United States–Canada border into the Great Lakes region, westwards towards the Rocky Mountains, northwards across the islands of the Canadian Arctic archipelago, and eastwards into the Canadian Maritime Provinces (Mayewski *et al.*, 1981). Glaciological modelling (Figure 4.3) suggests a major ice dome over Hudson Bay extending westwards across Keewatin, where ice thicknesses in excess of 3 km accumulated (Boulton *et al.*, 1985; Hughes, 1987). Total ice volume for the Late Wisconsinan maximum has been estimated at between $30–35 \times 10^6$ km^3 (Hughes *et al.*, 1981).

This is over four times the size of the Late Weichselian Scandinavian ice mass, and comparable with the estimated volume of the present-day Antarctic ice sheet. To the west, a major glacier complex extending from Alaska to southern British Columbia developed over the western cordillera which was coalescent at the Late Wisconsinan maximum with the Laurentide ice sheet in the plains and foothills of Alberta and eastern British Columbia. In the western USA, smaller mountain glaciers and minor glacier complexes formed as far south as California (Fullerton, 1986).

Between western Europe and North America, an ice sheet covered almost all of Iceland, while the Greenland ice sheet was slightly more extensive than at the present day (Andersen, 1981).

The ocean core record from the North Atlantic suggests that deglaciation from the last ice

Figure 4.2   The last glaciers in Britain and Ireland (after Bowen et al., 1986)

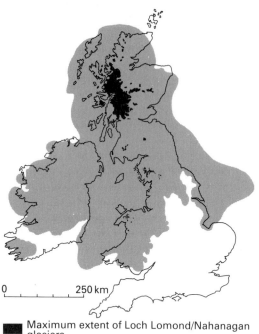

0   250 km

■ Maximum extent of Loch Lomond/Nahanagan glaciers

▨ Maximum extent of late Devensian glaciers

high latitude regions of the Northern Hemisphere from around 17 ka BP onwards. This, it is suggested, would be sufficient to inititate the process of deglaciation and the release of large volumes of meltwater, floe ice and icebergs into the North Atlantic. The subsequent fall in sea-surface temperatures and extension of winter sea-ice cover north of 50° latitude (resulting from lower winter insolation values) would, in turn, lead to a reduction in the poleward moisture flux, and hence a reduction in precipitation over the ice sheets. This hypothesis appears to have been confirmed by the occurrence in North Atlantic cores of large amounts of ice-rafted debris indicative of increased iceberg activity and biological indicators of a marked increase in volume of low-salinity surface meltwaters (Ruddiman and Duplessy, 1985). The evidence, therefore, implies that during the period 16–13 ka, the volume of the Northern Hemisphere ice sheets and glaciers was reduced partly because of a higher summer radiation balance, and partly through starvation of precipitation input. However, this relationship between ocean moisture and glacier mass balance would have been most significant in the wastage of the Fennoscandian and British ice sheets which lay downwind of the North Atlantic. In North America, by contrast, where the initial large-scale melting of the Laurentide ice sheet did not begin until around 14 ka BP, the timing and trend of ice volume loss during deglaciation is directly attributable to insolation forcing (Ruddiman, 1987).

Terrestrial proxy records, involving a range of geological and biological evidence, indicate that around 13 ka BP, a significant rise in temperature occurred throughout north-west Europe (Atkinson *et al.*, 1987), while oxygen isotope data from North Atlantic ocean cores point to a major glacier melting episode (due to warmer conditions?) between 13 and 11.5 ka BP (Ruddiman and Duplessy, 1985). This marked increase in the rate of glacier wastage around the maritime fringes of the Northern Hemisphere ice sheets may have led to the abrupt chilling of western North Atlantic surface waters resulting in the short-lived climatic deterioration of the European Younger Dryas Stadial (see above). During that episode (11–10 ka

maximum may have begun before 16 ka and that by 13 ka BP, over half of the Northern Hemisphere ice had disappeared (Ruddiman and McIntyre, 1981b; Ruddiman and Duplessy, 1985). Terrestrial proxy data indicate that cold and dry conditions existed to the south of the ice sheets during this period, which implies that a considerable volume of glacier ice disappeared while the climate of the Atlantic basin region remained cold. The British ice sheet, for example, which may have reached its maximum some time after 18 ka BP (Rose, 1985), had virtually disappeared *before* the climatic warming that occurred throughout western Europe at around 13 ka BP (Bowen *et al.*, 1986). One hypothesis that has been advanced to account for this apparent paradox involves feedback mechanisms between ocean and atmosphere. The influence of the Milankovitch variables would lead to an increase in summer insolation receipts in the

Figure 4.3   Computer model of the Laurentide ice sheet at its maximum extent (after Boulton et al., 1985; reproduced by permission of the Geological Society)

BP), marked re-advances occurred around the southern and western fringes of the Fennoscandian ice sheet (Lundqvist, 1986), and there was renewed glacier activity throughout the mountain regions of western Europe (e.g. Billard and Orombelli, 1986; Schlüchter, 1986). In Britain, a large mountain icefield developed in the western Highlands of Scotland (Thorp, 1986), while smaller glaciers and glacier complexes were to be found in the eastern and northern Scottish Highlands and on many of the Hebridean islands (Figure 4.2). Further south, cirque glaciers formed in the Southern Uplands, the Lake District, the uplands of North and South Wales and the hills of eastern and south-western Ireland (Gray and Coxon, 1991). The Loch Lomond Readvance was the last occasion when glaciers existed in the British Isles. In North America, the equivalent of the European Younger Dryas climatic oscillation appears to have affected only eastern parts of the USA and Canada (see above). Inland, the period from *c.* 14 ka to 10 ka BP saw the gradual contraction of the Laurentide ice sheet, albeit with frequent local re-advances

particularly in the Great Lakes region (Mickelson *et al.*, 1983; Andrews, 1987). In the mountains of western North America, deglaciation was well advanced by 14 ka BP (Porter *et al.*, 1983).

The global climatic warming which marked the beginning of the present interglacial was reflected in rapid deglaciation on a hemispherical scale. By 10 ka BP, or perhaps even before, all traces of glacier ice had vanished from the British Isles (Lowe and Walker, 1984), the Fennoscandian ice sheet had virtually disappeared by 8.5 ka BP, while by *c.* 7 ka BP the much larger Laurentide ice sheet had been reduced to residual ice masses in Labrador–Ungava, Keewatin and Baffin Island (Hughes, 1987), with only a remnant of the last-named surviving to the present day. Throughout the Holocene, however, there is abundant evidence for glacier activity in the mountain regions of Europe and North America. In Alpine regions of western Europe, glacier advances appear to have occurred intermittently over the past 10 ka (Grove, 1979), while in Sweden the available evidence suggests a sequence of glacier advances and

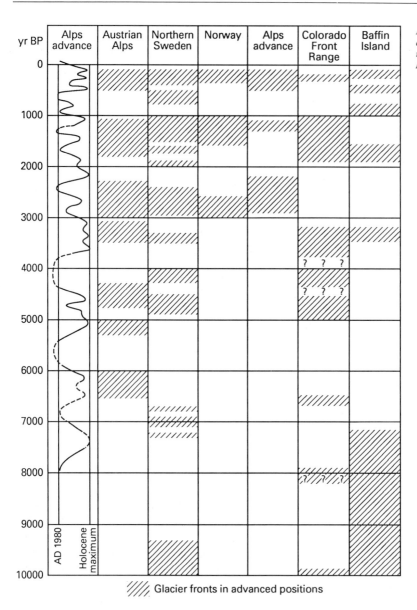

Figure 4.4   Summary of glacier expansion phases in Europe and north America during the Holocene (after Grove, 1988)

//// Glacier fronts in advanced positions

retreats at irregular intervals from 7.5 ka BP onwards (Karlén, 1988). In western North America, glaciers had retreated to near present limits by 10 ka BP, and while there are some indications of Early Holocene advances, glacier activity during the Altithermal/Hypsithermal interval (c. 8–5 ka BP) was minimal. Evidence for renewed glacier advances during the Neoglacial (post-5 ka BP) is found along the entire length of the western cordillera (Osborn and Luckman, 1988; Davis, 1988). In both Europe and North America, there is unequivocal evidence for widespread glacier re-advances during the Little Ice Age of recent centuries (Grove, 1988), and there are indications of widespread glacier expansion around 2.5 ka BP. Prior to that during the Holocene, hemispherical patterns of glacier activity are less easy to detect (Figure 4.4) partly,

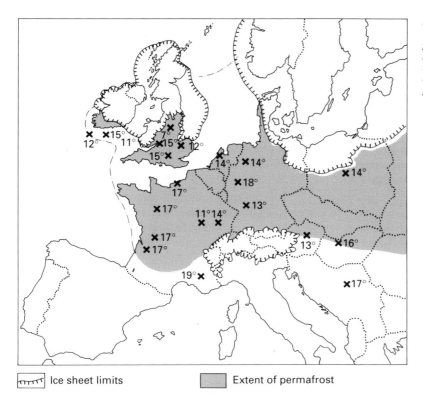

Figure 4.5   Distribution of Weichselian/Devensian permafrost in Europe (based on Maarleveldt, 1976) and some minimum temperature increases as estimated from relict periglacial features (based on Washburn, 1979)

Ice sheet limits   Extent of permafrost

✗   Sites of last glaciation maximum age

perhaps, as a consequence of chronological uncertainties relating to glacier behaviour.

Periglacial activity

The present periglacial domain covers three of the great vegetational zones of the world: the subarctic and northern forests, the arctic tundra and ice-free polar desert zones, and the birch–pine forests and mountain scrublands. Together these constitute around 25 per cent of the earth's land surface, but during the cold stages of the Quaternary, periglacial conditions characterized some 40–50 per cent (French, 1976).

The distribution of relict periglacial phenomena shows that at the height of the last cold stage regions beyond the ice margins experienced a range of periglacial conditions. In Europe, the limit of permafrost is believed to have extended from southern France around the foothills of the Alps and then eastwards approximately along the course of the River Danube (Figure 4.5). To the north of this line, mean annual temperatures of -2 °C or less have been inferred, with temperatures in many areas being depressed by 12–16 °C compared with the present day (Maarleveld, 1976). There is evidence, however, that the periglacial domain may have extended even further south, for fossil cold climate phenomena have been found in the mountains of the Balearic Islands (Butzer, 1964), while on Corsica periglacial activity down to sea level has been inferred during the last cold stage (Conchon, 1978).

Following the establishment of warmer North Atlantic waters around the coasts of Europe around 13 ka BP, more equable conditions replaced the periglacial climatic regime that had prevailed for much of the previous 100 ka. Thawing of the permafrost occurred over large areas, although some permafrost either remained throughout the Lateglacial or re-formed during the short-lived cold episode between 11 ka and 10 ka BP. In the southern Netherlands, for example, there is evidence for deep seasonal frost or localized permafrost development during the Younger Dryas (Vandenberghe *et al.*, 1987), while periglacial indicators of discontinuous permafrost have been found in Belgium (Pissart, 1987). In Britain, the distribution of relict periglacial phenomena of Loch Lomond Stadial age suggest active, if discontinuous, permafrost in Scotland, the Isle of Man and Wales, seasonally frozen ground in the Midlands and eastern England, and only intermittently frozen ground in the south-west where cool temperate conditions prevailed between 11 ka and 10 ka BP (Rose *et al.*, 1985). With the establishment of full interglacial conditions around 10 ka BP, permafrost disappeared almost completely from western Europe, and at the present day discontinuous permafrost is to be found only in the mountains of northern Fennoscandia (Seppala, 1987), on the Kola Peninsula, and in central and southern Iceland (Washburn, 1979). However, periglacial conditions (albeit in the absence of permanently frozen ground) exist in all mountain regions of central and northern Europe, and even in the British Isles where the climate of the uplands is characterized by extreme wetness and strong winds rather than severe cold, active periglacial geomorphological processes occur (Ballantyne, 1987).

In eastern and central North America, relict periglacial phenomena have been found in the Appalachian Mountains as far south as South Carolina, although there are no unequivocal indications of Late Wisconsinan permafrost. In the western cordillera, alpine permafrost bounded the mountain glacier complexes, with the permafrost zone perhaps continuing up to 100 km south of the main mountain ice sheet. To the east, the permafrost zone that fronted on to the Laurentide ice sheet (Figure 4.6) extended south for some 700 km in the high plains region of Colorado and Kansas, but narrowed eastwards towards the northern Appalachians (Péwé, 1983a). Proxy data indicate that at the Wisconsinan maximum, temperatures in the permafrost zone may have been depressed by over 16 °C in eastern areas (Walters, 1978), although further west values in the range 10–13 °C have been inferred (Washburn, 1979). Climatic simulation modelling suggests that around 18 ka BP, mean annual temperatures immediately to the south of the Laurentide ice sheet were up to 10 °C below those of the modern times, whereas in the southern United States temperatures were only a few degrees below those of the present day (Kutzbach, 1987). During the retreat of the ice sheet from around 14 ka BP onwards, the continental periglacial region migrated northwards. This zone, which ranged from 80–250 km wide over much of its length, was narrower than the belt in Europe, reflecting the more southerly position of the ice margin in the USA, and also the influence of extensive proglacial water bodies during the retreat phase of the ice sheet that reduced the permafrost area and helped to ameliorate climate (French, 1976).

Large areas of North America have remained under the influence of a periglacial climatic regime to the present day, a pattern that became established following the eventual disappearance of the Laurentide ice sheet during the Early/Mid-Holocene. Continuous or discontinuous permafrost characterizes 80–85 per cent of Alaska and 50 per cent of Canada, while over 100 000 km^2 of alpine permafrost have been mapped in the United States, mostly in the western cordillera extending as far south as the Mexican border (Péwé, 1983b).

Sea-level change

Few areas of the world have experienced any degree of coastal stability during the course of the Quaternary, but in those areas of the mid-latitudes that lay around the fringes of the great ice sheets, the effects on landscape of changing levels of land and sea have been particularly pronounced. This section examines the components of sea-level

Figure 4.6   Distribution of permafrost and periglacial features in the United States (after Péwé, 1983a)

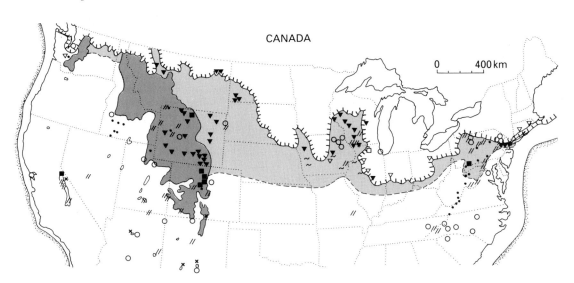

Explanation
⊤⊤⊤⊤ Edge of continental ice sheet in late Wisconsinan time
‾‾‾‾ Lowest shoreline in Wisconsinan time
▓▓ Areas of alpine permafrost
░░ Areas of continuous and discontinuous permafrost
- - - - Speculative southern limit of discontinuous permafrost

Features indicative of past permafrost
▼ Ice-wedge cast
▽ Doubtful ice-wedge cast
□ Relict pingo field
■ Inactive cryoplanation terrace
× Inactive rock glaciers (not in cirques)

Inactive features of past rigorous climate but not necessarily requiring permafrost
• Patterned ground
// Rock streams, block streams and fields, talus, rubble sheets.
○ Solifluction deposits and other colluvia
~ Involutions and frost-stirred sediments.

change in the Atlantic basin region and considers the effects of fluctuating sea levels on the coastlines of Europe and North America at the close of the last cold stage and during the present interglacial.

Changing levels of land and sea reflect the interplay of two major elements: global (**eustatic**) changes in sea level and localized tectonic activity which results in vertical displacement of the land. These are known as **isostatic** movements, the term **isostasy** referring to the state of balance that exists in the earth's crust so that depression in one locality will be compensated for by a rise in the crust elsewhere (Fairbridge, 1983). Over successive glacial/interglacial cycles, the dominant control on both components is the expansion and contraction of the continental ice sheets. Hence, global sea-

level changes that result from the repeated abstraction of water from, and subsequent return to, the ocean basins are referred to as **glacio-eustatic changes**; similary crustal deformation caused by loading of glacier ice is termed **glacio-isostasy**. Changes in sea level that take place through the interplay of these and other factors are known as *relative sea-level changes*, i.e. a change in the position of sea level relative to the land; such changes are essentially local in effect. In tectonically stable areas, evidence for sea-level change should, in theory, reflect only the eustatic component and it has often been suggested that in such regions a sequence of *absolute* sea-level changes can be established. Inductive reasoning has then been used to establish regional sea-level

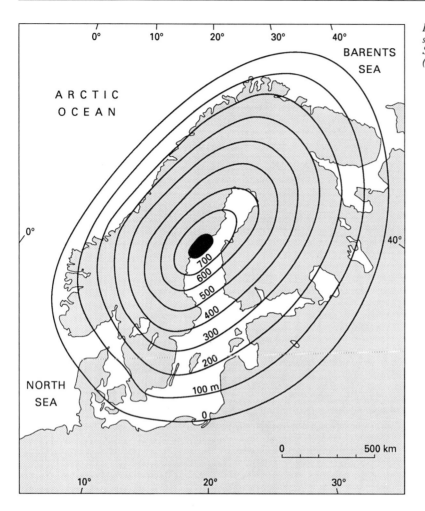

Figure 4.7   Isobases (in metres) showing absolute uplift of Scandinavia during the Holocene (after Mörner, 1980a)

histories (Shennan, 1987). However, in view of the complexity of land and sea-level movements it remains questionable whether estimates of sea-level change which are in any way meaningful at the global scale can be obtained from shoreline data alone (Peltier, 1987).

At the height of the last cold stage around 18 ka BP, sufficient volume of water had been removed from the ocean basins and stored in the continental ice sheets and glaciers to reduce global sea level by around 130 m (Bloom, 1983; Chappell and Shackleton, 1986). This glacio-eustatic lowering was accompanied by the isostatic depression of Fennoscandia, northern Britain and Canada through glacial loading. Following the wastage of

the Northern Hemisphere ice sheets, which may have begun as early as 16 ka BP (Ruddiman and Duplessy, 1985), global sea level rose steadily while thinning and eventual wastage of the continental ice sheets resulted in rapid glacio-isostatic recovery. Shorelines that formed around the margins of the melting ice sheets were progressively raised above sea level as isostatic uplift outstripped eustatic sea-level rise, these raised beaches being progressively tilted away from the centre of isostatic depression. Detailed analysis of raised shorelines and associated features provides evidence of the extent of isostatic uplift since deglaciation (Rose, 1981). In Scandinavia, for example, isobase maps (maps showing lines of equal uplift) indicate that in the

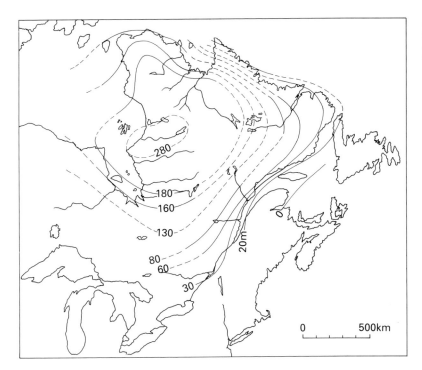

Figure 4.8   Isostatic uplift (in metres) in eastern Canada since 7.5 ka BP (after Hillaire-Marcel & Occhietti, 1980)

Baltic region, over 700 m of uplift has taken place during the Holocene (Figure 4.7), while in eastern Scotland over 250 m of uplift has occurred since deglaciation, the amount of uplift further west near the centre of the ice sheet being considerably greater (Sissons, 1967). In North America, over 200 m of uplift has occurred in the area to the east of Hudson Bay in the last 7 ka alone (Figure 4.8), and the total amount of depression near the ice sheet centre inferred from glaciological modelling (Denton and Hughes, 1981) may have exceeded 900 m.

In all of these areas, the process of land emergence has continued throughout the Holocene. In Scandinavia (which is the classic locality for the study of glacio-isostasy), the detailed chronology of shoreline displacement that has been developed for the Gulf of Bothnia area of the northern Baltic (Broadbent, 1979) shows a steady decline in uplift over the past 7 ka (Table 4.1), although land is still rising in this region at a rate of around 0.8 to 0.9 cm yr^{-1} (Figure 4.9). In other words, the shoreline that formed during the Viking

period is now almost 8 m above that of the present day! In Scotland, total glacio-isostatic uplift since the Loch Lomond Stadial is in the order of 40–50 m (Sissons, 1983), and is still incomplete. Raised shoreline data indicate that in the inner Forth, Clyde and Tay Valleys of Scotland, current rates of uplift range from 1.8 to 2.0 mm yr^{-1} (Shennan, 1989), and hence near the centre of isostatic recovery in the west-central Grampian Highlands uplift rates in excess of 2.0 mm yr^{-1} may be envisaged. Near the centre of the former Laurentide ice sheet where over 200 m of uplift has occurred (Figure 4.8), rates of isostatic recovery have declined from around 6 cm yr^{-1} at 8 ka BP to around 1.1 cm yr^{-1} at the present day (Hillaire-Marcel, 1980), and it has been estimated that at least a further 160–180 m of isostatic recovery remains to be completed in this area (Devoy, 1987). On the east coast of North America, emergence continued in the Maritime Provinces of Canada and in New England until at least 10 ka BP (Bloom, 1983b).

In the southern part of the North Sea basin,

Table 4.1 Shoreline displacement calculations for northern Vasterbotten, Gulf of Bothnia, Sweden, using varve chronology, archaeological and botanical data. Curve 2, which has two dendrochronologically-corrected points, probably provides the most accurate estimates of displacement rates above 49 m; between 49–24 m, the rates were probably intermediate between Curves 1 and 2; below 24 m, the rates of Curve 1 are considered to be the most accurate (based on Broadbent, 1979)

CURVE 1 (uncalibrated)			CURVE 2 (calibrated)		
Year BP	Metres above sea level	Uplift rate cm/year	Year BP	Metres above sea level	Uplift rate cm/year
9000	186.920	3.991	9000	141.778	2.752
8500	167.765	3.675	8500	128.482	2.569
8000	150.125	3.384	8000	117.070	2.398
7500	133.882	3.117	7500	104.484	2.238
7000	118.924	2.870	7000	93.668	2.090
6500	105.150	2.643	6500	83.572	1.951
6000	92.466	2.434	6000	74.147	1.821
5500	80.786	2.241	5500	65.349	1.700
5000	70.030	2.064	5000	57.136	1.587
4500	60.126	1.900	4500	49.469	1.481
4000	51.005	1.750	4000	42.313	1.383
3500	42.606	1.611	3500	35.632	1.291
3000	34.872	1.484	3000	29.396	1.205
2500	27.751	1.366	2500	23.574	1.125
2000	21.192	1.258	2000	18.140	1.050
1500	15.153	1.159	1500	13.067	0.980
1000	9.592	1.067	1000	8.331	0.915
500	4.471	0.983	500	3.911	0.854
0	−0.245	0.905	0	−0.216	0.797

isostatic depression has continued throughout the Holocene (Figure 4.9). There are indications that the North Sea has long been a focus for crustal downwarping, a trend which has been particularly pronounced following deglaciation and the collapse of the glacio-isostatic forebulge (the marginal displacement of the crust involving upbulging beyond the ice margin) associated with the Fennoscandian ice sheet (Fairbridge, 1983). This, along with continued sediment loading from inflowing rivers has caused up to 170 m of subsidence in the southern North Sea since regional deglaciation (Mörner, 1980b). Similar downwarping trends over the past 10 ka are apparent along the east coast of the USA, with maximal subsidence being recorded in the vicinity of Delaware where over 40 m of depression has

occurred since *c.* 10 ka BP (Newman *et al.*, 1980).

The pattern of absolute change in eustatic sea level at the close of the last cold stage has hitherto proved difficult to establish, principally because of the limitations of shoreline data outlined above. In recent years, however, new approaches involving, *inter alia*, oxygen isotope evidence from deep-ocean cores and uranium series dating of submerged coral reefs in low latitude regions, are beginning to provide an indication of global sea-level trends during the period of ice wastage (Bard *et al.*, 1989, 1990b). The data suggest that at 14.5 ka BP, sea levels stood at around −100 m, but a rapid rise of 40 m occurred over the next millennium (*c.* 3.7 m per century). A second major glacier melting phase around 11 ka BP (*c.* 2.5 m per century) raised eustatic sea level to around −40 m by the

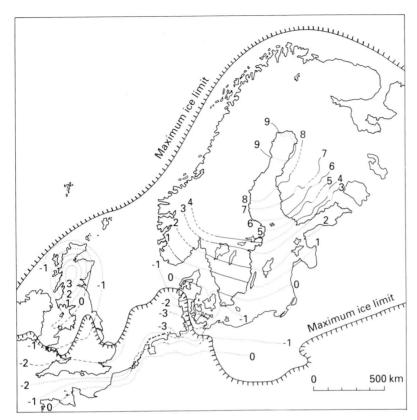

Figure 4.9   Present pattern of land uplift and subsidence in north-west Europe based on tide gauge data. The isobases show the rate of change in mm yr^{-1}. Dashed isobases are less certain, whilst dotted isobases are based on interpolation (after Devoy, 1987)

beginning of the Holocene, by which time global ice volumes had been reduced by over 50 per cent. During the Lateglacial period, however, rapid glacio-isostatic recovery in north-west Europe and North America outstripped eustatic sea-level rise and hence shorelines formed during that period now stand well above those of the present day. In Maine isostatic uplift has raised marine deposits up to 150 m (Bloom, 1984a); in Scotland the highest Lateglacial shorelines (c. 13 ka BP) now stand at 50 and 41 m OD on the east and west coasts respectively (Jardine, 1982), while in the Oslofjord area of Norway, Late Weichselian/Early Holocene shorelines have been raised by over 220 m (Hafsten, 1983). In the early Holocene, however, eustatic sea-level rise (at rates of around 1 cm yr^{-1}) began to exceed the rate of isostatic recovery in many areas. This resulted in a major marine transgression around the coastline of Scotland

between c. 8.5 ka and 6.5 ka BP (Sissons, 1976), in the flooding by the sea after c. 7.5 ka BP of the extensive freshwater lake that had occupied the Baltic following deglaciation (Eronen, 1983), and in the penetration of marine waters into Hudson Bay around c. 8 ka BP which led to the rapid disintegration of the remaining mass of the Laurentide ice sheet. The subsequent rapid isostatic recovery in this part of Canada resulted in shorelines that formed around 7 ka BP being raised by over 200 m (Peltier and Andrews, 1983).

In the North Sea area, the Dogger Bank was submerged beneath the rising Holocene sea by 8.7 ka BP, while further south the Straits of Dover were breached before 8 ka BP. Following a period of rapid marine inundation, the present configuration of the coastline of the southern North Sea basin was more or less established by 7.5 to 7.8 ka BP (Devoy, 1987). It is worth noting,

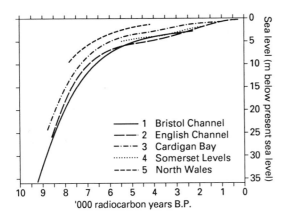

Figure 4.10   Curves showing Holocene sea-level rise in southern Britain: 1. Bristol Channel; 2. English Channel; 3. Cardigan Bay; 4. Somerset Levels; 5. north Wales (after Shennan, 1983)

Fennoscandia was so rapid that on parts of the Norwegian coast, for example, isostatic emergence kept pace with or even exceeded eustatic sea-level rise throughout the Holocene (Hafsten, 1983). In the southern North Sea, continuing crustal subsidence meant that the maximum of the Holocene marine transgression was delayed until after 4 ka BP (Jelgersma, 1979). Evidence from the Fenland region of eastern England shows that even in this peripheral area, subsidence over the past 6.5 ka has averaged 0.9 cm yr^{-1}, while sea level has risen at an average rate of 0.1 cm yr^{-1} since 3 ka BP, although there are short periods of enhanced and reduced rates (Shennan, 1986b).

In eastern North America, relative sea level around 10 ka BP stood at −25 to −32 m, after which rapid Early Holocene submergence led to rapid inundation of much of the coastal zone (Bloom, 1984). Deceleration in the rate of sea-level rise is apparent after around 7 ka BP (Figure 4.11), although since around 6 ka BP, the process of submergence has continued with a relative sea-level rise (reflecting both eustatic rise and crustal subsidence) of between 3 and 17 m along the coastline between New York and South Carolina (Newman *et al.*, 1980). Further north where

however, that the separation of Britain from Europe occurred some 2–3 ka after the severing of the last land bridge between mainland Britain and Ireland which resulted from eustatic sea-level rise during the Lateglacial (Devoy, 1985). Estimates of Holocene sea-level rise in southern Britain (Figure 4.10) suggest that in the Bristol Channel area sea level rose from around −35 m OD (Ordnance datum) at 9.5 ka BP to <−5 m OD at 5 ka BP at the following rates: 9–8 ka BP – *c.* 1.2 cm yr^{-1}; 8–7 ka BP – *c.* 0.9 cm yr^{-1}; 7–6 ka BP – *c.* 0.4 cm yr^{-1}; 6–5 ka BP – *c.* 0.2 cm yr^{-1} (Heyworth and Kidson, 1982). In the Thames estuary, broadly comparable figures of *c.* 1.3 cm yr^{-1} between 8.5 and 7 ka BP and 0.5 cm yr^{-1} between 6.5 and 5.0 ka BP have been obtained (Devoy, 1979).

After around 6 ka BP, marine incursions into coastal areas of north-west Europe took place more slowly. The configuration of the British coastline was comparable, in terms of its principal elements, with that of the present day, although considerably more indented due to the drowning of wetlands and estuaries which have subsequently silted up (Figure 5.12). Isostatic recovery in northern England and Scotland had once again outstripped eustatic sea-level rise, and the same was true of Scandinavia. Indeed, the rate of isostatic uplift in some areas of

Figure 4.11   Curve showing Holocene sea-level rise in eastern north America: 1. SE Massachusetts (Oldale & O'Hara, 1980); 2. Delaware coast (Belknap & Kraft, 1977); 3. Bermuda; 4. Florida (Newman et al., *1980)*

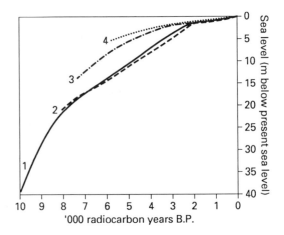

isostatic readjustment continued into the Holocene, slow marine transgression has been recorded on the Maine coasts between 4 ka and 3 ka BP (Bloom, 1984).

In conclusion it should be noted that eustatic sea level is still rising. Tide-gauge data from numerous localities around the world suggest that over the course of the last century, global sea levels have risen by 10–15 cm (Gornitz *et al.*, 1982; Barnett, 1984). As this may partly reflect the anthropogenically induced global warming of recent decades (Titus, 1987), present and future sea-level trends are considered more fully in Chapter 9.

Vegetational and pedological changes

Late Quaternary climatic change had dramatic effects on the terrestrial biosphere. In the temperate mid-latitude regions of the Northern Hemisphere, vegetational belts migrated over several thousands of kilometres in response to the fluctuating rhythms of successive glacial/interglacial cycles. These macroscale vegetational changes were accompanied by (and in some instances were partly in response to) changes in soil properties, and while these pedogenic variations were governed primarily by climatic changes, they also reflect the feedback effects from the regional vegetation cover. The ways in which soil–vegetation relationships have evolved over the course of the Late Quaternary, and particularly the response of vegetational and pedogenic processes to climatic change, are of fundamental importance in understanding the dynamics of contemporary ecosystems. This section outlines the major vegetational and pedological changes that have occurred over the closing stages of the last glacial and during the present interglacial cycle. The data base is provided principally by pollen analysis, augmented by evidence from plant macrofossils and buried palaeosols.

The pattern of biotic changes during the Late Quaternary can most easily be understood by reference to a simple model of ecological processes,

the so-called interglacial cycle. Such a model was outlined by Iversen (1958) to explain the sequence of vegetational changes that occurred over successive interglacials in north-west Europe, and this scheme has subsequently been further elaborated by Birks (1986). The model envisages vegetation and soils changing in response to the influence of climate over four successive stages:

1. *Cryocratic phase*: This is the glacial stage when climate is cold, dry and continental in the extra-glacial areas. Vegetation consists of sparse, species-poor, arctic–alpine or steppe herbaceous communities growing on skeletal mineral soils, often underlain by permafrost, which are disturbed by cryoturbation.

2. *Protocratic phase*: This is the first stage of an interglacial and, in response to rising temperatures, the steppe and tundra communities are replaced successively by species-rich grassland, scrub and open woodland. Soils are base-rich and fertile with low humus content: leaching is at a minimum.

3. *Mesocratic phase*: This corresponds with the climatic optimum of an interglacial, and is characterized by the expansion and establishment of closed temperate deciduous woodland. Tree immigration varies with species and so woodland diversification occurs throughout this episode. Soils are typically brown earths, initially at least of high base-status.

4. *Oligocratic phase*: Soil retrogression from brown earths to podzols and, in some areas, to acid peats leads to progressive elimination of some of the nutrient-demanding plants of the mesocratic phase. Climate begins to deteriorate towards the end of this stage, with vegetation changing from mixed woodlands to open conifer-dominated woods, ericaceous heaths and bogs. Some protocratic plants expand, however, with the opening up of the woodland. Major climatic deterioration heralds the arrival of the succeeding cryocratic phase as the woodlands disappear, the acidic soils are destroyed by gelifluction and cryoturbation, and herbaceous communities become established once again on the skeletal, mineral soils.

In Britain, pollen analytical evidence has shown that for the cold stages, it is possible to differentiate between full glacial and interstadial spectra, and that for interglacials the data can be resolved into a series of pollen assemblage zones reflecting sequential vegetation changes characteristic of successive interglacial cycles (Turner and West, 1968; West, 1970):

Zone I   Pre-temperate zone: This zone is characterized by rising values for arboreal pollen, particularly of the boreal trees such as birch (*Betula*), and pine (*Pinus*). Light-demanding shrubs and herbs are initially dominant, but rapidly decline.

Zone II   Early temperate zone: A zone dominated by trees associated with mixed-oak woodland such as oak (*Quercus*), elm (*Ulmus*), and hazel (*Corylus*). Non-arboreal pollen is minimal.

Zone III   Late temperate zone: This zone reflects the expansion of trees not present in zone II including hornbeam (*Carpinus*), alder (*Alnus*), and spruce (*Picea*) and the gradual disappearance of some of the previously established mixed-oak forest genera.

Zone IV   Post-temperate stage: Boreal trees are dominant once more, but there is a significant increase in non-arboreal pollen, reflecting opening of the woodland and the expansion, in particular, of heathland plants (*Ericaceae*).

Vegetational developments in the mid-latitude regions of the Northern Hemisphere over the past 20 ka will now be examined in the context of the framework provided by these two models.

At the end of the last cold stage the periglacial zone (see above) was characterized by a vegetation for which there are no known contemporary analogues (Birks, 1986). The vegetation cover of this crycocratic phase was frequently sparse, with large areas of bare ground and disturbed or moving soils. The flora consisted largely of herbaceous taxa of mixed ecological and geographical affinities, including arctic, alpine, steppe, southern, ruderal, marsh and halophytic taxa (Iversen, 1973; West, 1977b; Davis, 1984). Such diverse floristic elements reflect the polar desert environment that

existed beyond the ice margins, where a markedly continental climatic régime prevailed, but within which local variations in edaphic conditions, exposure, and slope stability provided microhabitats for a range of plant communities. Arctic structure soils, often with loessic materials incorporated into the upper horizons, developed above the permafrost table (Kemp, 1986; Rose *et al.*, 1985), these soils being recognizable today by the presence of such relict periglacial features as ice wedge casts, sand wedges and cryoturbation structures (Figures 2.21 and 2.22).

South of the European periglacial zone, steppe characterized by mugwort (*Artemisia*) and associated dryland herbs was to be found across much of southern and south-eastern Europe, although there are indications in the pollen records from the Iberian peninsula of pine and oak woodland in the south of Spain (Watts, 1980; Huntley and Birks, 1983; Huntley, 1990). In the eastern and central United States, palynological data suggest that boreal woodland was widespread throughout the last cold stage with *Picea* and *Pinus* as the dominant elements, although *Quercus* was locally present (Watts, 1983; Webb *et al.*, 1984). Extensive coniferous forests also blanketed the mountain slopes of the western cordillera (Spaulding *et al.*, 1983).

The protocratic phase of the Iversen model began in north-west Europe around 14 ka BP and ended shortly after 9 ka BP (Birks, 1986). As such it includes the Lateglacial climatic fluctuation and the first millennium of the Holocene. The thermal improvement that began around 13 ka BP saw the gradual replacement of steppe/tundra by open boreal woodland across large areas of Europe (Figure 4.12). By 11 ka BP, spruce forest was extensive in the east, while a belt of birch–conifer woodland extended from Poland south-eastwards into central France. Steppe vegetation still characterized much of southern Europe, although such areas were less extensive than during the Weichselian, while mixed deciduous forests were widespread in the south of the Iberian peninsula. In western Europe, the open tundra was invaded by shrubs, and in lowland areas extensive stands of tree birch became established. The uplands of

Figure 4.12   Isopollen maps showing patterns of vegetation change in north-west Europe during the Lateglacial period (after Huntley & Birks, 1983)

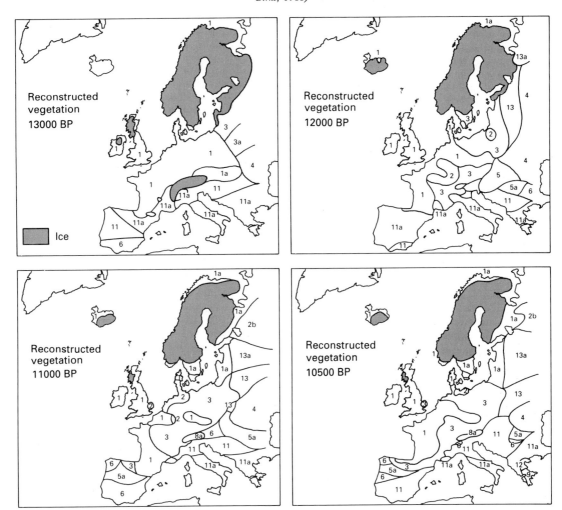

1 Tundra; 1a Tundra 'xeric' variant; 2 Birch forest; 2b Birch forest 'xeric' variant; 3 Birch-conifer forest; 4 Spruce-dominated forest; 5 Northern mixed conifer-deciduous forest; 5a Northern mixed conifer-deciduous forest: *Pinus* variant; 6 Mixed-deciduous forest; 8a Montane mixed-conifer forest: *Pinus* (Haploxylon)-*Larix* variant; 9 Mediterranean forest; 10 Xerix Mediterranean vegetation; 11 Steppe; 11a Steppe: treeless variant; 12 Grassland; 13 Mixed forest; 13a Mixed forest: xeric variant.

western Britain supported open grassland and acidophilous crowberry (*Empetrum*) heath (Lowe and Walker, 1984). Only limited pedological data are available for this episode, but buried palaeosols show accumulation of raw humus in wetter areas or shallow incorporation of organic matter in rendzinas (Catt, 1979). At a number of sites in southern Britain, a rendzina-type soil (the 'Pitstone

Soil') of Lateglacial Interstadial age has been identified (Kemp, 1986), and parallels have been drawn with palaeosols that may be correlatives in The Netherlands and in Belgium (Rose *et al.*, 1985).

The abrupt climatic deterioration of the Loch Lomond/Younger Dryas Stadial was reflected in significant changes in the vegetation pattern

Figure 4.13　Isopollen maps showing patterns of vegetation change in north-west Europe during the early and mid-Holocene periods (after Huntley & Birks, 1983)

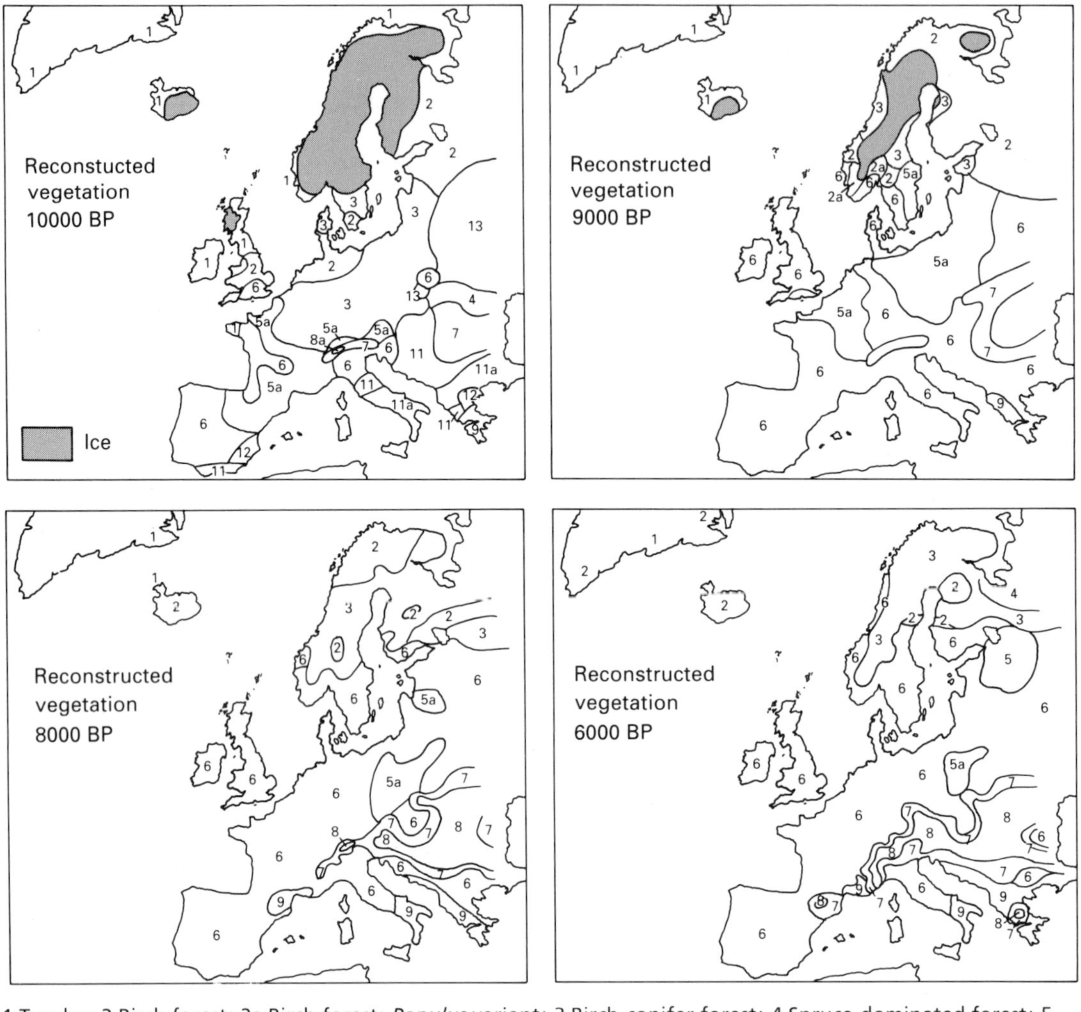

1 Tundra; 2 Birch forest; 2a Birch forest: *Populus* variant; 3 Birch-conifer forest; 4 Spruce-dominated forest; 5 Northern mixed conifer-deciduous forest; 5a northern mixed conifer-deciduous forest: *Pinus* variant; 6 Mixed deciduous forest; 7 Montane mixed conifer-deciduous forest; 8 Montane mixed-conifer forest; 8a Montane mixed conifer forest: *Pinus* (Haploxylon)-*Larix* variant; 9 Mediterranean forest; 11 Steppe; 11a Steppe: treeless variant; 12 Grassland; 13 Mixed forest.

throughout the extra-glacial areas of Europe, but especially along the Atlantic seaboard. Tundra extended from southern Sweden through the British Isles and across much of France, while in north-east Europe, xeric variants of tundra, mixed forest and birch forest were widespread. Steppe dominated southern Europe with deciduous woodland restricted to north-west Iberia.

Birch–coniferous forests were confined to a wedge trending north-eastwards from Iberia across central Europe (Huntley and Birks, 1983). Throughout north-west Europe, the process of soil maturation that began during the preceding interstadial was disrupted by cryoturbation processes as arctic structure soils developed once again in many areas in response to the prevailing periglacial climatic regime.

Table 4.2 Summary of the Holocene pollen stratigraphy of the Eastern United States (after Davis, 1984)

Age (years)	Northern New England	Southern New England	Central Appalachians	Coastal Plain	Central Florida
2000	Expansion of boreal elements		Oak, hickory sweetgum, pine		
				Pine, oak, hickory	Sand pine scrub
	Deciduous forest Hemlock decline		Hemlock decline		
5000	Expansion of temperate elements		Oak, hickory, sweetgum, chestnut	Oak, sweetgum hickory, pine	
	Mixed forest				
		Mixed forest			
10 000	Tundra	Open spruce woodland	Oak, hornbeam spruce	Mesic deciduous forest	Oak scrub
12 500		Tundra	Spruce forest with pine and deciduous trees	Pine and spruce forest	
	Ice				
15 000		Ice	Jack pine, spruce		?

Rapid climatic amelioration between 10.3 and 10 ka BP led to dramatic changes in the vegetation of Europe (Figure 4.13). By 9 ka BP, large areas of northern and western Europe had witnessed the replacement of tundra first by open grassland, and then successively by dwarf shrub heath with juniper (*Juniperus*), and willow (*Salix*), *Betula* forest and finally *Betula–Corylus* woodland. Birch forest also covered much of newly deglaciated Fennoscandia, with birch–conifer forests well established around the mountain ice sheet and glaciers in central Norway and Sweden (Huntley and Birks, 1983; Huntley, 1990). *Pinus*-dominated woodland covered much of the lowlands of northern Europe, while in the south mixed woodland extended from the Iberian peninsula to Greece. By this time,

steppe and tundra communities had been virtually eliminated from mainland Europe. This phase of vegetational development, the end of the protocratic stage of Iversen or the pre-temperate zone of Turner and West, was also characterized by the widespread development of basiphilous soils, typically brown earths (Catt, 1979).

In North America during the Lateglacial, the tundra belt along the ice sheet margin was progressively invaded by boreal forest trees (Jacobson *et al.*, 1987). In southern New England, where deglaciation occurred around 13 ka BP, open tundra was succeeded by scrub and then by *Picea* woodland, spruce becoming widely established by 12–11 ka BP (Table 4.2). Subsequently, the spruce forests were invaded by pine and larch and after around 10 ka BP *Picea* was progressively replaced as a woodland tree by such mixed woodland species as birch, oak, hemlock, pine and alder. Further north, the scrub tundra persisted until after 10 ka BP and was succeeded directly by mixed forest (Davis, 1984). Mixed woodland also replaced the boreal forests further south. To the west, the tundra vegetation of Minnesota was replaced by a *Picea–Larix–Betula* woodland around 12 ka BP, and this spruce forest dominated what is now the northern prairie zone during the Lateglacial phase. It existed in the Great Lakes region until an invasion by pine around 10.7 ka BP established forest similar to that of the southern boreal forest at the present day. In the Midwest, a deciduous forest of diverse composition succeeded the *Picea* phase and was established by 10 ka BP (Webb *et al.*, 1984). Further west, the *Picea* woodland was succeeded directly by prairie by *c.* 9 ka BP (Watts, 1984). Pedological data for the Lateglacial are limited, but basiphilous, nitrogen-deficient skeletal soils seem to have been widespread on newly deglaciated sites and throughout the tundra zone where gelifluction and frost heaving continued until the establishment of closed forest.

The mesocratic stage of the interglacial cycle (Zone II – early temperate zone in Britain) spans the period from around 9 ka to 5 ka BP. In north-west Europe (Figure 4.13), this stage is characterized by the immigration and expansion of temperate deciduous trees which formed dense mixed woodland dominated by *Corylus*, *Quercus*, *Ulmus*, and lime (*Tilia*), and by the widespread development of fertile brown earth forest soils with mull humus (Birks, 1986). The mixed forest zone extended from southern Spain to Poland and around the southern fringes of the Alps where montane deciduous and mixed deciduous–coniferous woodland occurred. To the south, Mediterranean forest developed in southern Italy and in Greece, while in Fennoscandia birch coniferous forests blanketed the mountains, flanked by mixed coniferous–deciduous woodland on the lower slopes (Huntley and Birks, 1983; Huntley, 1990). In terms of immigration and expansion, *Corylus* was the first mesocratic tree to attain dominance and was widely established by 9 ka BP.

By 8 ka BP, *Quercus* and *Ulmus* had expanded across much of north-west Europe, followed between 8 ka and 7 ka BP by *Alnus* and between 6 ka and 5 ka BP by ash (*Fraxinus*) (Birks, 1989). The expansion of alder, which probably replaced *Salix*, *Betula* and *Pinus* on damp or waterlogged sites, may have heralded a shift to wetter climatic conditions in western Europe which appears to have begun around 7 ka BP. Evidence in the upland regions of the beginnings of hydromorphism, including gleying and ombrogenous peat growth, and also podzolization (Guillet, 1982), are further evidence of this climatic shift. In some lowland areas also, the early stages of soil deterioration along the continuum argillic brown earth to brown podzolic soil to podzol (Anderson *et al.*, 1982) are apparent during the later part of the mesocratic stage. It is possible that these pedological changes may be due, at least in part, to Mesolithic anthropogenic activity (Chapter 6).

In the eastern United States throughout the mesocratic stage mixed forest existed (Table 4.2), its composition undergoing progressive changes as new species (oak, hemlock, hickory, chestnut) migrated northwards (Davis, 1984). With the establishment of mixed woodland in the Holocene, sterile, skeletal soils of the Lateglacial had been replaced by brown earths and forest soils. Further west, the spruce and pine woodland of the Lateglacial was succeeded by mixed woodland in the region of the Great Lakes and southwards towards the valley of the Mississippi. By 10 ka BP, *Quercus*, *Ulmus* and *Pinus* were well established, along with a range of mesic deciduous trees including hazel, ash and hickory (Jacobson *et al.*, 1987). As in the east, diversification of the woodland occurred with the subsequent immigration or expansion during the Mid-Holocene of other tree species including beech (*Fagus*), hemlock (*Tsuga*) and maple (*Acer*). To the west of the mixed forest zone in the Great Plains region, *Picea* woodland was succeeded at the opening of the Holocene by prairie. Movements of the prairie–woodland ecotone in the Midwest of the USA have occurred throughout the Holocene (Webb *et al.*, 1984), with the prairie reaching its maximum expansion north-eastwards into Minnesota around 7 ka BP (Ruhe, 1983). The transition from Lateglacial to Holocene conditions in the mountains of western North America has been reflected primarily in a retreat of the tundra belts to about their current altitudinal position, and the establishment of contemporary forest zones at lower altitudes (Baker, 1984).

In western Europe, the vegetation and pedological changes of the Late Holocene (Oligocratic phase; Late Temperate zone) become increasingly difficult to interpret, because of the widespread evidence for anthropogenic impact on landscape (Chapter 6). From c. 5 ka BP onwards, the European vegetation patterns were disrupted and a retrogressive sequence of vegetational changes resulted in the progressive replacement of deciduous woodland by an incompletely forested landscape with heaths, extensive grasslands and, in upland regions particularly, blanket mires (Huntley and Birks, 1983). The climatic shift to wetter conditions, which became especially marked after c. 2.5 ka BP led to accelerated leaching of soil nutrients, the change in the brown soils from mull to mor humus and the widespread development of a range of acid podzolic soils, including gley-podzols, stagnopodzols and podzols *sensu stricto* (Catt, 1979). Hydromorphic activity, which had previously been associated primarily with upland areas now became characteristic of lowland regions throughout western Europe, the process of soil degradation being further accelerated by anthropogenic clearance of woodland (Macphail, 1986). The results of soil deterioration and the progressive climatic shift to cooler and moister conditions led to a decline in those tree species requiring deep, fertile soils. Hence *Ulmus*, *Tilia* and *Corylus* were replaced progressively by *Quercus* and *Betula* from around 6 ka BP onwards, while in the mountains extensive blanket mire development led to a progressive fall in treeline altitudes (Birks, 1986). By 4 ka BP, the upland regions of the maritime fringes of north-west Europe were essentially treeless, and covered largely by acidophilous heathland, grassland and ombrogenous blanket mire.

The effect of human impact on the vegetation cover of North America is less marked than in western Europe, and hence the sequence of natural vegetational retrogression during the oligocratic stage of the present interglacial can be more readily discerned. The climatic cooling of the Late Holocene is most apparent in the reappearance of spruce in the pollen records from the northern United States from around 4 ka BP onwards (Webb *et al.*, 1983). In the Appalachians, the increase in *Picea* began even earlier around 5 ka BP (Watts, 1979). Over the past two millennia, the reappearance of spruce in parts of New England has been accompanied by an increase in fir and alder and a decline in beech and hemlock, trends which are attributed by Davis (1984) to a decrease in temperature or a rise in moisture or both. Increasing climatic wetness in the Midwest resulted in the development of the extensive patterned peatlands of Minnesota which date from around 4 ka BP (Griffin, 1975), the time at which *Picea* began to reappear in the pollen records, and in the westward retreat of the prairie/forest border (Webb *et al.*, 1984).

Palaeohydrological changes

River systems are extremely sensitive to climatic change and relict fluvial landforms throughout the temperate zone bear witness to the major climatic and environmental changes that have occurred over the course of the past 20 ka. The principal factor governing hydrological changes is precipitation, but the impact of rainfall variations on fluvial systems is modulated by other landscape components, notably soils and vegetation, which are themselves controlled, *inter alia*, by climate. Hence, variations in river incision, terrace development, channel change, sediment loading, etc., are determined either directly or indirectly by the course of climatic change. During the Late Holocene, however, it is now apparent that people exerted an increasing influence on hydrological processes (Chapter 7). This section is concerned with those areas of palaeohydrology that reflect climatic as opposed to anthropogenic control.

Although relict fluvial landforms and deposits provide clear indications of the climatic and landscape changes that occurred during the transition from the last cold stage to the present interglacial, reconstructing the chain of processes and responses that connects palaeoclimatic and palaeohydrological events to discernible field evidence is far from straightforward (Baker, 1984). In particular, the problems of equifinality (Chapter 2) loom large in palaeohydrological studies. A good example is provided by river terraces. Terraces represent former floodplains abandoned by the river as it becomes incised, and are major components of a fluvial landscape. Sequences of terrace fragments are found in most river valleys of the temperate zone and careful analysis of the altitude, morphology and stratigraphy of the terraces can form the basis for reconstructions of fluvial histories (Gibbard, 1985). Moreover, in so far as terraces are adjacent to, but raised above, the contemporary river channel, they provide potential sites for settlement and are, therefore, significant in an archaeological or historical context (Limbrey, 1983). Terrace formation clearly results from changes in river level, but this may be brought about by a range of factors, some of which are external such as fluctuations in sea level, tectonic processes (e.g. isostatic uplift), changes in precipitation, and periodic inputs of glacial meltwaters, whereas others (such as variations in nature and quantity of the sediment load) are internal to the system (Clayton, 1977; Dawson and Gardiner, 1987). Inferring the major controlling variable (process) from the terrace fragments themselves (form) requires a multiple-working hypothesis approach in which all possibilities are eliminated save one (Chapter 2), and this may not always be possible on available field evidence. Furthermore, understanding temporal variations in fluvial processes and the resulting landforms and sedimentary sequences is exacerbated by the spatial variation in fluvial activity that characterizes most river systems, with local changes in fluvial regime resulting from the transgression of key geomorphological thresholds (Schumm, 1979). However, despite these and other difficulties that have served to constrain palaeoenvironmental inferences from relict fluvial landforms and deposits, significant advances have been made in the field of palaeohydrology over the past two decades (e.g. Gregory, 1983; Gregory *et al.*, 1987). Consequently, it is now possible to begin to reconstruct the pattern of fluvial response to climatic change at the end of the last cold stage and during the present interglacial over large areas of the northern temperate zone.

It has been estimated that during the deglacial phase of a glacial cycle, rates of sediment yield may increase to about 10 times the geological norm (i.e. the natural long-term rate) and then rapidly subside (Church and Ryder, 1972). Hence enormous quantities of outwash material in the form of sandur (outwash plains) and valley trains (spreads of gravel material along valley floors) are characteristic fluvial features of newly deglaciated regions. In these environments, fluvial channels are initially braided, steeply graded and of low sinuosity, but these gradually change to deeper, more sinuous, low gradient and single-thread channels (Maizels, 1983b). A change is also apparent from an aggradational to an erosional regime with rivers becoming deeply incised into the unconsolidated fluvial sediments forming suites of outwash terraces leading away from the former glacier margin. Relict outwash terrace sequences dating from the last cold stage characterize river systems throughout the temperate zone (Schumm and Brakenridge, 1987). Meltwater channels that formed either beneath the ice or at the ice margin are also distinctive elements of former glacierized landscapes. In addition, in favourable topographic situations around the decaying ice margins proglacial lakes developed, some of which were of spectacular dimensions and hence major palaeohydrological phenomena. In north-west Europe, for example, the Baltic Ice Lake, impounded to the north by the Fennoscandian ice sheet (Figure 4.14), occupied a large part of the Baltic basin at an early stage in deglaciation and was over 1000 km in length at its maximum development (Eronen, 1983), while in North America, the proto-Great Lakes evolving along the margins of the wasting and thinning Laurentide ice sheet covered several hundred square kilometres (Teller, 1987). In some

Figure 4.14   Stages in the development of the Baltic during the wastage of the Fennoscandian ice sheet (after Eronen, 1983)

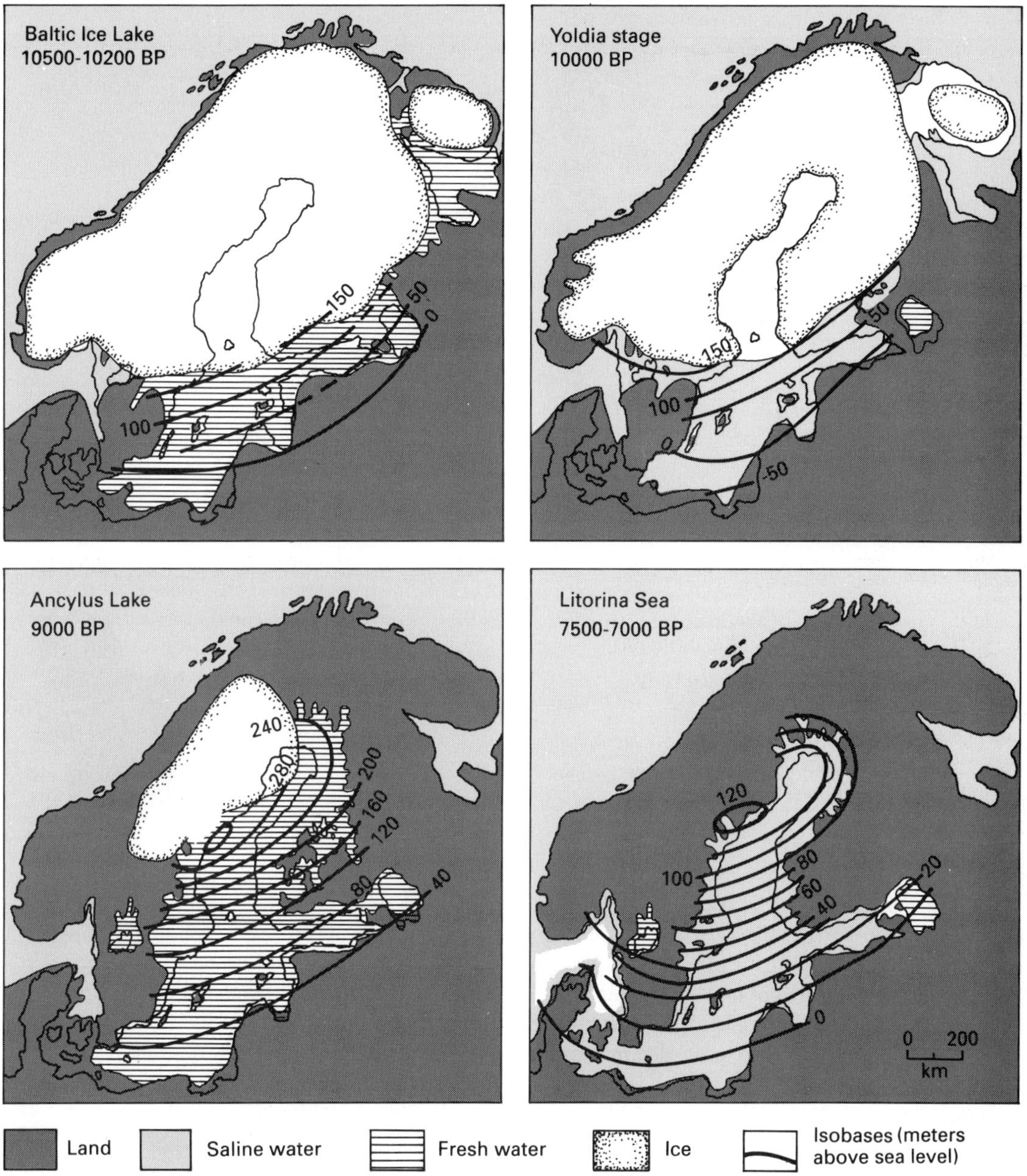

instances, outbursts from ice dammed lakes (jökulhlaups) exerted a major influence on the geomorphology of the proglacial area. Some of the most spectacular phenomena of the entire Quaternary period was the series of cataclysmic floods that occurred in the Columbia River system of the north-west USA between *c.* 16 ka and 12 ka BP (Baker, 1981). These may have been successive

Figure 4.15   Regions of the north-west United States affected by cataclysmic flooding during the late Wisconsinan (after Baker & Bunker, 1985)

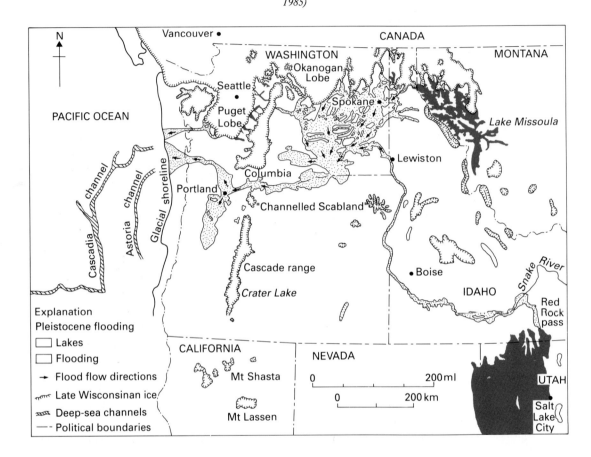

jökulhlaups from glacial Lake Missoula (Figure 4.15), and their effects can be seen in the network of deeply dissected river valleys that are incised into the loess-covered lava plateaux of eastern Washington State, an area of over 40 000 km² known as the Channelled Scabland. The magnitude of the flood events is reflected in the fact that their effects are also apparent on the abyssal sea floor off the mouth of the Columbia River some 700 km to the west. Palaeodischarge models for Lake Missoula suggest peak discharge hydrographs ranging from 2.7×10^6 to 14×10^6 m³ s⁻¹ over 8–20-day periods (Clarke *et al.*, 1984). Debate over the concept of cataclysmic flooding in the Columbia River region has continued throughout the century (Baker and Bunker, 1985), not least because it provides a classic example of the way in which

cataclysmic events can be reconciled within a framework of methodological uniformitarianism in the explanation of landscape evolution (Chapter 2).

Throughout western Europe and North America, there is also abundant evidence for former periglacial fluvial activity. Contemporary periglacial valleys are characterized by extensive aggradational sequences reflecting a combination of fluvial processes within the river channel and gelifluction activity on adjacent slopes (Washburn, 1979). Braided streamflow is a typical feature of such fluvial systems (Bryant, 1983a). Aggradational sequences relating to Late Weichselian/Wisconsinan periglacial fluvial regimes are to be found in numerous river valleys of Europe and North America (Starkel, 1987; Baker, 1984). A distinctive feature of contemporary periglacial rivers is a marked seasonality in annual discharge regimes, with peak flows in some rivers being concentrated in the spring following snowmelt (Arctic 'nival' rivers) while in others where there is a glacially derived component of discharge (Arctic 'proglacial' rivers) flows remains high throughout the summer (Bryant, 1983a). Evidence for both types of fluvial regime occurs in sequences of periglacial river sediments in lowland England (Bryant 1983b; Dawson, 1987). In addition, the characteristic forms of river valleys within present-day periglacial regions, notably the markedly asymmetric cross-profile which relates, *inter alia*, to slope aspect (Washburn, 1979), have been noted as relict periglacial features of contemporary river valleys within the temperate zone (French, 1976).

Some of the most dramatic palaeohydrological changes that have occurred in western Europe over the past 15 ka years accompanied the climatic fluctuations and associated vegetational changes of the Lateglacial (Starkel, 1983a, 1987). Braided or multi-thread channels that had characterized the extra-glacial areas at the close of the last cold stage gave way to deeper, meandering channels that developed following the climatic amelioration around 13 ka BP and the increasing stabilization of the surrounding landscape that resulted from the closing of the vegetation cover. Subsequently, the climatic deterioration of the Younger Dryas/Loch Lomond Stadial saw an increase in fluvial activity reflected in the coarsening of channel material, widespread aggradation with an increased supply of sediment geliflucted into the channel from adjacent hillslopes, and a return to multi-thread channel form. In the Gipping Valley of south-east England, Rose *et al.* (1980) have analysed the Lateglacial fluvial sequence in great detail employing a range of sedimentological and biological evidence (Table 4.3). In particular, they noted that the gradual deterioration in climate at the close of the Lateglacial Interstadial (11.3–11 ka BP) was accompanied by a marked erosional episode. They attribute this to very high discharge through the channel resulting from spring and early summer melt of winter snow over seasonally frozen ground, accompanied by restricted infiltration and high surface runoff. This episode is followed by a major

aggradational phase which is associated with the marked climatic deterioration of the Loch Lomond Stadial. Increasing continentality led to a reduction in snow-melt runoff and hence a decrease in channel discharge, and this was accompanied by increased instability on the hillslope above the channel. The change from erosion to sedimentation reflects the fact that increased runoff and the initial high seasonal discharge would be an immediate response to climatic deterioration, whereas slope instability, which is a function of soil moisture conditions, weathering processes and vegetation, would exhibit a delayed response to the time required to loosen and break up surface materials. The Gipping Valley evidence, therefore, shows very clearly the linkages between climate, vegetation, soils and fluvial activity that determine the pattern of temporal change in river systems.

The early millennia of the present interglacial saw a marked reduction in discharge in most rivers of the temperate zone. In western Europe, there is evidence from lake basins of lower water levels during the period 10–8 ka BP (Barber and Coope, 1987), and Lamb (1977) has estimated precipitation values for this period in Britain as being around 92–95 per cent of the early twentieth century average. Reduced precipitation, therefore, accompanied by the establishment of mixed woodland across much of the European landscape led to an abrupt decline in stream loading, a gradual cessation of alluviation and the development of single-thread meandering river channels which progressively incised into the Lateglacial valley fills. Characteristic features of the river valleys of the European lowlands during the Early Holocene were, therefore, stable river channels with small discharge variation and limited bedload transport. Gradual backswamp sedimentation and extension of the floodplains were also characteristic features of these well-vegetated low relief environments (Rose *et al.*, 1980). Wetter conditions after *c.* 7 ka BP, however, prompted an increase in fluvial activity most notably in flood frequency, the effects of which were increased stream loading and episodic phases of aggradation and incision (Starkel, 1983b). From around 5 ka BP onwards, however, with the increasing

Table 4.3 Late Devensian and early Flandrian events in the Gipping Valley, south-east England (after Rose et al., *1980)*

Deposits and landforms	Geomorphic processes	Inferred environment	Inferred age
Present floodplain Clay mud	Overbank sedimentation	Temperate climate with anthopogenic deforestation and ploughing	Present
Peat with lenses of sand and gravel Palaeosol on bars	Backswamp fen behind levees Soil formation	Temperate climate with extensive forest cover	9500 BP
Sand and gravel Small-scale cross sets with interdigitating chalk and flint head ground ice Large-scale cross sets	Braided river sedimentation and contemporaneous hillslope gelifluction Meandering river sedimentation	Arctic climate with seasonal development and possibly permafrost	11 000 BP
Channel of discontinuous gully	River incision	Arctic climate with extensive winter snow cover	11 300 BP
Laminated calcareous silt	Shallow eutrophic lake development	Deteriorating climate	
Palaeosols on chalk mud and chalk head Chalk mud and nodular flints	Soil formation and slope stability Mudflow development	Cool temperate climate with grasses and shrub vegetation	12 200 BP
Chalk head	Hillslope gelifluction	Periglacial climate	

anthropogenic impact, particularly woodland clearance and the establishment of a settled agricultural economy, the effects of climate on river regimes become increasingly difficult to identify (Chapter 7).

In North America, the response of river systems to Holocene climatic changes has been discussed by Knox (1984) and is summarized in Figure 4.16. The data indicate that most areas were warmer and drier between *c.* 10 ka and 8 ka BP and valley alluviation was dominant, especially in the drier western regions of the USA. An erosional episode between 8 ka and 7.5 ka is apparent in the Midwest and the Great Plains, and especially in the south-west reflecting, it appears, warmer and wetter conditions in those regions from around 8 ka BP onwards. There was a general decline in alluviation up to 6 ka BP as large-scale vegetation communities became established, but this relatively quiescent phase was succeeded after 6 ka BP by widespread erosion of Early Holocene fills. According to Knox, this is a reflection partly of decreased stream loading due to the more extensive vegetation cover, and partly of an increase in flooding as a consequence of higher precipitation levels and frequency of storms. In more arid areas, however, alluviation continued as the dominant geomorphological fluvial process throughout most

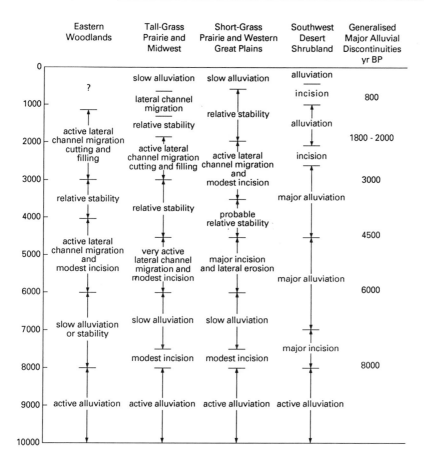

Figure 4.16   Generalised alluvial chronologies in different areas of the United States during the Holocene (after Knox, 1984)

of the Holocene. Both erosional and depositional intensity appears to have subsided between *c.* 4.5 ka and 3 ka BP, but there is evidence in most regions of renewed fluvial activity after 3 ka BP although the type of process varied from region to region. In the south-west, where fluvial sediments can be correlated with dendroclimatological data for the last two millennia, phases of erosion and deposition relate to changes in water-table and precipitation (Chapter 5).

5

The Impact of Environmental Change on People

Introduction

Chapters 3 and 4 have reviewed environmental changes and their causes over a range of temporal scales. This chapter considers the various ways in which environmental change impacts on human activity. Examples range in time from the early hominids to the Little Ice Age. Broad-scale and long-term changes undoubtedly exerted a range of influences on human evolution and cultural development but it is equally important to consider human response to short-term and localized changes right down to single events. Some of the changes involved are of a cyclical nature and reflect extra-terrestrial factors, ranging for example from tides to Milankovitch cycles, while others result from more localized stochastic events.

Some environments are subject to major changes, for example, tectonically unstable continental plate margins where earthquakes and volcanism occur, or low-lying areas liable to flood. Human activity in some areas is delicately balanced

and particularly vulnerable to the effects of environmental fluctuations. Such areas can be regarded as marginal in the context of particular economic strategies and social conditions. This chapter includes several examples of marginal situations such as the fringes of uplands and arid zones, coastal regions and the subarctic. These areas often contain well-preserved archaeological and palaeoenvironmental records, due partly to a lack of later disturbance but also to climatic factors such as aridity and coldness, and to waterlogging in coastal and upland situations, all of which help to preserve organic remains.

Where the archaeological and palaeoenvironmental record is of sufficient quality, it may be possible to identify contemporary changes in the two data sets. Even so it is not a simple matter to infer from the observed result the process which brought it about. One problem is equifinality (Chapter 2, p. 47), in other words various causes, natural as well as cultural, giving rise to similar results. The difficulties are especially acute in the case of human beings who have a capacity to act independently of narrowly prescribed environmental parameters. That capacity has increased through time with technological developments and increasing social complexity. People are affected not just by their environment but by social factors internal to human societies which also need to be considered.

Human adaptation occurs over various timescales. In the long term there is the role of environment in helping to determine the course of biological evolutionary change. The timescale here is such that people would have been unconscious of the overall trajectory of change, though it may well have represented the outcome of a multiplicity of conscious decisions. These shorter term adaptive responses are of a social/cultural nature and give rise to ways of life which represent adjustments to a given set of environmental parameters. In addition, there is the human response to particular environmental changes. Some will be gradual changes observed over a period during which knowledge about former environmental conditions was passed from one generation to the next until it constituted the basis of a decision to adapt. Generally, however, decisions would have been based on observed changes over periods much less than an individual lifetime. The trigger may have been a single event, a volcanic eruption, a flood or a failed harvest. Here a key concept is perceived recurrence interval (Chapter 1, p. 7), in other words, was the risk of repetition sufficiently great to justify moving away? Increasingly through time, human communities when faced with environmental stress, hazard or demographic pressure have opted to change not themselves but their environment. This they did by burning, clearing, irrigating, draining, etc., which ultimately produced a cultural landscape (Chapters 6 and 7).

Through a combination of environmental modification and technological advance, human beings have become increasingly insulated from the effects of all but the most severe environmental change. For archaeologists this constitutes a paradox, since just as the technological ability to reconstruct past worlds improves so the investigators become more remote from the concerns of the people they are studying. Prehistoric communities would have been much more aware of regular cycles of change in the cosmos than most westerners today. Former preoccupation with such matters may be indicated by the astronomical alignment of prehistoric megaliths and other sites. Ecological anthropology has provided valuable insights into the range of human responses to environmental phenomena among societies of various types, especially in North America and Australasia. Anthropology offers a glimpse of how particular groups perceived their environment and its changing nature. The mythology of Eskimo communities in north-east Alaska associated famine with changed climate and reflected the alternating availability of marine and terrestrial fauna during different periods (Minc and Smith, 1989). These fluctuations occur today as a result of the changing extent of sea-ice, which can be correlated with evidence for periodic climatic change from dendrochronological sequences that extend back to the tenth century AD. Thus the mythological evidence from ethnohistory tells us about social attitudes to environmental perturbations and can be correlated with the physical palaeoenvironmental record. Such instances play a key role in linking scientific and social perspectives, as they do in examples from Papua New Guinea (p. 132) and southern California (p. 139).

Anthropology also aids understanding of the various ways in which people cope with change. Many develop forms of cultural behaviour which act as a buffer between them and natural fluctuations in their environment. Buffering, or coping, strategies may take the form of mobility, diversification, food storage or sharing and reciprocity (Halsted and O'Shea, 1989). Potlatch ceremonies in the American north-west involved the giving of enormous feasts and gifts in exchange for prestige. This practice meant that there was always sufficient surplus in the economy for exchange with other communities during times of stress, which occurred periodically as a result of the effect of ocean current changes on salmon runs (Hardesty, 1977). Tribal gatherings such as the *fandangos* of the North American Great Basin were clearing houses for environmental information. People discovered where good and bad nut harvests were likely in the following year and on this basis made decisions about future scheduling and population levels.

This chapter does not attempt a general evaluation of human response to the established sequence of Late Quaternary environmental changes. The pattern is too diverse, the factors too complex and the archaeological record too patchy. The approach adopted is to select examples of both short- and long-term changes in situations which mostly have good archaeological records and chronological precision. The examples include: evolution; the origins of agriculture; coastal wetlands; volcanism; the American south-west; and the Little Optimum and Little Ice Age, and these are discussed in a loosely chronological order. Most examples take a critical look at situations where it has been claimed that environmental change caused major social repercussions. It has already been noted that for many millennia people have not just been responding to environmental change but initiating it themselves. Inevitably this has feedback effects for human societies. Consequently, the effect of environmental change on people is a topic which will also recur in Chapters 6 and 7.

Figure 5.1 Simplified reconstruction of African vegetation patterns. (a) In the late Miocene and Pliocene, circles indicate the main early hominid localities: 1. Ethiopia; 2, Lake Turkana Basin; 3. Lake Baringo Basin; 4. N. Tanzania/S. Kenya; 5. Transvaal. (b) Under glacial conditions (c) under interglacial conditions (after Foley, 1987)

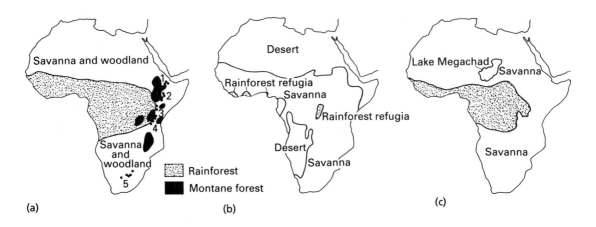

Evolution and environmental change

The relationship between hominid evolution and long-term (10^4–10^6 ka) environmental change has long excited the interest of natural scientists. It raises fundamental questions about the place of people in nature and it forms a background to the shorter timescales and largely Holocene themes considered in the remainder of this book.

Environment played a key role in Darwin's (1859) ideas about evolutionary change. Varieties of organisms able to adapt to solve environmental problems were seen as enjoying a greater level of reproductive success, and therefore increasing at the expense of less well-adapted varieties (natural selection). This led to gradual genetic change giving rise to better adapted, higher forms of life with time. More recently it has been suggested that some traits are neutral in adaptive terms, their frequency being fixed by stochastic factors (Dunnell, 1980). Some scientists such as Gould and Eldredge (1972) have questioned the idea that evolutionary change was indeed gradual, and have reinterpreted the fossil record in terms of punctuated equilibrium with species remaining

stable for long periods and then undergoing what, on a geological timescale, are relatively sudden bursts of evolutionary change. Whichever model is preferred, the evidence for global cycles of environmental change reviewed in Chapter 3 is clearly very pertinent to the study of evolution.

The early stages of hominid evolution took place in Africa, the first hominids appearing 5 ma BP at Tabarin and Lothagam in Kenya (Foley, 1987). They are classified as *Australopithecus afarensis*, which was already bipedal as confirmed by the discovery of footprints at Laetoli, Kenya dated to around 3.6 ma BP and the 'Lucy' skeleton from Hadar, Ethiopia dated to *c.* 2.9 ma BP. Since the time of Darwin one explanation for bipedalism has been the abandonment of the arboreal habitat, occupied by most primates, in favour of an open environment. Early Miocene Africa had been covered by a vast undifferentiated rainforest, but around 10 ma BP conditions became more arid (Brain, 1981). As the extent of the equatorial rainforest diminished, competition between species would have intensified at a time when the adjacent savanna environment was coming into being (Figure 5.1a). Foley (1987) noted that bipedalism is a radical departure for primates and must reflect severe selective pressure: 'we are what we are because we are an arboreal animal which has taken up the challenge of a terrestrial life'.

Further important climatic changes accompanied the expansion of the Northern Hemisphere ice sheet some time around 2.4 ma BP. From then onwards clearly defined cold/warm cycles are attested in deep-ocean cores (Chapter 3). In Africa interglacial optima were, contrary to earlier opinion, apparently marked by pluvial episodes with high levels of precipitation and an extensive tropical rainforest belt flanked by savanna (Roberts, 1984). This contrasts with the phases which saw glacier expansion in mid- and high latitudes; these were drier, leading to the reduction of rainforest to small refugia with more extensive spreads of savanna and desert (Figure 5.1b). Isolation of groups of biota in refugia would stimulate speciation just as recurrent stress resulting from cycles of climatic change would have led to a need for frequent adaptation. The result was accelerated evolutionary change in many groups of plants and animals including hominids. Between 2 ma and 1.5 ma BP there were at least five hominid species (as presently recognized – Foley, 1987) in the Transvaal and East Africa. Three were *Australopithecus* spp.; then there was *Homo habilis*, the first stone tool user dating from around 2 ma BP at sites such as Koobi Fora and Olduvai Gorge, Tanzania; then at around 1.6 ma BP *Homo habilis* was replaced by *Homo erectus*.

Palaeoenvironmental evidence shows that nearly all early African hominid finds were associated with savanna environments ranging from woodland to open grassland. Savanna has pronounced dry seasons with stress induced by plant food shortage to which the various hominid species may reflect differing responses (Foley, 1987). *Australopithecus robustus* had dentition indicating dietary diversification towards poor quality plant foods which, in the event, proved to be an evolutionary dead end. *Homo habilis*, on the other hand, adopted a more omnivorous diet, perhaps exploiting concentrations of animals around waterholes during dry seasons. Stone tool use would have given this species a major competitive advantage, whether for hunting or scavenging. The replacement of *Homo*

habilis by *Homo erectus* at *c.* 1.6 ma BP occurred at a time of major faunal change within the Olduvai sequence. *Homo erectus* had the capacity to adapt to a wider range of environments spreading to South East Asia between 1 ma and 1.5 ma BP and to Europe around 0.9 ma BP. Stress during a glacial phase may well have been the initial trigger for colonization. Cold/warm changes on a regular 100 ka cycle became progressively marked from 900 ka to 450 ka BP (Chapter 3, p. 65). Despite these changes and the range of habitats colonized by *Homo erectus*, including parts of Europe experiencing recurrent glaciation, artefact assemblages give little hint of corresponding social change. The hand axes, which are so characteristic of Acheulean assemblages, change little over the 1 ma prior to 200 ka BP (Butzer, 1982). It appears that, during this period at least, climatically induced stress did not lead to major changes of material culture. Around 200 ka BP *Homo sapiens* appeared in Africa and Europe, followed between 100 ka and 50 ka BP by distinctive populations of *Homo sapiens neanderthalensis* in Europe and South West Asia. These were heavily built with large brow ridges, prognathous jaws and limb proportions which suggest cold adaptation, perhaps the result of evolutionary change among a community isolated during a glaciation (Butzer, 1977a).

Homo sapiens sapiens appeared in Africa and South West Asia at *c.* 100 ka BP, much earlier than elsewhere. Once again the effects of glacial/interglacial cycles on the African savanna are seen by some as a key factor in promoting speciation (Foley, 1989). The sub-species then spread to Europe during the last cold stage, where it appeared at 35 ka BP, perhaps coexisting with Neanderthal populations which were then absorbed or replaced (Stringer and Grün, 1991). That was part of a rapid colonization which took people to nearly all parts of the world's major land masses. Australia was reached around 40 ka BP by boat (Thorne, 1980). The Americas were colonized at a still disputed date, possibly as early as 20 ka BP or as late as 12 ka BP, by peoples who had adapted to the Arctic zone and crossed the Bering Land Bridge during the low last glacial sea level (Fagan, 1991).

Colonization by *Homo sapiens sapiens* marks something of a technological and cultural explosion. 'Tool kits' (artefact groups associated with particular activities or communities) became more complex and regionally diverse, blade-based lithic technology, bone tools and art all appeared. Early art consists of engraved slabs from 32 ka BP, followed by cave painting from 25 ka BP (Champion *et al.*, 1984). The rich and lively art of the French Dordogne, suggests that from the time of initial colonization *Homo sapiens sapiens* was as artistically, and perhaps therefore as intellectually, able as people are today. In the Upper Palaeolithic the Dordogne exhibits particularly dense settlement and art, the range of animals depicted suggesting to some that the area was especially rich in environmental terms. A converse proposition by Mithen (1990) regards the main development of the art as relating to a time of declining hunting yields during the glacial maximum 20–17 ka BP. It is proposed that hunters were forced to turn away from co-operative game drives to the tracking of individual animals for which accurate observation of tracks, trails and signs, and the dissemination of this knowledge through art, played a key and adaptively decisive role in the decision-making processes of individual hunters.

The most vigorous scientific debate of the nineteenth century centred on the conflict between the Biblical account of human origins and the Darwinian theory of biological evolution, in which adaptation to environmental conditions played a key role. Although Darwin's theory has come to be almost universally accepted there remains debate about whether humanity can be seen as part of nature, or is in some way separate. Foley (1987, 1989), for example, regards humans as unique but points out that so is every other species! Human uniqueness is seen as the product of biological processes common to all living matter and can only be fully understood in the context of the evolution of the ecological communities of which hominids were a part. This interpretation emphasizes the effect of environmental factors on selection from existing genetic diversity, and is essentially deterministic (Chapter 1, p. 8). An alternative view might be that the role of human choice has been

overlooked and that several evolutionary outcomes may have been possible, the result depending on cultural or social preference. Durham (1978) saw culture as evolving in a complementary way to biology by the selective retention of non-genetic traits that enhance the ability of human beings to survive and reproduce. If, on the other hand, culture is seen as something quite separate, the question arises as to the point in the long history of human evolution when cultural factors eclipse the role of biological evolution. Was it the first step, the development of bipedalism, or the comparatively recent emergence of *Homo sapiens sapiens*? For those who emphasize the separateness of people from nature the fundamental question remains, as it did before Darwin, when and why did this happen?

The origins of agriculture

The transition from hunting and foraging to agriculture was an event of major significance in human terms. It made possible the feeding of much larger populations in one place, led to the development of complex societies in which individuals could specialize in activities other than food production, and fostered the emergence of civilization. It was also via the process of domestication that humans transformed the face of the earth by taking species to new areas and creating conditions for domesticates to flourish. By intervening in the breeding process of plants and animals, people have both consciously and unconsciously selected certain traits favoured by humans or by the agricultural regime which they brought about. Eventually genetic changes led to morphological differences which distinguish wild and domestic forms. In the wheats of South West Asia, for example, mutant cereal ears with a tough rachis (connecting axis of the ear) were preferentially, and largely unconsciously, collected in the harvesting process because the natural brittle rachis tended to shatter at a touch. People also selected in favour of larger forms, of maize cob for instance, and those with a better food value. So too with domestic animals, humans bred from those

Figure 5.2   The regions, sites and dates at which some of the most important domestications took place (after Simmons, 1989)

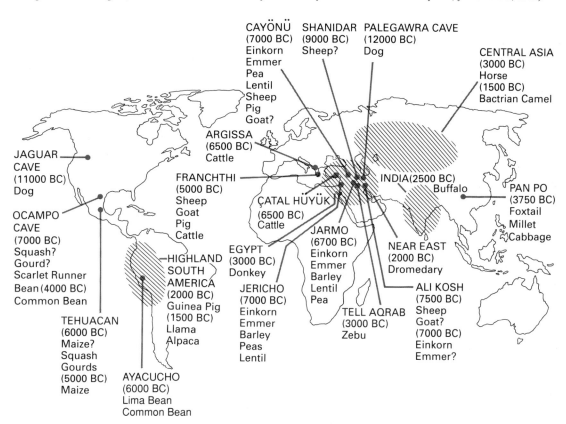

with favoured traits of size, docility, coat, or for the retention of particular juvenile characteristics which made animals easier to control. This led to recognizable changes in bone morphology. Many early domesticates, such as sheep, cattle and pigs are smaller than their wild ancestors (Davis, 1987).

Domestication was not a one-off event, for it occurred independently at many times and several places; some examples are shown on Figure 5.2. Three major centres of domestication which have been investigated in particular detail are considered here: South West Asia, Meso-America and Andean America. Others were also important, particularly South East Asia, and it has long been suspected that tropical 'vegeculture' involving crops such as yams, taro and manioc is likely to have early origins,

although the evidence would seldom survive (Harris, 1969).

A distinction needs to be made, however, between domestication and environmental manipulation, such as cultivation, which can be carried out to encourage undomesticated species. Australian Aborigines lacked domesticated plants but none the less had a remarkable awareness and knowledge of their biotic environment which was manipulated by the practice of burning (Chapter 6, p. 154) and the replacement planting of yams. Increasing awareness of the complexity of relationships between hunter–foragers and both animals and plants has led to the view that domestication should be seen in terms of a gradient of gradually intensifying relationships between people and biota through time (Jarman *et al.*, 1982; Harris and Hillman, 1989). Sedentism is also likely to increase awareness of the life cycles and conditions of the growth of plants. The 'dump heap' theory holds that domestication came about because waste plant material was discarded and multiplied in the artificial, disturbed and nitrogen-rich areas around settlements. Cohen (1977) describes how around hunter–gatherer camps there may be 'gardens' of favoured species growing naturally from seeds germinated from dumped foods and faeces.

The critical question is why some groups became fully agricultural, for it was not a simple passport to an easier lifestyle. Many foragers spend less time getting food and have more leisure than agriculturalists (Lee and DeVore, 1968). Furthermore, vegetable resources, particularly cereals, are regarded by hunter–gatherers as low preference foods. As Flannery (1973) has observed 'since farming represents a decision to work harder and eat more third choice food, I suspect that people did it because they felt they had to, not because they wanted to'. Jones and Meehan (1989) have shown the importance for Aboriginal societies of curation of the countryside and its spiritual essence which leads to sanctions against innovation

or disturbance of the natural social order. The reasons why some communities eschewed such conservatism and became particularly responsive to change and innovation are far from clear. Cohen (1977) has argued that the only advantage of agriculture over hunting and gathering is that more calories can be produced per unit of land, thus a denser population can be supported. Population growth in several parts of the world at the Pleistocene/Holocene transition is seen as triggering the need to exploit an increasingly broad range of resources, particularly plants and coastal resources leading to domestication.

A very different view by Hodder (1990) emphasizes the role of the opposition which exists in people's minds between the wild and that which is created or controlled by people. Domestication is seen not in purely utilitarian subsistence terms, but as a social and symbolic process by which the wild was brought within the control of a social and cultural system. Whichever model is favoured, the context of these changes in many parts of the world is provided by the dramatic environmental changes around 10 ka BP. The extent to which these changes influenced decisions to domesticate remains a subject of active debate.

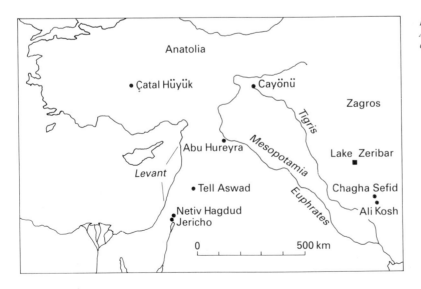

Figure 5.3   Main South West Asian sites associated with the early history of domestication

South West Asia

This area is important, both because it is the earliest known and most intensively studied centre of domestication and because of the remarkable cultural manifestations associated with the earliest farming communities. These developments also form the background to the emergence of civilization in Mesopotamia in the fourth millennium BC. The area concerned (Figure 5.3) has traditionally been known as the fertile crescent, and comprises the Levant, southern Turkey and Mesopotamia. Here are found the wild progenitors of domestic cereals, emmer wheat (*Triticum dicoccoides*), einkorn wheat (*Triticum boeoticum*) and barley (*Hordeum spontaneum*) (Figure 5.4). Five companion domesticates also occur naturally within this area, pea (*Pisum humile*), lentil (*Lens orientalis*), chickpea (*Cicer reticulatum*), bitter vetch (*Vicia ervilia*) and flax (*Linum bienne*), as also do the first domesticated animals: sheep, goats, cattle and pigs.

The prelude to domestication was a shift away from reliance on hunting of a few species of large animals, which characterized many Palaeolithic communities, towards the exploitation of a broad spectrum of food sources including fish, crustaceans, molluscs, birds and plant foods (Flannery, 1969). This change is a characteristic feature of Late- and Post-Pleistocene adaptation in many parts of the world and occurs in South West Asia between 22 ka and 12 ka BP. During the Early Holocene, Natufian communities in the Levant occupied small seasonal settlements equipped with grinding stones and storage pits. They exploited wild cereal stands which can produce yields comparable to those under cultivation. During the Pre-Pottery Neolithic, the earliest evidence for large-scale crop growing appears at Netiv Hagdud, Jericho and Tell Aswad (Hillman, 1989). Jericho exemplifies the remarkable nature of initial Holocene adaptation in this area, for a flourishing township of perhaps 2000 to 3000 people surrounded by walls and with a stone tower had been established by 10 ka BP. In the period after 9.5 ka BP farming villages appeared throughout the area.

A particularly detailed picture of Early Holocene

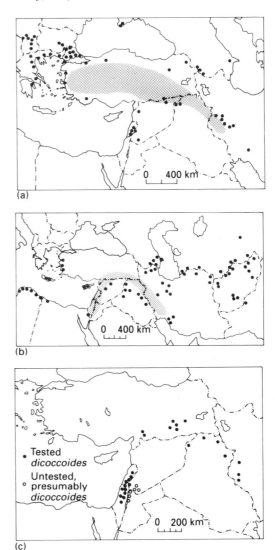

*Figure 5.4   Present day distributions of (a) wild einkorn wheat (*Triticum boeoticum*) (b) wild barley (*Hordeum spontaneum*) and (c) wild emmer wheat (*Triticum dicoccoides*); shaded area shows main distribution; dots indicate isolated population (after Zohary, 1989)*

economic change has been obtained from Tell Abu Hureyra in the Euphrates Valley where the Epipalaeolithic (transitional Palaeolithic/Postglacial) economy was characterized by a broad spectrum of plant resources including wild einkorn; animal exploitation was concentrated on gazelle which contributed 80 per cent of the excavated bones (Hillman *et al.*, 1989; Moore, 1989). Already at this pre-agricultural stage, the environmental evidence suggests sedentism. The transition to agriculture occurred suddenly in a few centuries around 9 ka BP. Sheep and goats then made up 70 per cent of the excavated bones and there is macrofossil evidence of crop plants which had been introduced from elsewhere.

To the east in Mesopotamia and the Zagros it is evident that farming began at least as early as in the Levant. The sites of Ali Kosh and Chagha Sefid show that sheep and goats were herded and emmer, einkorn, barley and lentils were cultivated on the De Luran Plain from 9 ka to 10 ka BP (Hole and Flannery, 1967; A. M. T. Moore, 1985). In Anatolia farming was established at Çayönü by 9 ka BP and sometime before 8.7 ka BP the site of Çatal Hüyük was settled (Mellaart, 1967). This was the largest Neolithic settlement in the Near East, with the earliest evidence of cattle domestication which probably occurred locally on the fertile grasslands of the Konya Plain. Parts of the settlement were decorated by a remarkable series of wall paintings and reliefs reflecting cult practices. Many of these feature cattle, suggesting that ritual and symbolic aspects may also have played a part in domestication (Hodder, 1990).

In an early attempt to account for the origins of agriculture, Childe (1928) put forward a deterministic model which saw domestication as a response to environmental change. This was based on the prevailing view of the time that glaciations in Europe had been accompanied by wetter pluvial episodes throughout the Near East. Drier conditions during the Holocene were seen as leading to concentrations of human and animal populations in remaining wetter 'oasis' areas which became increasingly isolated by desert tracts. Such enforced juxtaposition of people and other biota, it was suggested, promoted symbiosis leading to domestication. However, testing of this hypothesis has revealed that the environmental changes were virtually the reverse of those envisaged by Childe (Wright, 1977). A series of pollen diagrams from lake basins, most notably Lake Zeribar in Iran, show that prior to 10 ka BP the area experienced

cold, very dry conditions which gave rise to treeless steppe over most of South West Asia (van Zeist and Bottema, 1982). The onset of a less harsh climate began in the southern Levant around 14 ka BP and by 11 ka BP that area was covered by continuous forest (Figure 5.5), although there was a drier episode with reduced tree cover between 11 ka and 10 ka BP. Further north conditions sufficiently moist for tree growth began rather later, about 10.5 ka BP. Between then and 5 ka BP open oak/pistachio forest spread and became more dense in southern Turkey and down the Zagros mountains. Zohary (1989) contended that by 11 ka BP the vegetation formations containing the early plant domesticates were established in at least the western part of the area. However, as Figure 5.5 shows, this was a period of considerable vegetational change, so the distribution of cereals in the critical period before 10 ka BP remains uncertain.

On present evidence this area does not seem to have experienced the sudden and marked environmental change at the Lateglacial/Holocene transition that occurred around the North Atlantic (Chapter 3, p. 69), although it must be acknowledged that the South West Asian pollen sequences are fewer and less well dated. Although farming settlements with domesticates appear suddenly in a number of places in the Levant and Mesopotamia in the period 10–9 ka BP, there is no obvious relationship with natural environmental change. A drier episode with reduced tree cover in the southern Levant between 11 ka and 10 ka BP has been seen as prompting Natufian communities to domesticate cereals (Henry *et al.*, 1981; Bar-Yosef and Kislev, 1989), but at approximately the same time communities to the north and east were experiencing a wetter climate than previously and an increasing tree cover, yet they too were experimenting with farming. Much depends on the date of initial domestication. It has recently been suggested, for example, that the origins of domestication should be sought in the Epipalaeolithic between 18 ka and 8 ka BP (Moore, 1989). This hypothesis envisages an experimental period corresponding to the closing phases of the last cold stage followed by adoption of an

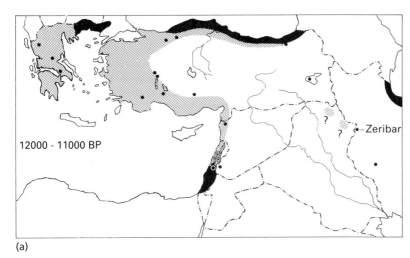

(a)

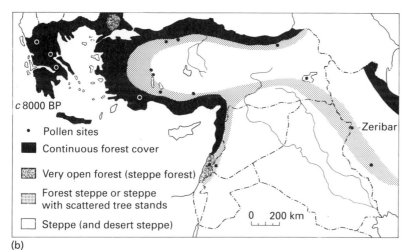

(b)

Figure 5.5   Palaeo-vegetation map of South West Asia (a) 12 000–11 000 BP (b) c. 8000 BP (after van Zeist and Bottema, 1982)

agricultural way of life close to the beginning of the Holocene, in other words a much earlier prelude to domestication than generally envisaged.

The environmental changes at the Pleistocene/Holocene transition created the communities of plants and animals within which domestication took place. They also contributed to the sedimentary context of domestication. Alluvium and sediments were exposed by the shrinkage of Pleistocene lakes as temperatures rose. This exposed rich soils exploited by farming communities, for instance the grassland around Çatal Hüyük (Roberts, 1989). The changes created opportunities but do not of themselves explain why those opportunities were taken up. One argument

against environmental determinism has been that similar opportunities would have presented themselves during earlier cold/warm shifts. Such changes may not necessarily, however, have been as sudden as in the Early Holocene and they certainly did not affect such large human populations. Furthermore, the emergence of *Homo sapiens sapiens* during the last cold stage means that there are likely to have been significant biological and behavioural differences which might have led to particular reproductive success once climatic amelioration allowed extensive settlement in the temperate zone (Binford, 1983). Population is a key variable (Cohen, 1977). In areas which were especially favoured ecologically, for instance by dense stands of wild cereals or concentrations of animals, populations may have increased, and this in turn would have led to 'packing' of communities and thus have reduced options for seasonal movement, resulting in increased sedentism in favoured habitats. As people ceased to be mobile so they would have relaxed the regulatory practices (contraception, abstinence, prolonged lactation, abortion, etc.) which ethnographic studies show maintain hunter–forager populations below the levels at which they would deplete the local food supply (Hardesty, 1977). Differences in the rate of population growth between optimal areas and those beyond would give rise to emigration and attempts to replicate optimal conditions artificially by domestication (Binford, 1968). In South West Asia there is indeed evidence for population growth and increased sedentism in the archaeological record (A. M. T. Moore, 1985, 1989). However, these conditions did not obtain to the same extent in other areas where domestication occurred at roughly the same time.

America

In the Americas the best documented early centre of plant domestication is in Meso-America, particularly the Tehuacan Valley and the Tamaulipas and Oaxaca areas of Mexico (Figure 5.6). Some of the earliest domesticates were non-staple species. Squash (*Cucurbita* spp.) were domesticated by 9 ka BP and there is also evidence of possibly domesticated bottle gourds (*Lagenaria siceraria*) and peppers (*Capsicum* spp.) (Heiser. 1989). Maize (*Zea mays*) first appears in the cave sequences of Tehuacan 7 ka BP in the form of tiny cobs (Wilkes, 1989). It was probably domesticated from the wild grass *Teosinte* although some specialists favour domestication from an extinct progenitor. Even at this stage it has been estimated that the domesticates only provided 5 per cent of food and were essentially supplements to a diet still largely based on foraging and hunting and involving seasonal movements (MacNeish, 1967). The size of maize cobs gradually increased until, sometime prior to 3.5 ka BP, its genetic changes crossed a threshold which gave rise to a highly productive

food plant (Flannery, 1973; Wilkes, 1989). Henceforth Indian communities cleared native vegetation and planted maize on a much larger scale, and this move to cultivation registers in pollen sequences (McAndrews, 1988). This relatively sudden change led to the emergence of permanent villages which were widespread in southern Mexico and Guatemala by 3.3 ka BP. Ultimately these changes led to the development of cities based on irrigation agriculture in the highland valleys of Mexico.

Domestication in the central Andes of Peru may have occurred as early as in Meso-America. Domesticates appear around 10 ka BP and by 8 ka BP include peppers, beans (*Phaseolus*) and squash (Bonavia and Grobman, 1989). Maize appears on several sites 6 ka BP, although there is dispute as to whether it represents a local domesticate or derives from contact with Meso-America for which there is little other evidence at this early date. Even so it is generally accepted that the two areas were independent centres, since some different species, of squash for instance, were involved (Pickersgill, 1989). Domestic animals in the Andes were the llama, alpaca and the guinea pig, all of which were domesticated by 6 ka BP (Davis, 1987).

North America was not a major centre of domestication but there is growing evidence of local small-scale domesticates, possibly squash, from 7 ka BP, sumpweed (*Iva annua*) and grain chenopods (*Chenopodium* spp.) from 3 ka BP in Illinois and Kentucky (Watson, 1989). Between 3 ka and 4 ka BP a number of cultivars appeared, and Eastern Woodlands hunter–gatherer communities became part-time horticulturalists which provided a basis for year-round settlement. Maize was introduced from Meso-America into south-west North America by 3 ka BP (Fagan, 1991) but agriculture based on the nutritionally balanced triad of maize, beans and squash, though widely practised south of a line from the Great Lakes to the Gulf of California (Figure 5.6), only dominated the economies of selected areas, such as some of the main river valleys. Elsewhere crop growing remained a supplement to hunting and foraging activities (Delcourt, 1987). The only domestic animals in North America were the turkey

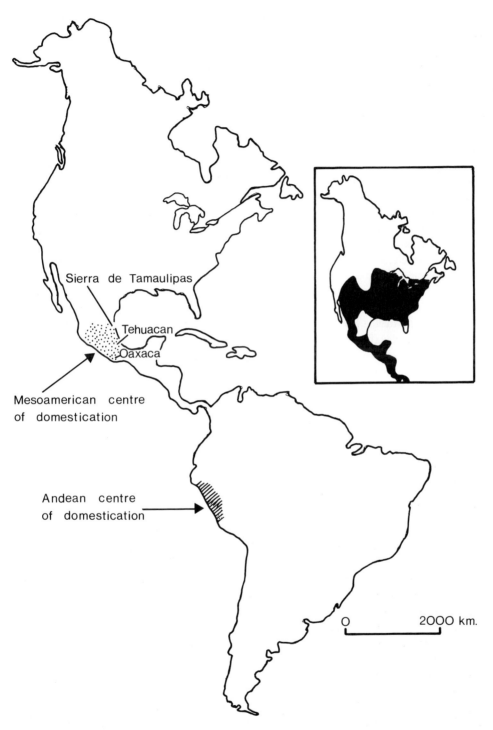

Figure 5.6   Main American areas associated with the early history of domestication. Inset shows the area of North America where Indians grew maize before European contact (after McAndrews, 1988)

and the dog (Harris, 1989).

Although in both Meso-America and the Andes initial small-scale domestication took place between about 10 ka and 9 ka BP, there is no clear link with environmental changes relating to the global warming at the end of the last cold stage, *c.* 10 ka BP (Chapter 3). Certainly the old South West Asian model of desiccation as the trigger is no more satisfactory in Meso-America, because domestication did not take place in areas which experienced drying during the Early Holocene (Roberts, 1989). In the Tehuacan Valley, conditions after 10 ka BP were slightly less dry than during the Lateglacial, but throughout the Holocene plant macrofossil assemblages contain no evidence of significant climatic change (Byers, 1967). One reason why people took greater control over their plant resources may have been marked seasonal and annual variations in their availability in this area, as suggested by the evidence from Guilá Naquitz Cave in the Oaxaca Valley (Flannery, 1986). The extinction of Pleistocene megafauna (Chapter 6, p. 148) may also have encouraged use of a broader spectrum of food resources particularly plants (Wright, 1977; Cohen, 1977) which eventually led to domestication. It is not, however, simply a case of the megafauna being replaced by domesticates, because the extinctions were mostly 1–2 millennia before the earliest evidence of domestication and many millennia before domesticates made a major contribution to the diet. In North America it is likely that the Mid-Holocene Climatic Optimum or Hypsithermal (Chapter 3, p. 70) encouraged greater use of optimal floodplain sites which led ultimately to local domestication (Watson, 1989).

There are some interesting contrasts between the American and South West Asian evidence. Flannery (1973) has argued that in both Meso-America and Peru there is little evidence that demographic pressure led to domestication (although Cohen, 1977, argues otherwise). Furthermore, in the Andes the shift to full domestication and sedentism was not sudden, but occurred 4–5 millennia after initial domestication. This may be because Meso-America lacked the animal domesticates that could be integrated within

a fully fledged agricultural system. Consequently, relatively advanced societies had to rely on hunting.

Despite these contrasts, it is apparent that the earliest domestication seems to have taken place at similar dates (9–10 ka BP) in South West Asia, Meso-America and Peru, and arguably in the less intensively studied area of South East Asia. In New Guinea, for example, there is evidence for swamp clearance for agriculture as early as 9 ka BP (Golson, 1989). In view of the apparent contemporaneity of initial domestication in totally unconnected areas of the world and in such different environmental situations, it seems unlikely that the transition to agriculture occurred as the result of any single factor, such as sedentism, population growth or any one environmental change. It must, however, be seen in the context of gradually intensifying relationships between people, plants and animals through time (Harris, 1989), and as a particular consequence of the range of post-Pleistocene adaptations by anatomically modern hominids in several geographical areas.

Coastal wetlands

A knowledge of Quaternary coastal and sea-level change is central to understanding patterns of prehistoric colonization, for example, of the Americas and Australia (Masters and Fleming, 1983). Coastal wetlands are also valuable archives because of the range of environmental and organic artefactual evidence which they preserve (Coles and Coles, 1989). Ecotones, especially coasts and estuaries, are among the most productive ecosystems and often contain particularly rich and stable cultural manifestations, such as the Ertebølle of Denmark (Figure 6.6). Even in the extreme north of Norway, the ameliorating effects of Atlantic currents and high plankton populations gave rise to settlements of largely sedentary hunter–fishers by 4 ka BP. These remained stable for long periods and showed little evidence of cultural change (Renouf, 1988). Further south the ameliorating effects of the Atlantic Gulf Stream create more favourable conditions for plant growth

Figure 5.7   Climate and prehistoric settlement in Wales (a) The number of days into the year when selected plant species leaf or flower (after Lockley, 1970) (b) The distribution of later Mesolithic sites in Wales (after Jacobi, 1980) (c) The distribution of Neolithic tombs in Wales (after Lynch, 1980)

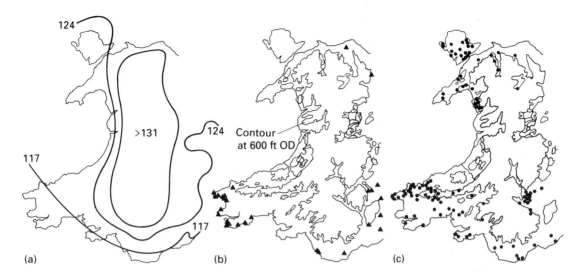

on the south and west coasts of Wales and that may be one reason why those areas contain the main concentrations of both later Mesolithic sites and Neolithic tombs (Figure 5.7).

Although attractive for settlement because of the resources they offer, coastal areas with gentle relief are particularly vulnerable to environmental change, which can be both sudden and large scale. Such areas are valuable laboratories for examining the impact of natural environmental change on human communities. The eustatic rise in sea level that followed the wastage of the last ice sheets (Chapter 4, p. 88) was rapid in the Early Holocene, with a gradually diminishing rate of rise from around 6 ka BP onwards. In The Netherlands the Early Holocene sea level rise of up to 2 m a century would have encroached rapidly across the relatively flat North Sea Plain and forced Mesolithic communities inland, making them adapt to a constantly changing geography (Louwe Kooijmans,

Figure 5.8   A submerged forest at Borth, Wales which has radiocarbon dates around 5300 BP (photo A Heyworth)

Figure 5.9   A sediment block from Westward Ho!, England with (a) fen peat (b) fragmentary shells from a Mesolithic midden dated 6–7 ka BP and (c) clay (photo A Philpott)

1987). In the earlier Mesolithic a basically similar Maglemosian culture existed across the land connecting Britain and Denmark, but divergent cultural traditions emerged after the isolation of Britain from the continent around 8 ka BP. The isolation of Ireland from Britain 2–3 ka earlier (p. 94) accounts for its impoverished fauna and for the contrasting nature of Ireland's Mesolithic culture (Sleeman *et al.*, 1983). On the east coast of North America, eustatic sea-level rise around 5 ka BP transformed the Gulf of Maine, USA from a tideless relatively unproductive system to one in which the exploitation of shellfish represented an important resource (Yesner, 1988).

Perhaps even more significant, in human terms, than the general rise in global sea levels, were smaller scale marine regressions and transgressions. Regression episodes are, for instance, represented by the occurrence, within marine and estuarine sequences, of terrestrial peats. Many are now exposed as submerged forests in the intertidal area (Figure 5.8). Such regressions are frequently associated with evidence of human activity, one example being the Mesolithic shell midden at Westward Ho!, England (Figure 5.9) which today is only exposed at low tide (Balaam *et al.*, 1987). People may have concentrated their activities in these areas (Figure 5.7b) because of coastal resources and a milder climate but also perhaps because the periodic nature of coastal change created broad ecotones with mosaics of subclimax plant communities.

Regressions are sometimes seen as reflecting short-term eustatic falls in sea level or, alternatively, they may reflect more local factors such as the development of coastal bars. Recognizing short-term sea-level changes in proxy records is not easy, however, for the error margins of the data are often greater than the supposed oscillations (Kidson, 1982). Detailed examination of sequences in The Netherlands shows that the evidence for falls in relative sea level is frequently equivocal, although there are clear indications of fluctuations in the rate of sea-level rise at intervals of 0.5–1 ka (van de Plassche, 1982). In view of the importance to human communities of small-scale, often local, coastal changes, two case studies relating to The Netherlands and the Somerset Levels/Severn Estuary in Britain will be considered in detail.

The Netherlands

Much of The Netherlands is a cultural landscape *par excellence*. About 50 per cent is around or below present sea level and has been subject to Holocene sedimentation; this area owes its very existence to human endeavour in the form of dikes and drainage. The sedimentary record along the Dutch coast is known in considerable detail and is closely correlated with the archaeology (Louwe Kooijmans, 1974, 1980). Figure 5.10a is a reconstruction of the landscape of The Netherlands at around 4.3 ka BP. The main sediment zones from the coast inland are:

1. Dune barrier broken by tidal inlets.

2. Tidal flats extending inland to saltmarsh.

3. Peats – fen peat towards the coast and in river channels with raised bog behind.

4. Fluvial minerogenic sediments in river valleys.

The boundaries between these sedimentary regions varied spatially through time, partly as a

Figure 5.10   The Netherlands (a) palaeogeographic reconstruction of the western Netherlands c. 2400 BC (b) east-west section through the Rhine/Meuse delta (after Louwe Kooijmans, 1987)

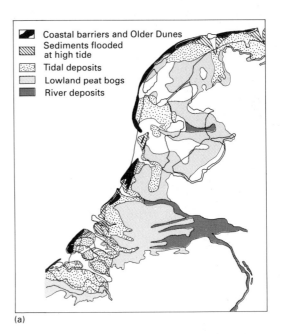

(a)

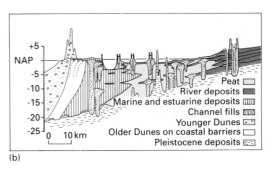

(b)

result of transgressive and regressive marine episodes, producing an interleaving of deposits (Figure 5.10b). The contrasting landscape regions and the sedimentary cycles which they underwent have been important influences on human activity. A number of Early Neolithic sites are on old river dunes and levees just above the surrounding marsh and backswamps (Louwe Kooijmans, 1987). The economy was only semi-agrarian with some

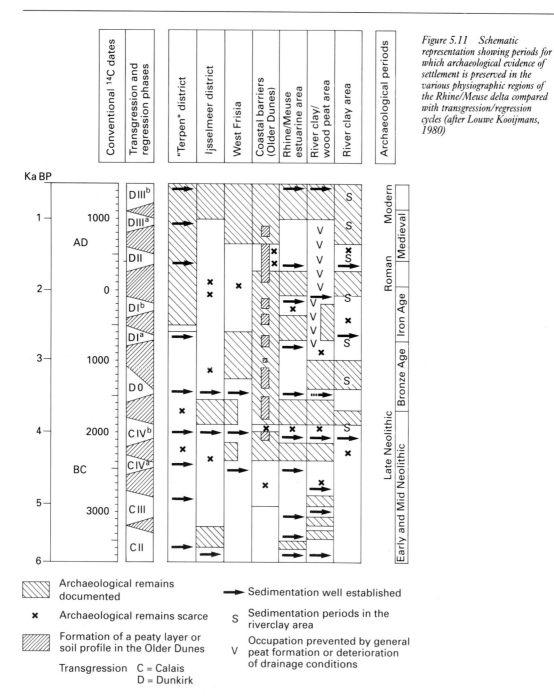

Figure 5.11 Schematic representation showing periods for which archaeological evidence of settlement is preserved in the various physiographic regions of the Rhine/Meuse delta compared with transgression/regression cycles (after Louwe Kooijmans, 1980)

domestic animals and cereals. Hunting, fishing and fowling were still important elements of the economy; some sites only seem to have been occupied seasonally, and their occupants only accepted those aspects of the Neolithic economy

(p. 158) appropriate to their own unusual ecology.

The sequence of later prehistoric settlement envisaged in many areas involves cyclical patterns with occasional visits, settlement and abandonment in parallel with alternating wetter and drier phases

(Brandt *et al.*, 1987). This accords with the evidence for widespread transgressive and regressive phases which are broadly contemporary in the major landscape regions (Figure 5.11). Archaeological evidence of settlement is well preserved in the regression episodes, which are separated by phases of sedimentation. While settlement was certainly more extensive during regressions this picture has also been exaggerated by the destructive effects of transgressions on settlement evidence (Louwe Kooijmans, 1985). Sometimes even transgressions created local conditions suitable for settlement, for example the Assendelver Polders were initially settled during an Iron Age marine transgression, when the peat surface became drier as a result of improved drainage by the landward extension of tidal creek systems.

Current Dutch thought tends to be critical of so-called 'wet feet hypotheses' which assume abandonment during transgressions (Abbink, 1986; Therkorn and Abbink, 1987). Instead the degree of human adaptability is emphasized and there are indications that this varied temporally (Louwe Kooijmans, 1991). During the Middle Bronze Age, for instance, there was a very rigid subsistence strategy with fully agrarian settlements on higher claylands and sandy creek ridges. This is in contrast to the more diverse settlement locations and economic strategies found in the Neolithic and Iron Age when wetter habitats were used to a greater extent. This must reflect temporal contrasts in people's perception of the environment and its possibilities and differences in the behavioural margins within which particular communities were adaptable (Louwe Kooijmans, 1991). Logically, the narrower the margins of adaptability the greater the repercussions of environmental change are likely to have been.

Increasingly with time, adaptation took the form, not just of changing human lifestyles, but fundamental modification of the environment itself. In Assendelver Polder during the Iron Age this led to the creation of a more homogeneous cultural landscape created by digging ditches, the changing of plant communities by grazing and the building of mounds (Brandt and van der Leeuw, 1987). A

current debate about the purpose of low mounds which carried structures in this area mirrors contrasting environmental and social perspectives on Dutch prehistory. Some see them as a response to rising water-tables, others stress the need to consider social explanations such as the enhancement of the visibility and status of structures in a very flat landscape (Therkorn and Abbink, 1987).

Much more substantial settlement mounds were constructed in the salt marshes of Friesland where a rich pastoral resource became dry enough for settlement of some higher areas by 500 BC. Even so, this landscape was still subject to regular flooding as shown by the macrofossils of salt-tolerant plants on settlement sites (van Zeist, 1974). Transgressive phases forced the abandonment of some sites, but other communities in north Holland, Germany and Denmark adapted by constructing terps (settlements raised on sods). The fortunes of these settlements were closely linked to coastal change and in the Groningen area four separate transgressive phases can be identified, each of which was followed by an episode of terp construction (Louwe Kooijmans, 1980). The success of terps is demonstrated by the very high density of settlements, particularly during the Roman Iron Age. A marked reduction in the evidence for settlement in the terp areas in the fourth and fifth centuries AD has often been correlated with the effects of the Dunkirk II marine transgression (Figure 5.11). This may have helped to prompt the migrations of the fifth century AD from Friesland. It must also be acknowledged, however, that these migrations were part of a much more widespread, virtually pan-European social and economic collapse, consequent upon the disintegration of the Roman Empire. The terps would have lost their markets in the Roman frontier to the south and the collapse of trade also makes sites difficult to ascribe to this period because of the paucity of datable artefacts.

Terp construction ceased in the eleventh century AD when the strategy for avoiding inundation took more ambitious forms. Dike construction began outside the terpen area in the eighth century AD and by the thirteenth century

AD the completion of ring dikes surrounding major areas made settlement possible throughout The Netherlands (Berendsen and Zagwijn, 1984). Such a significant increase in the extent of environmental manipulation implies social organization on a regional scale, the emergence of which may have been partly encouraged by substantial loss of land elsewhere. Of particular significance were flooding which formed the Zuiderzee, partly as a result of a storm surge in AD 1170, and early Medieval dune encroachment near the coast.

The earliest surviving coastal dunes had formed by 4.8 ka BP. The prehistoric Older Dunes (Figure 5.10b) are characterized by an alternating sequence of stabilization phases with peaty organic bands separated by aeolian sands which form coeval blankets over large areas. The dunes were settled throughout prehistory, albeit with interruptions during sand blow phases, and well-preserved settlements and field systems are widespread (Jelgersma et al., 1970). The stabilization episodes are regarded as relating to phases when the water-table in the dune belt was sufficiently high for the surface to support vegetation. These stabilization episodes can be shown to be generally contemporary with the previously noted marine transgression phases recorded in the clays and peatlands behind the dune bar (Figure 5.11).

A dramatic change in depositional regime occurred in the coastal dune belt between AD 1000 and 1600 when tall parabolic dune systems known as the Younger Dunes formed (Zagwijn, 1984). That so much material was available for aeolian transport suggests that the main cause was a change in the pattern of coastal erosion. Migration of this sand inland covering settlements and fields may have been exacerbated by the destruction of woodland which had developed on the dune belt during a stabilization phase over the preceding millennium. An additional factor may have been a particularly dry period between AD 1000 and 1180 at the beginning of Younger Dune deposition (Berendsen and Zagwijn, 1984), perhaps corresponding with the warmer and drier period of the Medieval Climatic Optimum (Chapter 3, p. 73).

Somerset Levels and Severn Estuary, Britain

The Severn Estuary in south-west Britain (Figure 5.12) is funnel shaped with extensive estuarine clays round its margin. To the south in the Somerset Levels a coastal barrier of sand dunes is succeeded inland by estuarine clays and finally peat deposits. From an archaeological point of view these peats are the most intensively investigated wetlands in Britain (Coles and Coles, 1986).

Holocene sea-level rise (Figure 4.10) meant that by the end of the Mesolithic the whole area was a great tidally inundated expanse in which silts and clays were accumulating. Between 5400 and 5000 BP the Somerset Levels were colonized first by *Phragmites* reeds then, as peat accumulated, fen woodland developed. It was at this stage that the first of many wooden trackways was constructed between bedrock ridges and drier islands. The Sweet Track is the earliest and most sophisticated, a raised walkway built across fen peat to maintain communications at a time when plant macrofossils and insect evidence show that conditions were temporarily rather wetter with areas of standing water along the route. This track is dated dendrochronologically to winter/spring 3807–6 BC (Hillam et al., 1990). That corresponds to a period between 4000 and 3800 BC when other proxy data also point to increased wetness and possibly marked coastal change (Morgan et al., 1987). From about 2500 radiocarbon years BC, raised bog developed over parts of the fen peat.

A number of trackways were constructed during the later Neolithic period (Figure 5.13). Some were major structures like the Abbot's Way, a corduroy track (baulks of timber laid at right angles to the line of the track) dating from around 2000 BC laid across the spongy raised bog surface between two islands. Some tracks were made of hurdles (Figure 6.12) and others were bundles of brushwood laid down to enable crossing of particularly wet areas of the bog surface or edge.

In the earlier Bronze Age few tracks were constructed and there is less evidence of human activity on the large area of raised bog which had by this time covered nearly all of the area. Trackway

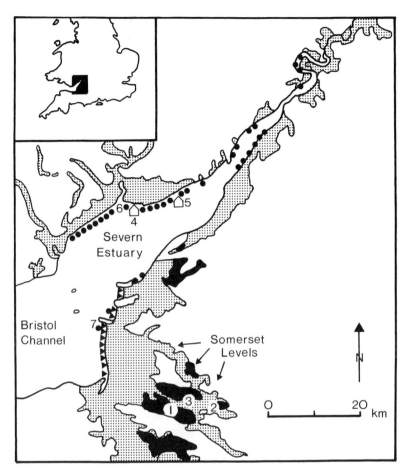

Figure 5.12   The Somerset Levels and Severn Estuary showing the main sediment types and archaeological sites

construction increased again in the Late Bronze Age (Figure 5.13) and there is further evidence for flooding around 700 BC. This wetter episode occurs at a time of well-attested climatic deterioration (Chapter 3, p. 71) although it has also been hypothesized that other environmental factors such as the activity of beavers and the possible effects of increasing clearance on runoff could also have contributed to increased wetness (Coles and Coles, 1986). In fact, it now appears that increased waterlogging was caused by a major marine incursion between 850 and 550 BC which

Figure 5.13 The radiocarbon dates of trackway construction on the Somerset Levels. Uncalibrated, upper axis; calibrated (according to Clark, 1975), lower axis. It should be noted that this picture may to some extent be affected by the pattern of trackway survival and recovery (after Orme, 1982)

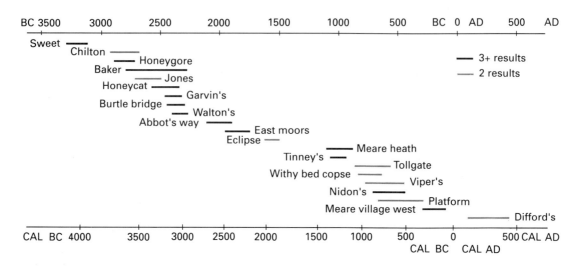

led to the deposition of clays up the Axe Valley as far as Glastonbury (Housley, 1988; Figure 5.12). Most Somerset Levels' trackways represent responses to wetter conditions; however, the substantial Meare Heath Track was constructed a century before the onset of widespread flooding and may partly reflect changing communication needs.

Before the Late Iron Age, human activity on the Levels is attested by pollen diagrams, trackways and a few artefacts left by people hunting, grazing animals or gathering raw materials. No evidence of settlements has yet been found: these must have been on intervening ridges and islands, their locations suggested only by the convergence of trackways and pollen evidence for arable fields and pasture just off the wetland. What survives is very much biased towards those aspects of human activity which took place on, or adjacent to, wetland. One way in which people responded to environmental change was by making trackways, rather than allowing existing patterns of communication to be disrupted. Wetter or drier conditions would have had economic effects, but arguably only a limited impact on settlements and the activites carried out on neighbouring dry ground.

In the Late Iron Age wetland settlements began to appear. The best preserved is Glastonbury lake village, a crannog (artificial island) surrounded by shallow open water and reed swamp. It was a large permanently occupied settlement located, perhaps for defensive reasons, at some distance from its dryland arable resources. Meare Village East was different; it was on the dried-out raised bog edge and only seems to have been seasonally occupied by communities exploiting wetland grazing (Coles, 1987). Increasingly during the later Iron Age and Romano-British period the environment was dominated by human activity, an open agricultural landscape was created and a concomitant increase in alluvial clay deposition may relate to increased runoff from agricultural land.

The Severn Estuary to the north (Figure 5.12) is a very different environment with related problems in terms of human settlement history. It contains an estuary-wide sedimentary sequence (Figure 5.14) with a succession of transgression and regression episodes (Allen, 1987). Rising sea level in the earlier Holocene is represented by a thick clay deposit (the Wentlooge Formation). Clays exposed on the foreshore, datable to the Mesolithic period, are covered with the footprints of people (Figure 5.15), wild animals and birds. Peat bands within the

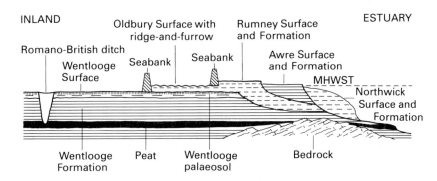

INLAND
Oldbury Surface with
ridge-and-furrow
Rumney Surface
and Formation
ESTUARY

Figure 5.14   Schematic section of the sedimentary sequence in the Severn Estuary (after Allen, 1987)

Romano-British ditch
Wentlooge
Surface
Seabank
Seabank
Awre Surface
and Formation
MHWST
Northwick
Surface and
Formation

Wentlooge
Formation
Peat
Wentlooge
palaeosol
Bedrock

Wentlooge clay testify to marine regressions, the most important of which occurred between 4 ka and 2.5 ka BP. Waterlogged wooden buildings are associated with the peat bands and date to the Middle Bronze Age at Chapel Tump and the Iron Age at Goldcliff, this activity may relate to seasonal or specialized wetland exploitation (Whittle, 1989). Subsequently, the peat was covered by the Upper Wentlooge clay during a marine transgressive phase around 2.5 ka BP, which seems likely to correspond to that previously noted in the Somerset Levels.

By the Romano-British period marine influence had declined and the flats were sufficiently dry for very large-scale and probably Roman military-inspired drainage operations marked by alignments of ditches which are now exposed in the intertidal area. These probably lay behind a now lost sea defence (Allen and Fulford, 1986, 1987). There is also evidence of extensive Romano-British settlement and drainage of the clayland of the Somerset Levels at this time. In the Severn Estuary following Roman activity there was a series of transgression phases which cut low cliffs into earlier sedimentary formations and were each followed by an episode of sediment deposition (Figure 5.14). Periodically, areas of saltmarsh were reclaimed for agriculture behind seabanks. The cycles of erosion and deposition within the Severn Estuary are most clearly defined and closely dated during the Post-Medieval period when the timescale of instablity which they represent was as short as a single century (Allen and Rae, 1987).

There is evidence of much earlier episodic environmental instability in the Severn Estuary on the coastal dunes at Brean Down. Five distinct

Bronze Age occupation horizons, two with well-preserved circular houses, are separated by phases of sand blow (Figure 5.16). Between periods of occupation people moved away for sufficiently long to change the fabric, form and decorative motifs on their pottery (Bell, 1990). Sand blow from the intertidal area is thought to relate to regressive phases and, therefore, the cessation of sand blow and final abandonment of the settlement around 2.6 ka BP may tie in with the previously noted marine transgression.

Coastal change: causes and consequences

These case studies reveal evidence for short-term marine transgressive and regressive episodes, the reasons for which are not always apparent. One hypothesis is that they relate to sea-level fluctuations within the overall rising trend of the Holocene. However, there are indications that sea-level falls (or reductions in the rate of rise) have been largely obscured by the effects of both regional and local coastal changes, which may, *inter alia*, be climate- or weather-related. In the Severn Estuary, for example, offshore shoal movements, changing sediment input, and short- or medium-term climatic change, specifically the frequency of westerly and south-westerly winds, have all been suggested as possible causal factors of short-term coastal change (Allen and Rae, 1987).

In The Netherlands, a key process has been the periodic breaching of coastal barriers during episodes of increased storminess. Storm surges, in particular, can be catastrophic in human terms. On 31 January 1953 sea level in the south-western

Figure 5.15   Human footprints of Mesolithic date in inter-tidal sediments in the Severn Estuary (photo D Upton)

Netherlands was 3 m above the predicted tide level and major flooding occurred all round the North Sea. In the Severn Estuary catastrophic flooding occurred in AD 1606–7 and the available historical records show that tides up to 1 m above HAT (Highest Astronomical, i.e. predicted, Tide) recur every 100–200 years (Bell, 1990). There are indications that a greater frequency of such events occurred during periods of climatic deterioration and there is also evidence to show that marine transgressions and regressions affected widely separated areas at approximately the same time (Tooley, 1985). Quasi-cyclical transgressions and regressions in these coastal areas *may* reflect

Figure 5.16   Sand dune sequence at Brean Down, Somerset showing Bronze Age occupation layers separated by blown sand and colluvium (photo A Philpott)

cyclical climatic change, particularly over shorter timescales (Chapter 3).

In terms of the archaeological record, The Netherlands and Somerset Levels present interesting contrasts. In the Levels dry ground is never far away and could be used for settlement. Even so, prehistoric communities had to adapt by constructing tracks so that routes could be kept open during wetter periods. As early as the Neolithic they were starting to construct a cultural landscape of clearings, communication routes and managed woodland which took account of the natural environment and its changes, but was by no means utterly constrained by them. Hence, when people eventually settled within the wetland itself, as they did in the Late Iron Age, it was apparently from choice, perhaps reflecting such imperatives as defence and grazing requirements. In The

Netherlands the low-lying area is dramatically larger; if prehistoric communities were to exploit it, they had to occupy slight rises such as levees, or slightly higher peat areas and even these were sometimes vulnerable to flood. The result was a Neolithic culture uniquely adapted to the local ecology, and Bronze Age and Iron Age societies which were, in a sense, opportunistic, exploiting areas when they were sufficiently dry, or because they offered particular resource opportunities. Eventually, by physically altering the landscape, digging ditches and constructing terps, people became less vulnerable to the vagaries of sea-level change. Large-scale dikes and drainage works were undertaken from around AD 1100, about a millennium after the extensive draining of many areas of estuarine environment in Britain under the Romans.

Volcanism

Volcanic eruptions are environmental changes which are precisely definable in time and space, and this makes them of particular interest in the evaluation of human response. Major eruptions eject many cubic kilometres of ash (tephra) which blankets the immediate area, parts of which are also inundated by lava flows and by mud flows resulting from the mobilization of ash by rainfall. Areas at a distance of 100 or more kilometres, which may be covered by only thin layers of ash, may also suffer serious but short-lived damage to vegetation and livestock. Accompanying seismic sea-waves (tsunami) can devastate coastal areas at considerable distances from the volcanic centre, as they did when Krakatoa, erupted in 1883 killing 36 000 people in coastal Java and Sumatra. Major volcanic eruptions eject such large quantities of sulphur-rich dust into the atmosphere that a globally detectable signal is produced, and thus past volcanic events register as peaks in the acidity trace from polar ice cores (Figure 3.17). Moreover, volcanic eruptions have short-term climatic effects which can be identified in tree-ring sequences (Chapter 3, p. 79). Recent recognition of these global signals has prompted renewed interest in the effects of volcanism on past societies. The local effects are obvious enough in the form of buried sites like Pompeii, Herculaneum and Akrotiri. Eruptions have also been claimed to have had widespread social repercussions over much larger geographical areas.

The effects of volcanic eruptions on present-day complex and literate societies have been documented in great detail (Blong, 1984), while ethnographic studies are of particular value to the interpretation of prehistoric hazard response. In parts of Papua-New Guinea, oral legend tells of a 'time of darkness' lasting several days in which plants and animals were seriously affected and in about 30 per cent of communities people died mostly in collapsing houses or of starvation. One group even turned briefly to cannibalism as a consequence (Blong, 1984). Interpretation of these legends represents a fascinating integration of evidence for human perception from ethnography and data from the earth sciences (Blong, 1982). The legends correspond closely with tephra distributions and are found to constitute a reliable historical record for a previously unknown eruption on Long Island in the Bismark Sea probably between AD 1630 and 1670. This produced 30 km^3 of ejecta which covered 80 000 km^2 at depths of between 1.5 and 16 cm. The legends record differing attitudes to the eruption, some communities regarded it as very harmful and felt they had never been the same since. Surprisingly, by contrast, about 20 per cent of legends recorded beneficial effects, a time of great plenty brought about by the fertilizing properties of the ash. Some saw the eruption as leading to great cultural achievements and one group even invoked earth magic to secure a repetition! Positive and negative memories of the eruption do not correlate with the evidence for its purely physical effects and are thought to relate partly to social and cultural variation. A much earlier eruption c. 3.5 ka BP by Mount Witori in the same general area has been shown to have caused abandonment of settlements and obsidian workshops and is apparently also correlated with changes in tool type (Torrence *et al.*, 1990).

Archaeological evidence for the beneficial effects of tephra comes from the Sunset Crater eruption, Arizona (Figure 5.20) in AD 1066. After short-term disruption the Indian population and cultural diversity of this densely settled area actually increased in the 150 years after the eruption. This is thought to have resulted from a combination of the beneficial mulching effects of the ash on water retention and a coincidentally more favourable rainfall regime (Pilles, 1979; Hevly *et al.*, 1979). Similarly in America, the initially devastating effects of the eruption of Mount Mazama, Oregon c.. 6.8 ka BP produced 30 km^3 of ejecta covering a large area (Figure 2.6) and led to the massive caldera collapse of Crater Lake, but this event produced only negligible long-term effects on the fauna in cave sequences in the area (Bacon, 1983; Grayson, 1979).

One area where it is possible to compare recent and archaeological events is the Aleutian area of Alaska. There Katmai erupted in 1912 producing

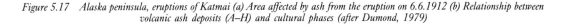

Figure 5.17   Alaska peninsula, eruptions of Katmai (a) Area affected by ash from the eruption on 6.6.1912 (b) Relationship between volcanic ash deposits (A–H) and cultural phases (after Dumond, 1979)

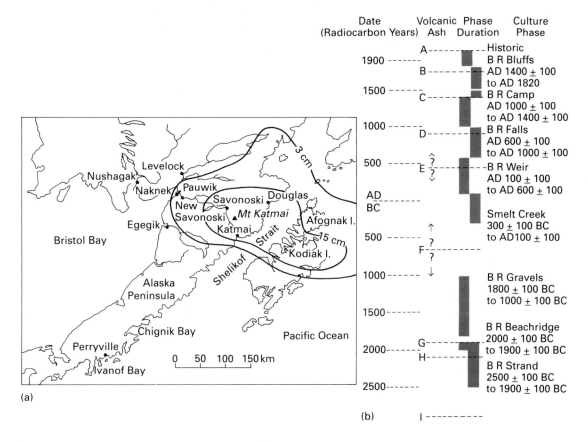

(a)

(b)

28 km³ of ejecta. Even today the immediate surroundings are a devastated lunar-like landscape used for the training of space crews (McCracken Peck, 1990). Short-term effects where tephra was deposited (Figure 5.17a) were serious but a return to previous ways of life was possible in most of the area within 20 years. It had been sparsely settled and the key factor was that kin networks and exploitation territories of the Aleuts extended beyond the devastated area (Dumond, 1979). People simply moved away as refugees, later to return. Nine distinct ash layers have been identified in sediments that have accumulated over the last 7.4 ka. Figure 5.17b compares these volcanic events with the timescales of cultural phases for the area identified on the basis of assemblages of artefact types. Some thin ash layers (B,D,H) occur within cultural phases without apparent disruption whereas others (C,F,G) occur between phases. The latter could imply either a cultural response or hiatus sufficiently long for cultural change to have taken place. The cultural changes are not, however, restricted to the areas directly affected by eruptions and it is concluded that despite the severity and frequency of volcanic activity around the Aleutians this had a minimal effect on cultural change.

For the eruption of Vesuvius in AD 79 there is a considerable amount of direct information regarding its effects, including an eye witness account by the Younger Pliny. Pompeii and villas in its vicinity were buried by ash and Herculaneum by a mud flow (Jashemski, 1979). The towns were never rebuilt although there is some evidence for a few later burials and buildings by the third century AD in the area. Of the population, 90 per cent escaped and had time to remove many of their valuables. About 2000 people died, some by fume asphyxiation, and are represented by evocative casts of their dying bodies in the solidified ash.

The island group of Thera (Santorini) in Greece formed a single island until an eruption which produced an estimated 30–50 km^3 of ejecta and

was followed by collapse of the island's centre as a vast caldera (Figure 5.18a). Ash buried the settlement at Akrotiri in which buildings containing magnificent frescoes survive in places to a height of 6 m (Renfrew, 1979). Traditional dating of the eruption, based on the associated Late Minoan 1a pottery was c. 1500 BC. Radiocarbon dating suggested an earlier date, however, and this now seems to be confirmed by an acidity peak at 1644 BC in the Dye 3 Greenland ice core (Figure 5.18b)

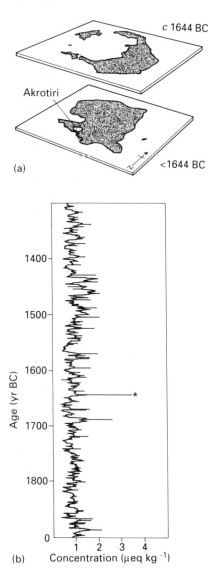

(a)

(b)   Concentration (µeq kg^{-1})

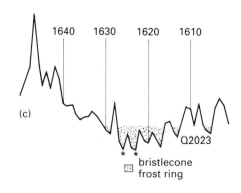

(c)

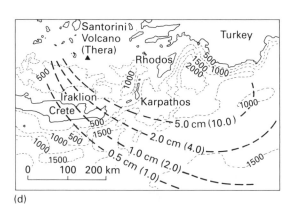

(d)

Figure 5.18   The Thera eruption (a) The island of Thera before and after the eruption. (b) Acidity concentrations in Dye 3 core, Greenland. The peak with the asterisk is that correlated with the Thera eruption (a and b after Hammer et al., 1987) (c) Tree ring width in Northern Ireland. Shaded areas are particularly narrow rings, those with an asterisk the narrowest in the tree's life. The shaded box marks the Californian bristlecone pine frost rings, which are tentatively correlated with the Thera eruption (after Baillie and Munroe, 1988) (d) The distribution of Thera ash in deep sea cores. Figures in brackets are estimated depths on dry land (after Watkins et al., 1978)

and by exceptionally narrow tree rings in Ireland and California at 1627/8 (Figure 5.18c). These new dates place the Thera eruption at around 1625–45 BC. The correlation of these signals with the Thera eruption and the conflict with the traditional archaeological dating continues to excite controversy (Pyle, 1989; Aitken *et al.*, 1988).

The date is of considerable archaeological interest because several writers have followed Marinatos (1939) in arguing that this massive eruption caused the collapse of the Minoan civilization on Crete, 120 km away, where up to 4 cm of tephra were deposited by the blast (Figure 5.18d). They point to the sudden abandonment and destruction of Cretan palaces and settlements and the possibility that a tsunami of Krakatoan proportions accompanied caldera collapse. The problem is that the abandonment layers contain Late Minoan 1b pottery at least half a century later than the Late Minoan 1a pottery buried by ash at Akrotiri. The Thera eruption seems unlikely to be the direct cause of Cretan abandonments, which may be attributed to social and economic tensions inherent within this early state society (Renfrew, 1979).

Another major acidity peak in the Greenland ice cores occurred in *c.* 1100 BC and is almost certainly associated with the Hekla 3 eruption in Iceland. This is correlated with a particularly narrow group of Irish tree rings beginning in 1159 BC (Baillie and Munro, 1988). The eruption clearly registered over a very wide area and traces of ash from one of the eruptions of Hekla (possibly Hekla 4 *c.* 4 ka BP) have recently been found in a bog in northern Scotland (Dugmore, 1989b). Recognition of the scale of volcanic signals has led to recent suggestions that eruptions had serious repercussions in Bronze Age Britain. Archaeologists have long been puzzled by the extent of settlement pattern changes, particularly the abandonment of the uplands between the earlier and later Bronze Age around the twelfth century BC. Burgess has long favoured a catastrophic explanation, originally seeing climatic change as the main factor (Burgess, 1974, 1980), although subsequently arguing that it may have been exacerbated by plague (Burgess, 1985). Most

recently the Hekla 3 eruption has been suggested as the cause of these changes (Burgess, 1989; Baillie, 1989). It has also been suggested that a succession of population declines seen in the prehistoric record from Britain, and elsewhere in Europe, correlate with tree-ring evidence for major volcanic eruptions. Baillie (1989) has suggested that other changes, including the Justinian plague and the beginning and end of the Chinese Shang Dynasty, can be attributed to the effects of individual eruptions.

The implication that volcanism has been a major cause of cultural change requires critical examination of the spatial and temporal scale of volcanic effects. The sizes of the areas over which social repercussions have been claimed are much larger than the area of clearly attested effects from recent documented eruptions. Britain is 1000 km from the nearest centre of contemporary volcanic activity in Iceland, but it has been suggested that Hekla 3 led to widespread British upland abandonment and 'population loss', in fact, more severe effects than those documented in many volcanic areas themselves! In Iceland large numbers of sheep have died as a result of fluorine toxicity arising from recent Hekla eruptions but most of the losses were within 100 km of the source of the eruption. There are only occasional records of geographically much more widespread effects such as the 1783 eruption of Laki (Lakagigar), ash from which damaged crops in Caithness (Thorarinsson, 1979).

Burgess (1989) has argued that in the later Bronze Age, the British uplands were abandoned for several hundred years. Sometimes, in areas unfavourable for agriculture, abandonments resulting from volcanic activity may be long term, for instance the eruption of Hekla in AD 1104 caused the desertion of a number of farms but these were mostly within 25 km of the eruption although one was 70 km away; none was ever resettled (Figure 5.19). Generalizing, however, from the effects of the six most recent Icelandic eruptions since AD 1104, Thorarinsson (1971) has shown that in all but the most severely affected areas the effects were short term; 10 cm of tephra caused abandonments of up to 1 year, 15 cm for

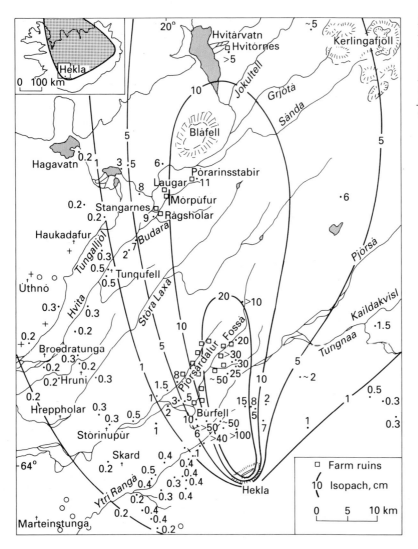

Figure 5.19 *The eruption of Hekla in AD 1104 showing isopachs of tephra thickness and the locations of abandoned farms, inset shows the distribution on Iceland as a whole. Key: 0.3 thickness of layer (cm); + layer just visible; ○ layer not found (after Thorarinsson, 1971)*

1–5 years and 20 cm for periods of some decades.

More problematic is the effect on the global climatic signal produced by major volcanic eruptions. Severe winters follow certain types of eruption but, as noted in Chapter 3 (p. 77), these effects are essentially short term, being typically less than ten years. Furthermore, the extent of their effect on agriculture and human activity remains to be established. The Papua-New Guinea and Sunset Crater evidence further challenges catastrophic hypotheses by showing that sometimes short-term disaster may even be counterbalanced by longer term benefits. Ethnohistorical and archaeological evidence highlights the rapidity of recovery and the flexibility of human response. It is acknowledged that much does depend on the scale of the eruption, population density and the capacity of a society to adapt. The balance may have been tipped by volcanism in societies under tension and at the margins of survival, but to regard a succession of major cultural changes as a direct consequence of volcanic activity is, on present evidence, unduly deterministic.

Arid zone margins in the American south-west

The physical variables affecting past human communities have been particularly well established in the American south-west (Figure 5.20). Here pueblo settlements (Figure 5.21) with organic artefacts are remarkably well preserved and dendrochronology provides a highly accurate chronological framework for both settlement archaeology and environmental change. In addition, there is a valuable ethnohistorical record. Cultural groups in the area have often been regarded as adapted to specific biogeographical regions (Martin and Plog, 1973). The Anasazi occupied the Colorado Plateau between the later first millennium BC and AD 1300 and their environmental relationships have been investigated in particular detail (Gumerman, 1988). By comparing the

Figure 5.20   The American South West showing the extent of Anasazi communities and palaeoenvironmental sites noted in the text. The Sunset Crater and Mount Mazama volcanos and temporary Lake Cahuilla are also shown

Figure 5.21   Pueblo Bonito, Chaco Canyon, New Mexico a ceremonial and population centre of the eleventh and twelfth centuries AD. An arroyo in the background was incised after AD 1100 and may have impeded floodplain cultivation (photo Harold D. Walter, Courtesy Museum of New Mexico, Negative No. 128725)

archaeological evidence with that from alluvial sequences, pollen and dendroclimatology (Figure 5.22), environmental change on two main wavelengths can be identified. Low frequency processes, longer than one human generation of *c.* 25 years, were probably not perceptible to contemporary communities but are registered in the alluvial sequences. High frequency processes with a timescale of less than 25 years are detected by dendroclimatology.

Alluvial sequences show four main phases of degradation represented by stream entrenchment and arroyo formation beginning AD 200, 750, 1250 and 1850 (Karlstrom, 1988). Degradation is associated with lower water-table and precipitation, while aggradation (sediment deposition) is associated with more frequent inundation. The regular sequence of alluvial changes points to underlying primary and secondary cycles lasting

550 and 275 years. Episodes with proportionately more arboreal than non-arboreal pollen are believed to indicate higher effective moisture and correspond with aggradation phases. In the south-west a close correlation has been established between tree-ring width, precipitation and temperature (Hughes *et al.*, 1982). Figure 5.22c summarizes the main dendroclimatic trends, with periods of particular variability, mostly corresponding to degradation phases (Dean, 1988).

Population in the area rose steadily from AD 1 to 1000 (Figure 5.22e) peaking during a moist episode between AD 950 and 1150 when settlement was more widely distributed than it is today. There was then a marked population decline, the exact date of which varies regionally. Much of the Colorado Plateau was totally abandoned and people concentrated in the better watered river valleys. Drought is widely acknowledged as the

Figure 5.22   Environmental and demographic change on the Colorado Plateau (a) Aggradation/degradation cycles in flood plains (b) Effective moisture reflected by pollen data, solid line primary fluctuations, dashed line secondary fluctuations. (c) Dendroclimate tree growth departures in standard deviation units from the long-term mean for each decade. Positive values reflect higher than average rainfall. Rectangular blocks show periods of particular variability (d) Spatial variability indicated by dendroclimate (e) Postulated population trends AD 1–1450 (after Plog et al., 1988)

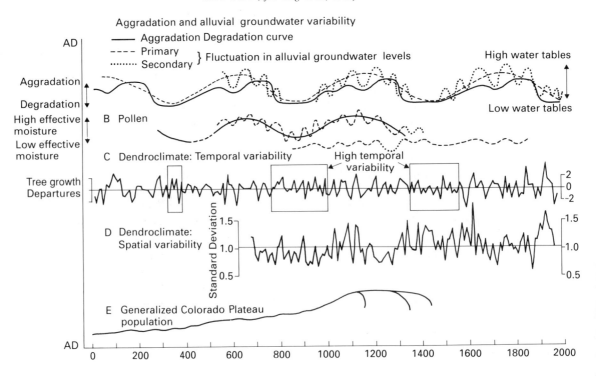

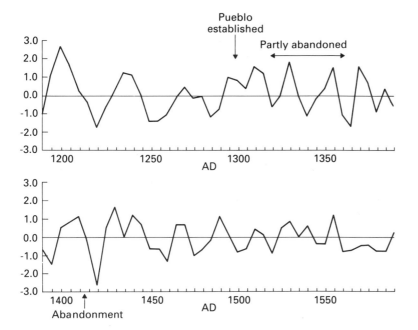

Figure 5.23   Reconstruction of 12 month precipitation for the Arroyo Hondo area, AD 1190 to 1590, expressed as departures from the mean (after Rose et al., 1981; Wetterstrom, forthcoming)

major trigger of these changes. Earlier pre-AD 600 droughts may only have had limited social repercussions because low populations left plenty of opportunities for mobility and diversification. By AD 1100 the dense population was already beginning to strain the capacity of the habitats concerned. At first this led to innovation, for there is increasing evidence of water control (e.g. irrigation), erosion control walls and food storage in the period after AD 1000. The spatial variability of precipitation, which is demonstrated by comparison of dendroclimatic sequences from various parts of the plateau, may have given rise to a social organization based on welfare and sharing as a way of spreading risks in a hazardous environment (Gumerman, 1988). Ceremonial structures known as kivas may have played a part in cementing the mutually supporting links within and between communities. A quite different response to stress is represented by increasing evidence for defensive sites.

Arroyo Hondo pueblo to the east of the Colorado Plateau in a tributary of the Rio Grande, New Mexico provides a particular instance of the relationship between settlement history and a dendroclimatic sequence (Figure 5.23;

Wetterstrom, forthcoming). This community was one of the last to be established in the area around AD 1300 and quickly grew to a population of 400–600 during an episode of wetter climatic conditions. A drought began in AD 1335 and led to intensification of hunting. Between AD 1340 and 1370 the pueblo was partly abandoned but its population grew again in the 1370s only to be finally abandoned in a drought in AD 1425. Skeletal evidence testifies to a community under stress with malnutrition and high infant mortality.

The role of water in the archaeology of the semi-arid south-west is reflected in the remarkable history of temporary Lake Cahuilla in southern California (Wilke, 1978; Aikens, 1984). From time to time this formed as a result of changes to the Colorado river delta (Figure 5.20) which periodically debouched northwards to fill the lake before returning to its old course south to the Gulf of California. The most recent occasion on which the lake formed was AD 1150–1550 (Waters, 1983). It was 185 km long and a considerable Indian population was attracted to its shores. Of particular interest, with regard to perception of environmental change, is that after almost half a millennium oral tradition records the lake and the abundance of its aquatic resources. These formed the economic basis of year-round hunter–gatherer activity as shown by the seeds and bones in human coprolites from lake shore settlements. Once input from the Colorado had ceased it would have dried up quickly with catastrophic ecological effects. Freshwater mussels, the main constituent of lakeshore middens, vanished after about 10 years. Although fish weirs and settlement on the retreating shores testify to attempts to maintain the old way of life, it is calculated that within 30 years the entire aquatic plant and animal community would have disappeared and the lake itself would have vanished in about 60 years. The people, it is thought, retreated to the upland, where their arrival is likely to have had repercussions that may explain the endemic intertribal hostility which characterizes the early historical record for the area.

The south-west provides some of the clearest evidence for the effects of climatic and other environmental change on human communities. Surprisingly, in view of the population levels attained on the Colorado Plateau in the twelfth century AD, there is little evidence here for environmental change induced by human communities. Conversely, in southern California manipulation occurred to the point where little or no natural landscape remained by the time of the first Spanish descriptions (Shipek, 1981, 1989). The range and quality of palaeoenvironmental evidence from the Anasazi makes it clear that each data set provides information about differing timescales of change. It is, therefore, desirable to think not just in terms of periods of favourable and unfavourable climate, but rather to consider the amplitude, frequency, temporal and spatial aspects of environmental variability as summarized in Figure 5.22 (Plog *et al.*, 1988). Short-term (high frequency) variation may have exacerbated or mitigated the effects of longer term (low frequency) variation. Effects may have been particularly severe when both coincided, as perhaps in the period after AD 1150. By that time population was at a high level and large, complex pueblo communities had fewer opportunities than during earlier crises for diversification and adaptation.

The Little Optimum and Little Ice Age

The warm episode of the Little Optimum between AD 700 and 1300, followed by the Little Ice Age between AD 1550 and 1850 is recorded in a wide range of proxy data sources (Grove, 1988), while the subsequent warmer phase is reflected in both proxy and instrumental records (Chapter 3, p. 73). What makes these episodes of particular interest is the opportunity for comparative studies of proxy, instrumental and historical sources in order to achieve a more detailed understanding of the effects of climatic change on human society. This demands a critical appraisal of the historical source material itself which has only recently become the norm in studies of climate and history (Le Roy Ladurie, 1972; Wigley *et al.*, 1981). The climatic changes of the last millennium will be considered in terms of their effects on human communities in Greenland, Iceland and parts of north-west Europe.

Greenland

Greenland was first settled by Norse communities in AD 985 (McGovern, 1981) during a marked warm stage of the Little Optimum (Figure 5.24a). During the same period the Norse explored as far as Newfoundland, settling at L'Anse aux Meadows, on the northernmost point of the island, though it is uncertain whether this is more than a temporary settlement. There are Norse artefacts but no certain evidence of agriculture (Ingstad, 1977). In Greenland the Norse established an Eastern Settlement which had, at its maximum, a population of 4000–5000 and a Western Settlement of 1000–1500 (Figure 5.24b). Settlement is reflected in pollen diagrams by a pronounced 'landnam' (land occupation) which shows an increase in weeds (some introduced) which are typically associated with pastoral activity, while charcoal horizons occur at the same time in lakes and settlement profiles. Cereals could not be successfully cultivated in this harsh climate and the

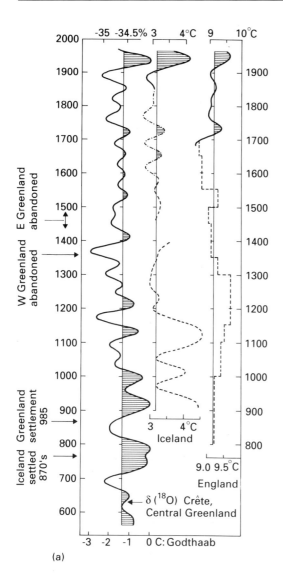

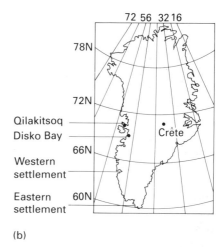

Figure 5.24   (a) Oxygen isotope curve (^{18}/0 from Crête, central Greenland compared with temperatures in Iceland and England. Dashed curves based on indirect evidence, the broken Iceland curve on sea-ice data and the solid Iceland and England curves on instrument records (after Dansgaard et al., 1975) (b) Location of Norse settlement in Greenland and the Crête core

economy was based on pastoralism involving sheep and cattle which needed to be kept in byres through the long winter and were, therefore, highly dependent on the hay harvest. Agriculture was supplemented to a major extent by the hunting of seals, caribou and seabirds. The Norse maintained contact with their homeland by limited trade, particularly in walrus ivory and polar bear skins obtained in hunting grounds at Disko Bay 800 km north of the Western Settlement.

At first the Norse colonies flourished. They had a bishop from AD 1125 to 1378 and a major church building programme took place between AD 1125 and 1300. Then a marked decline took place resulting in the abandonment in the mid-fourteenth century AD of the Western Settlement in mysterious circumstances, with domestic animals left wandering around. Official contact with Norway ended in 1408 and the Eastern Settlement died out in the late fifteenth century. Many hypotheses have been advanced to explain this decline. There is evidence for conflict with Basque fishermen and perhaps with Inuit communities who were at that time moving down into the Norse hunting grounds from further north. Changes in the orientation of Danish and Norwegian trade would have left the Greenland communities increasingly isolated, at the same time that increasing sea-ice was making communications more difficult. Environmental degradation caused by Norse activity may have been a further contributory factor. Re-establishment of sheep farms in the twentieth century has led to overgrazing and erosion of this fragile ecosystem and comparable episodes of sand blow are attested during the Norse period (Fredskild, 1988).

Modern analogues in both Greenland and Iceland (see below) emphasize the extent to which agriculture at this latitude is susceptible to climatic change; during the bad years of 1966–67, for example, the number of sheep in Greenland was reduced by more than half. Colder conditions would also have led to reduced seal populations in the settled areas and increased snow cover can lead to catastrophic declines in caribou numbers. There is certainly some archaeological evidence for deteriorating climate in the fourteenth century. Bodies buried between AD 1350 and 1500, for example, are better preserved than earlier interments due to the extension of permafrost (Hovgaard, 1925). The final days in the life of a farm at Nipáitsoq in the Western Settlement were marked by increasing hardship, for animal bones indicate that both dairy cattle and hunting dogs were slaughtered (Buckland et al., 1983; McGovern et al., 1983) probably during an exceptionally severe winter. The Crête ice core (Figure 5.24a) shows that the settlement abandonments occurred at a time when temperatures were for a prolonged period 1–2 °C lower than those which had obtained at the time of original settlement.

Greenland is of considerable interest as the only example of a developed European society which (so far!) has been extinguished. Given the marginal nature of the area there can be little doubt that the main factor that put the Norse settlers under stress was the climatic downturn of the Little Ice Age, which on the Crête oxygen isotope timescale (Figure 5.24a) began rather earlier here than in Europe. McGovern (1981) has suggested that not only did the Norse fail to adapt to these changes, but they did not fully exploit coastal resources or adopt Inuit technology and clothing (Hovgaard, 1925). Inuit clothing, preserved on frozen bodies at Qilakitsoq, demonstrates a remarkably sophisticated adaptation to a harsh environment (Hansen et al., 1985, 1991). The Norse lack of adaptability may have been brought about by the concentration of wealth and decision making in the hands of a small elite (McGovern et al., 1988), including the Church, the emergence of which is well attested by the archaeological record. The Norse communities were, therefore, economically and socially ill-adapted to a ecosystem that was (and still is) so highly susceptible to the effects of short-term as well as secular episodes of climatic change.

Iceland

Iceland was first settled by the Norse in the 870s and within about 10 years all the coastal and valley land suitable for farming had been taken up. In Norse times some barley was grown but this declined in the thirteenth century and ceased during the sixteenth century almost certainly for climatic reasons. At that time the Icelandic glaciers were expanding and several Norse farms were engulfed (Grove, 1988). There is abundant historical evidence for particular social hardship in the eighteenth century, including official returns sent back to Copenhagen, from where Iceland was ruled at the time (Ogilvie, 1981). The story is one of failed harvests, declining fish catches and an impoverished sometimes starving population with an increasing dispossessed underclass retreating from the most severely affected areas in the north.

Particular periods of socio-economic crisis resulted from a number of exceptionally severe winters but other environmental factors also played a part. There was a succession of major volcanic eruptions of which the most severe was that of Laki (Lakagigar) in 1783 (Thorarinsson, 1979). This poisoned the pastures with ash and, together with the effects of severe weather conditions, led to the death of half the cattle and three-quarters of the sheep and horses. One-fifth of the human population died of famine and the abandonment of Iceland was seriously considered. The effects of volcanism and climatic change are not easy to separate in Iceland where many of the glaciers are on volcanoes and eruptions also cause glacier surges and catastrophic floods (jökulhlaups) from glacier-dammed lakes (Chapter 4, p. 102). In addition, short-term climatic changes of 1–3 years' duration appear to follow volcanic eruptions (Chapter 3, p. 79), and acidity profiles from the Greenland ice sheet (Figure 3.17) indicate that much of the period of the Little Ice Age is marked by increased volcanic activity. As if the effects of severe winters and volcanism were not enough, the weakened population suffered a smallpox outbreak in 1707 which wiped out one-third of the population (Ogilvie, 1981).

A proxy climatic record relating to these changes is provided by data on the extent of sea-ice (Bergthórsson, 1969; Lamb, 1972) which has been calibrated by instrumental records since 1846 (Figure 5.24a). This record compares closely with the oxygen isotope curve from Crête, Greenland and clearly indicates a period of lower temperatures between AD 1600–1900. The effects of climatic change on the restricted economic base of Iceland are emphasized by the consequences of well-documented twentieth-century changes. A dramatic decline in cod catches accompanied lower temperatures in the, 1960s and led ultimately to the 'cod war' with Britain. Comparing the period 1873–1922 with the warmer episode 1931–60 the latter had a hay yield 16 per cent higher (allowing for fertilizers and technological change). The crucial factor, the period of winter feeding, was 13 days shorter enabling at least 20 per cent more stock to be kept (Ogilvie, 1981). In the Little Ice Age comparable, or greater *reductions* would have been suffered by farmers with very limited options for diversification. A fall in summer temperatures of 1 °C would have resulted in a 15 per cent reduction in yield (Grove, 1988). Inevitably the decision-making landowners were less severely affected than peasants (McGovern *et al.*, 1988). Other interlinked social factors included debilitation of the economy

by a Danish trade monopoly between 1602 and 1789 and the rule of Iceland from Denmark, an area much less severely affected by the climatic changes.

Norway

In southern Norway glaciers from Jostedalsbreen, the largest icecap in mainland Europe, advanced over agricultural land reaching their most advanced positions in 1740–50. Tax surveys before and during the main glacier advance (in 1667 and 1723 respectively) reveal widespread evidence of declining prosperity and a substantial drop in the number of cattle in each parish (Grove and Battagel, 1983). The social effects of ice advance

Figure 5.25   The incidence of mass movement, flooding etc., revealed by requests for tax relief (avtak) in parishes on the margins of the Jostedalsbreen in southern Norway (after Grove, 1988)

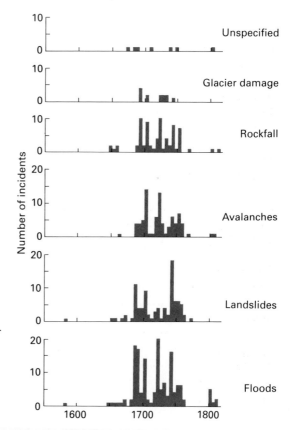

are demonstrated by requests for tax relief known as avtak which show considerable evidence of damage to farm property by glaciers, flooding and various forms of mass movement (Figure 5.25). Similar glacier advance, avalanches and flooding are widely reported during the Little Ice Age in the Alps (Grove, 1988).

British Upland

Throughout the upland and moorland areas of Britain there is evidence for the extension of Medieval agriculture, in the form of deserted farms, fields and cultivation ridges, well beyond present altitudinal limits. Opinions vary as to the relative importance of climatic and social factors in causing these desertions. An example is the deserted

settlement at Houndtor, Dartmoor (Figure 5.26) where Beresford (1981) saw climatic change as the principal cause of desertion. Austin (1985), on the other hand, pointed to the existence on Dartmoor of deserted and surviving settlements at the same altitude arguing that legal contrasts in the nature of landholding and individual manorial policies may also have been important factors. In an attempt to examine the issue of climatic causation in the Lammermuir Hills of south-east Scotland, Parry (1978, 1981) developed analogue models based on present-day crop–climate relationships. Cultivation ridges which are now abandoned had been used to grow oats, and the number of degree days, wind speed and water surplus necessary for successful ripening of oats was calculated. Manley's (1974) central England temperature record from AD 1659 was used to graph (Figure 5.27a) the predicted

Figure 5.26   The deserted Medieval settlement of Houndtor, Dartmoor, England, beyond the margins of present-day enclosed farmland seen lower down the valley (photo M J C Walker)

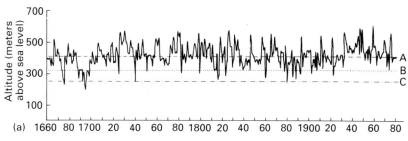

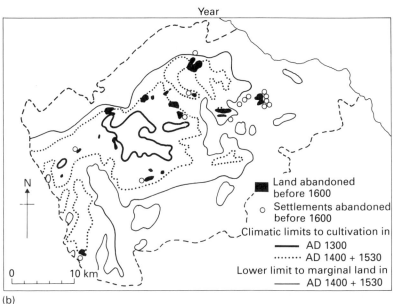

(b)

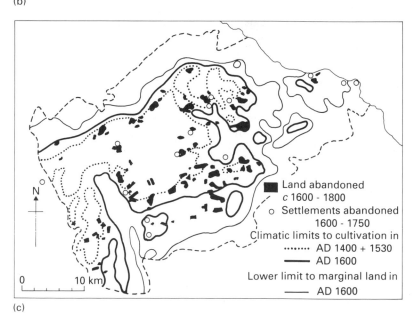

(c)

Figure 5.27   (a) Hypothetical shift in oat crop failure with altitude in southern Scotland between 1659 and 1981. A = mean altitude of crop failure, B = 1 in 10 failure frequency, C = 1 in 50 failure frequency (after Parry and Carter, 1985). (b and c) Lammermuir Hills, south-east Scotland showing the lowered climatic limits of cultivation and abandoned farmland, (b) between 1300 and 1600 and (c) between 1600 and 1750 (after Parry, 1975)

changing frequency of crop failure with altitude in southern Scotland (Parry and Carter, 1985). The results reveal a good correspondence with historical records of agriculturally bad years; for instance, the so-called 'seven ill years' of the 1690s, which had a devastating effect on Scottish agriculture, show up as a pronounced trough. Farms above 340 m would have suffered failure in 11 successive years between 1688 and 1698. For individual farmers the probability of crop failure would have been the key factor in decisions to abandon arable which may in turn have made farms in some parts of the upland unviable.

Parry (1981) has mapped the changing Medieval and Post-Medieval limits of cultivation which correlate closely with the distributions of abandoned settlements and fields (Figure 5.27b and c). This supports the hypothesis that abandonment is explicable in climatic terms, although of course a variety of other factors such as famines, the Black Death, dissolution of the monasteries and successive English invasions in the thirteenth to sixteenth centuries may have been more important in relation to some individual holdings.

European Lowlands

During the Little Optimum vineyards extended 3–5 degrees of latitude further north and 100–200 m higher up hillsides, and the Domesday Book (AD 1086) records 38 vineyards in southern England. The dates of vine harvests from the late fifteenth century provide a valuable proxy climatic source and correlate with glacier fluctuations (Chapter 2, p. 43). Even so, Le Roy Ladurie (1972) has shown that a decline in vine growing in the later fourteenth century north of Paris is not explicable in purely climatic terms but reflects a more general economic decline during which cultivation became unprofitable because of labour costs.

Desertion of Medieval villages in lowland, as in upland, areas has often been associated with the Little Ice Age. Beresford (1981) has argued that abandonment of the clayland settlement at Goltho, Lincolnshire, for example, was a result of increased climatic wetness which he also correlated with architectural changes in the later phases of the site. Such conclusions, however, seem overly deterministic in the light of other evidence which appears to show that deserted and non-deserted settlements occur alongside one another and hence that climatic factors may have been of relatively little importance. The consensus is that desertions in the lowland areas of Britain, and perhaps elsewhere in Europe, were mostly caused by more general economic change and decline, beginning before the Black Death in AD 1348 which reinforced the trend. Specific factors were agricultural changes in individual manorial holdings, such as clearance for sheep and parks, and the effects of soil exhaustion.

This picture contrasts with those examples of marginal situations discussed above. There the Little Ice Age temperature decline of 1–2 °C would have had serious consequences reducing the economic base to the point where abandonment may have been inevitable. Opportunities for adaption would have been very limited and may have been discouraged by tenurial considerations. It is also difficult to avoid the conclusion that some marginal holdings may, given the natural variability of climate, have simply been unsuitable for long-term agricultural activity. There is a limit as to how far the boundaries of cultivation can be pushed even with modern technology, and the longer timescales provided by detailed analysis of past situations will, as Parry's work has clearly shown, help to define that limit more precisely.

6

The Impact of People on Environment

Introduction

Human beings are not just responsive to natural environmental change but play an active role in the creation of their own environment, to the extent that many parts of the world have been totally transformed by long histories of human activity. This chapter considers the problem of establishing the relative importance of human and natural factors in causing particular environmental changes.

It is appropriate to begin by considering some of the ecological processes and problems concerned. It was noted in Chapter 1 that the factors of the environment (geology, climate, soil, flora, fauna, disease and people) are interlinked, so that an effect on one has repercussions on others (Figure 1.2). Contrary to some early ecological texts, people are specified as a separate ecological factor in acknowledgement that their effects are so much greater than those of any other faunal species. Theoretically, under a given set of environmental parameters, a stable ecological community will

develop in time: this is the **climax community**. If one of the factors changes significantly then that may set in motion succession towards a new climax. The timescale of a **succession** varies; a primary succession, on a rock surface newly exposed by glacial retreat, might take millennia for soil formation and the succession to a full climax. By contrast, secondary successions following abandonment of cleared (i.e. agricultural) land may reach climax woodland in 100 or 200 years.

The concept of the natural climax of an area is a useful yardstick in evaluating the effects of human activity, although it does assume stable conditions. Previous chapters have emphasized that few if any environments can be regarded as static, being subject to cyclical changes on various wavelengths and to stochastic events. It follows, therefore, that we need to adopt dynamic models in which human impact is seen as affecting ecosystems undergoing various degrees and types of change rather than static climax communities (Nicholas, 1988). In some cases anthropogenic changes may have occurred early in the Holocene before a natural climax had time to develop. Also problematic is whether climax communities, under theoretically stable conditions, experience change as a result of their own maturation (ageing) processes. Iversen (1964) used the term **retrogressive succession** to describe the effects of soil leaching and concomitant vegetational changes which characterize the later (oligocratic) part of the present and previous interglacials (Chapter 4). The implication is that some environmental changes of the Late Holocene would have occurred eventually, although human agency accelerated the process.

Human impact on a climax community returns it to a **subclimax** stage in the succession. This may be more productive in terms of biomass but it is less stable and conserving of nutrients than the climax and, therefore, more susceptible to the changes associated with retrogressive succession (Dimbleby, 1976). Apart from the activities of people, subclimax communities may be created by natural environmental changes described in Chapter 4. Hence problems of equifinality again confront the historical scientist in trying to infer cause from observed effects. In Europe the problem is acute because of the long history of intensive human impact, whereas in North America, where impact has been more limited, it is easier to distinguish the effects of, for example, climate and disease.

In all parts of the world people's ability to change the environment has increased through time. This is partly the result of an increasingly sophisticated and ecologically demanding economic basis, i.e.: hunter–gatherer > agriculturalist > early complex societies > urban > industrial societies. This has been accompanied by increasing technological capacity to change environments and an inexorable growth in human population levels. As a consequence, people have become increasingly out of balance with their environment. It must also be appreciated that people have been domesticating (i.e. manipulating) their environment over a very long timescale, in some cases before they domesticated individual plant and animal species.

So many habitat types once assumed to be natural are now seen as artefacts of long histories of human activity, e.g. moorland, heathland and some grasslands. These changes have come about largely as the long-term result of particular land-use regimes. People also transform their environment with non-utilitarian social objectives, exemplified by particular arrangements of monuments, for instance, the Neolithic tombs of the Boyne Valley, Ireland, or the immense stone heads of Easter Island. The concept of the 'cultural landscape' articulated by palaeoenvironmentalists in Scandinavia and Germany (e.g. Birks *et al.*, 1988) acknowledges the extent to which people have transformed landscape. Moreover, these landscapes should not be regarded simply as fossils of their past; they are evolving entities, the maintenance of which, if it is seen as conservationally desirable (Chapter 8), necessitates an understanding and continuance of traditional agricultural practices which can be elucidated by palaeoenvironmental and ethnohistorical research.

This chapter examines the contribution of human activity to environmental changes in selected periods and environment types. The themes include animal extinctions, the impact of hunter–gatherers, the effects of early farmers and the role of people in the formation of mires, heathlands and grasslands.

Late Pleistocene and Holocene extinctions

Geological time is characterized by major waves of extinction, the best known being the disappearance of the dinosaurs at the Cretaceous/Tertiary boundary. That extinction episode affected both terrestrial and marine organisms and was so dramatic that some have sought an explanation in extra-terrestrial causes (Alvarez *et al.*, 1980). Since the emergence of the genus *Homo* there have been a number of waves of extinction, the most marked of which occurred at the end of the last glaciation. Among the animals affected were the mammoths, mastodons and woolly rhino. These Pleistocene extinctions were largely of big game animals and avian herbivores over 50 kg in weight. One favoured hypothesis is that these extinctions resulted from pronounced climatic changes at the end of the last glaciation. However, similar climatic changes occurred at the end of previous glacials, yet no comparable extinctions are apparent in the fossil record.

Since the days of the pioneering geologist Charles Lyell, people have also been identified as a possible cause of the extinctions. A fully comprehensive theory of human causation only emerged, however, with the work of Paul Martin whose ideas, together with the counter-arguments of those who favour climatic hypotheses, have been rehearsed in three symposia volumes (Martin and Wright, 1967; Martin and Klein, 1984; Mead and Meltzer, 1985). Martin (1984) showed that the extinctions were concentrated in areas where people had settled late, such as the Americas and the Antipodes, and on remote islands where extinctions occurred at approximately the time of initial human settlement (Figure 6.1).

The Pleistocene megafauna has survived to a greater extent in Africa and Eurasia where there has been a long evolutionary development of people/animal relationships. Lower and Middle Pleistocene sites such as Olduvai Gorge, Tanzania, have associations between extinct megafauna and human artefacts but the dates and causes of the early extinctions are very unclear. Two later, minor African extinction episodes have been identified in

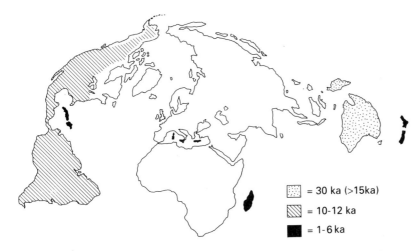

Figure 6.1   Simplified global pattern showing the dates of the main waves of extinctions (after Martin, 1984)

= 30 ka (>15ka)

= 10-12 ka

= 1-6 ka

the Cape 12–9.5 ka BP and in North Africa 4–5 ka BP, both of which occurred during periods of climatic change, but in both instances human activity is also regarded as playing an important role (Klein, 1984).

European extinctions include the mammoth, woolly rhinoceros, giant deer, cave bear, cave lion and cave hyena. The mammoth seems to have survived widely across Eurasia until 12 ka BP when it finally succumbed during the rapid climatic fluctuations and associated habitat changes of the Lateglacial (Vereshschagin and Baryshnikov, 1984). Even so mammoths had survived earlier, probably comparable, climatic changes. Perhaps the emergence of anatomically modern hominids in Eurasia around 30–40 ka BP enabled permanent settlement of the hostile environments in the north-western USSR and Siberia which in earlier episodes may have served as refugia for the mammoth. In northern Europe there is evidence for mammoth hunting on a grand scale, the remains of 516 mammoths being used in huts dated to 20 ka BP in the Dneiper Desna region of Poland. Equally intriguing is the extinction of the giant deer (Figure 6.2) which was abundant in Ireland during the Woodgrange Interstadial 12–10.6 ka BP but disappeared during the suceeding Nahanagan (Younger Dryas) Stadial, apparently because this brief cold spell led to a drastic reduction in the quality of its browse (Barnosky, 1986; Stuart and

van Wijngaarden-Bakker, 1985). None survived to the Holocene and the animal was extinct before people reached Ireland around 9 ka BP (Mitchell, 1986).

The smaller islands of the Mediterranean experienced the extinction of some bizarre endemic species including pygmy hippos and elephants, and large rodents, the evolution of which reflects the long isolation of these islands even during periods of low glacial sea level (Davis, 1987). The extinctions themselves are imprecisely dated, although the first people attested on the islands around 8 ka BP did exploit the extinct species on Mallorca and Corsica. Many of the extinctions were probably brought about by competition with, and predations by, people and their domestic animals.

Extinctions were especially dramatic in areas of the world only settled during, or since, the last glaciation. North America lost 35 genera at the end of the Wisconsinan glaciation, representing perhaps 70 per cent of the big game, and in South America the losses may have been even more severe. Martin (1967, 1984) hypothesized a wave of extinctions resulting from overkill by newly arrived human groups, the users of Clovis points, which are distinctively shaped flaked stone projectile points dating between *c.* 11.5 ka and 11 ka BP. A 'blitzkrieg' model was proposed whereby animals totally unaccustomed to human predation were very quickly exterminated. Speed was fundamental to

Figure 6.2   Giant Deer (photo National Museum of Wales)

the hypothesis because of the limited number of archaeological sites with animal bone evidence for predation on the extinct fauna. Only in the case of mammoths is there evidence of widespread exploitation by Clovis hunters (Fagan, 1991). It has often been assumed that the extinctions were coeval with human arrival in the Americas, but the date of this has long been debated. There is no universally accepted evidence for human presence before about 12 ka BP (Dincauze, 1984), which accords with evidence that entry to the continent from the Bering Land Bridge via Alaska would have been blocked in Canada by the coalescence of the Laurentide and Cordilleran ice sheets (Chapter 4, p. 83) during the Wisconsinan maximum (c. 18 ka). An ice-free corridor was available along the eastern slopes of the Rocky Mountains by 12 ka BP (White et al., 1985), although by this date people were apparently present as far south as Chile (D. R. Harris, 1987). This implies almost impossibly rapid migration, and may point to an earlier entry before the glacial maximum. Chronology is a key issue in the extinction debate and in recent years there have been a number of critical studies of the available radiocarbon dates (Mead amd Meltzer, 1984, 1985; Grayson, 1987, 1989). All the dates regarded as reliable imply that extinctions occurred by 10 ka BP and possibly by 10.8 ka BP (Figure 6.3a). However, of the 35 extinct North American taxa only seven have sufficient reliable dates indicating extinction between 12 and 10 ka BP; they include camel, horse, mastodon, mammoth, Shasta ground sloth (Figure 6.3b) and sabre-toothed cat. Other taxa have few reliable dates but the balance of evidence suggests that some of them became extinct, or suffered major population reductions, before 12 ka BP.

The period from c. 13 ka BP was one of major climatic change (Chapter 3, p. 68) and the extinctions episode does correspond to the most pronounced vegetational changes (Graham and Mead, 1987). The extent of these changes is highlighted by the presence in the Lateglacial of 'disharmonious assemblages' of animals – those not found today in association (Lundelius, 1989). That points to a reduction in habitat diversity at the Lateglacial/Holocene transition which may have been especially detrimental to certain species.

Australia represents a comparable case to the Americas. There the extinction of about 35 megafaunal species occurred, including a giant kangaroo, a Diprotodon and a koala. Accurate dating evidence for these extinctions is even less satisfactory than in North America but it appears that the main extinctions took place between 26 ka and 15 ka BP (Figure 6.4a). This is well after the colonization of Australia by humans which seems to have occurred about 40 ka BP; within 10 ka they had occupied all ecological regions (Figure 6.4b; Jones, 1989). Thus in Australia there is no evidence in support of a 'blitzkrieg' model. No major megafaunal kill sites have been found and there are only small numbers of archaeological sites with the bones of extinct species. Horton (1984) concluded that the extinctions resulted from an arid episode during which the megafauna were ecologically tethered at a declining number of waterholes which became exhausted. Vegetational reconstructions, inferred from pollen and other records, indicate a major episode of aridity between 30 ka and 10 ka BP, when coastal rainforests and woodlands were drastically reduced, but scrub and grassland habitats expanded (Dodson, 1989). If climatic change was the reason for the extinctions, the question remains as to why the megafauna survived earlier well-attested aridity phases. Perhaps the arid episode between 30 ka and 10 ka BP was more severe (Horton, 1984); alternatively, the presence of people and the practice of aboriginal burning (p. 154) leading to landscape modification from an early date may have tipped the ecological balance (Jones, 1989).

The dates and pattern of extinctions are very different on more remote islands. New Zealand has been isolated for the last 80 ma leading, in the absence of predators, to the evolution of a remarkable endemic bird fauna exemplified by the flightless moa. Thirty-four species of bird, living at the time when the Polynesians arrived c. 1 ka BP, had become extinct before European arrival in AD 1642 (Trotter and McCullock, 1984). Unlike the areas previously considered there are large numbers of sites on South Island where middens containing abundant moa bones are associated with

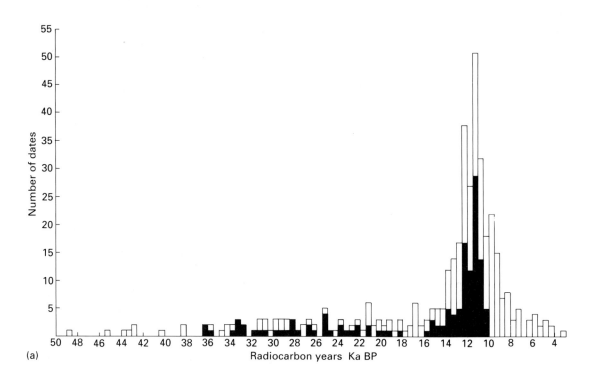

Figure 6.3　North American extinctions (a) chronological distribution of 307 radiocarbon dates (after Meltzer & Mead, 1985) (b) the youngest radiocarbon dates on dung of the Shasta ground sloth Nothrotheriops shastensis *(after Martin* et al.*, 1985). In both (a) and (b) the dates regarded as reliable are in solid black*

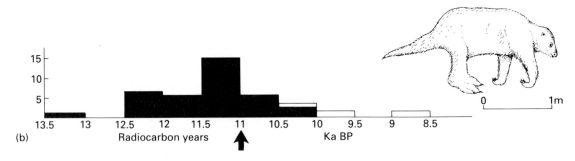

Maori sites. These have typically been radiocarbon dated between 1 ka and 0.5 ka BP. Overkill is apparent in that sometimes only the most palatable parts were eaten and there is the possibility of conspicuous consumption by groups competing for social supremacy. Even so, the single most important factor is likely to have been loss of habitat, arising from woodland destruction by the

Maori practice of burning, for preserved gizzard contents show that moa fed in the forest and forest margin (Anderson, 1988, 1989).

Madagascar also had a remarkable native fauna before 800 BP when it was first settled by people; this included a dwarf hippopotamus, extinct crocodile, large lemurs and *Aepyornis*, a giant flightless bird some 3 m high. Bones of the extinct

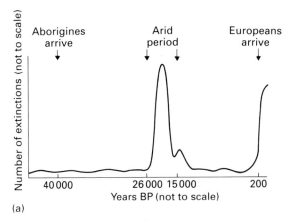

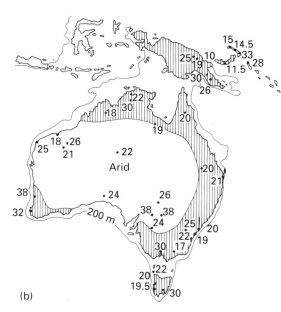

(b)

Figure 6.4   The colonization of Australia and megafaunal extinctions. (a) Diagramatic representation of the timing of extinctions in relation to climatic change and human settlement (after Horton, 1984) (b) The earliest dated occurrences of human occupation in Australia and New Guinea in thousands of years. Vertical shading distinguishes humid margins from arid interior (after Jones, 1989)

species are found on archaeological sites and the animals are recorded in oral legends. Hunting of the highly vulnerable *Aepyornis*, together with the plundering of its eggs, was almost certainly the reason for its extinction. Other losses are thought to

have resulted from extensive forest destruction and accelerated soil erosion (Battistini and Verin, 1967). Nearby on Mauritius the flightless dodo became extinct within 174 years of discovery in AD 1505.

Many island groups in the Pacific were colonized between 6 ka and 1 ka BP and there are numerous instances of extinctions of highly endemic species brought about by people and the animals they introduced, particularly rats, pigs and domestic fowl. Coupled with this is evidence for human habitat destruction on the Solomon Islands, Fiji and the Chatam Islands (Cassels, 1984).

The simplified global pattern of extinction dates (Figure 6.1) certainly shows a broad correlation with the spread of people. This map does, however, conflate two very different types of event. On the one hand there are the extinctions of the Holocene when human groups arrived on remote islands and encountered animals which had evolved without any predators. In these cases the extinct species are often well represented on early archaeological sites, and there is clear evidence that extinction resulted from a combination of hunting and anthropogenically induced environmental change. In none of these Holocene examples is there evidence of contemporary natural environmental change of an appropriate scale to explain the extinctions.

On the other hand, there are the major continental waves of extinction which affected the Americas and Australia, the latter rather earlier than the former, but both apparently coinciding with the major climatic changes at the end of the last cold stage. In both cases the evidence for direct associations between extinct fauna and people remains tenuous. However, if a climatic hypothesis is favoured, then it is necessary to explain why certain species and not others were affected, and why comparable waves of extinctions did not occur during earlier glacial/interglacial transitions. Furthermore, in some areas, such as New Zealand, major climatic and vegetational changes at the end of the last cold stage appear to have had little impact on the fauna (Diamond, 1989). There are some puzzling features in the pattern and scale of extinctions. In north-west Europe, for example, major fluctuations in climate and in vegetation

cover occurred during the Lateglacial period
(*c.* 13–10 ka BP) as the result of rapid changes in
the extent of North Atlantic sea-ice (p. 68). Much
of North America did not experience such abrupt
climatic and habitat change. Yet the scale of faunal
extinction is far greater in North America than in
Europe. The explanation may be that the native
American fauna was totally unaccustomed to
hunting by humans and was easily decimated. The
last glacial witnessed the dispersion of anatomically
modern hominids to the last of the world's great
land masses. Hence, even if climatic and natural
habitat changes were the predominant factors in
megafaunal extinctions at the end of the last cold
stage, hominids may have helped deliver the final
coup de grace to particular species under stress. The
'overkill' debate has considerable relevance to the
whole question of people's relationship with nature,
as Spaulding (1983) has written: 'Implicit in its
acceptance is the idea that Upper Palaeolithic man
was inherently as capable of disrupting his
environment as modern man'.

Burning by hunter–gatherers

Whatever our conclusions regarding the causes of
Late Pleistocene extinctions, there is no doubt that
hunter–gatherer communities can have a significant
environmental impact. Chroniclers of early
expeditions to North America, such as those of
Drake and Hudson recorded widespread burning
by Indians, not just for the purpose of growing
crops such as maize, but also on a more widespread
basis by hunter–gatherers. Burning has clearly
modified the forests of the north-eastern United
States (Day, 1953). More problematic is the extent
to which the character of the parkland and prairie
grassland vegetation (p. 183) in central North
America has been determined by burning
processes. Stewart (1956) noted some areas which
had been colonized by woodland once Indian
burning ceased. Similarly, in Australia and
Tasmania early explorers reported extensive
burning. Mellars (1976) reviewed the reasons for
hunter–gatherer burning, especially in America and
Australasia, and suggested that it aided hunting by

increasing mobility in the forest and was used to
drive game. It was also sometimes used to favour
food plants, particularly those bearing nuts. The
ecological effects of incendiarism included the
release of nutrients for recycling, producing new
growth and increasing species diversity. Studies of
burning for nature conservation purposes in
America show that the quality of browse can
increase by over 400 per cent and there are similar
qualitative increases so that the carrying capacity
for herbivores may be increased by between 300
and 700 per cent. The result is that animals
increase their reproductive rate and other animals
migrate in from adjacent areas, such that the
number of herbivores may increase to three or four
times their former level. These ecological effects
last for 10 to 15 years and so, if the more
productive regime is to be maintained, there is a
need for repeated cycles of burning.

Evidence for anthropogenically induced burning
in the palaeoenvironmental record is not always
easy to detect, for some fires must have been
naturally caused (Chapter 2, p. 29). Lightning fire
incidence varies geographically, being common in
the central United States and Canada, but rare in
the north-eastern United States (Patterson and
Backman, 1988). Combustibility also varies,
conifers burning more readily than deciduous
woodland. Estimates of charcoal abundance
associated with pollen diagrams from America
reveal considerable evidence of fire, though there is
seldom a clear indication of its cause in prehistory
(Dincauze, 1987; Patterson and Backman, 1988).
Even regular burning in some parts of North
America has been ascribed to natural causes. An
example is Cwynar's (1978) study of annually
varved sediments in Greenleaf Lake, Ontario where
there was evidence of burning and concomitant soil
erosion every 80 years between AD 770 and 1270.
Despite the wealth of ethnohistorical evidence for
burning by Indians in the context of
hunter–gatherer activities, the clearest evidence for
landscape manipulation by American Indians in the
pollen record follows the spread of maize
cultivation (McAndrews, 1988). In New England
the evidence comes largely from southern coastal
areas where agriculture was practised and the

greatest population density occurred (Patterson and Sassaman, 1988). Some areas, such as the Pacific North West, with high populations but no maize, show little sign of environmental modification.

Australian archaeologists have placed considerable emphasis on the extent of Aboriginal burning, adopting the phrase 'fire stick farming' to highlight the importance of environmental manipulation to the Aboriginal way of life (Jones, 1968, 1969; Yen, 1989). Many writers attribute evidence for burning in the palaeoenvironmental record largely to human activity (Bowdler, 1988; Jones, 1989) although others take a more limited view, pointing out the contribution of natural fires and the presence of charcoal pre-dating evidence of human settlement (Horton, 1982; Clark, 1983). Fire-adapted vegetation is widespread and as a result of burning, forests have been replaced in some areas by sedge fields and heaths. Some regenerated when burning was reduced after European settlement.

The extent to which natural factors, in addition to lightning fires, may have created openings in woodland has been little considered. Major storms can fell huge tracts of woodland (Figure 1.5). Although such events are infrequent, their effects last for decades and are especially concentrated in vulnerable topographic situations such as coasts where the effects of wind are exacerbated by periodic marine inundations (p. 122). Animals have also been underestimated as agencies of ecological change. The beaver fells trees and creates dams behind which ponds and lakes form; some Canadian examples have persisted for millennia. When sedimentation and vegetation colonization occur a 'beaver meadow' is created and attracts grazing animals. Beavers were once widespread in Europe and wood with their toothmarks is increasingly turning up on prehistoric archaeological sites. It is now appreciated that the vegetation changes brought about by their felling and flooding activities could be mistaken for anthropogenic effects (Coles and Orme, 1983). Open areas are also likely to have been created, or maintained, by the activities of wild herbivores, such as deer and wild cattle in European forests (Bottema, 1988) and bison in North American

prairies (Buckland and Edwards, 1984). Such effects are likely to have been particularly concentrated around watering places and to have increased the width of the transitional (ecotonal) zone between woodland and open conditions along coasts. Hippos in Africa can create treeless belts up to 1 km wide along a river's edge. Openings in the forest created by tree-throw, animal activities or natural fires would have been more ecologically productive than their surroundings and attracted both grazing animals and human settlement. Naturally created openings may well have been maintained artificially by burning, thus making it more difficult, than generally acknowledged, to establish the cause of the original ecological change.

People have long had the capacity to change the environment with fire from an early date. Early hominids in East Africa used, if not created, fire for domestic purposes as early as 1.5–1 ma BP (Clark and Harris, 1985) and there is evidence for hearths from some of the earliest sites beyond Africa, such as Zhoukoudian (China), Vértesszöllös (Hungary) and Terra Amata (France). During the British Hoxnian interglacial three sites, Hoxne, Marks Tey and Swanscombe, 120 miles (190 km) apart show broadly contemporary deforestation and grassland episodes (Turner, 1970; Evans, 1975). The nature of the vegetational change does not suggest climatic causation and fire seems likely in view of the presence of charcoal at Marks Tey and Hoxne. If this was a catastrophic natural fire then open conditions must have been maintained by grazing animals because varved sediments at Marks Tey show the episode lasted some 300 years. At Hoxne the phase coincides precisely with Acheulean occupation of the lake margin, suggesting that the episode may represent the earliest major fire clearance by people (Evans, 1975). Pine charcoal horizons are also widespread in Allerød Interstadial (11.8–10.8 ka BP) layers in the coversand of The Netherlands and north-west Germany, although it is unclear whether they are the result of lightning strikes or anthropogenic fires (Behre, 1988).

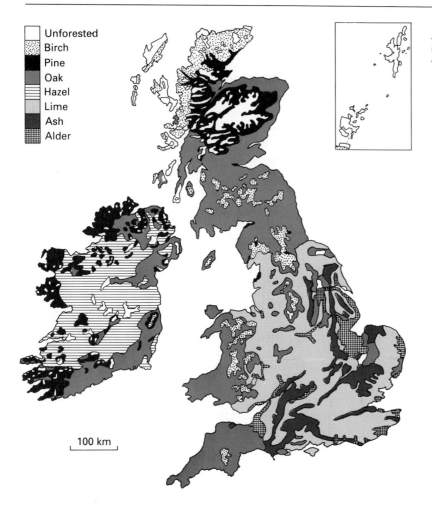

Unforested
Birch
Pine
Oak
Hazel
Lime
Ash
Alder

100 km

Figure 6.5   Map of dominant woodland types in the British Isles c. 5 ka BP (after Bennett, 1989)

Mesolithic forest clearance in the British Isles

During the Early and Middle Holocene, nearly all of the British Isles was covered by woodland, successive stages in forest development being shown in Figure 4.13. By 5 ka BP a complex woodland mosaic had developed (Figure 6.5), for which Rackham (1980) coined the name 'wildwood'. Unwooded areas were very limited: the upland summits of Wales, the Pennines, the Lake District and the more extensive areas of Scottish upland. Parts of central Ireland were unforested because they were already covered by raised bog.

Elsewhere there would have been localized unwooded areas in unstable landscapes, such as river banks and coasts subject to fairly frequent episodic environmental changes. This is particularly the case with the most exposed areas of the Outer Hebrides, Orkney and Shetland, together with parts of Caithness which were largely unforested. Even on the Outer Hebrides, however, some pollen diagrams, particularly those from more sheltered localities, show the presence of woodland and there are hints that people, as well as climatic factors such as changes in the frequency of westerly gales, may have been responsible for fluctuations in the extent of woodland from before the Neolithic (Bohncke, 1988; Bennett *et al.*, 1990).

The period between 10 ka and 5 ka BP represents the Mesolithic period of hunter–gatherer activity, which ends with the appearance of the first farmers. Recently, the clear distinction which earlier authors made between Mesolithic and Neolithic ways of life has become more blurred (Zvelebil, 1986). The view of Mesolithic communities as mobile hunter–gatherers constrained by the environment and its seasonal cycles but having minimal effect on it, has been modified in the face of increasing evidence of Mesolithic burning. Charcoal horizons occur in the stratigraphic records with evidence in the accompanying pollen diagram for a reduction in areas of woodland and an increase in plantain, sorrel and grasses at a time when lithic evidence on some sites shows that people were active. The North York Moors site of North Gill has episodes of vegetational disturbance and charcoal alternating with undisturbed conditions (Innes and Simmons, 1988) and at Bleaklow on the Pennines there were regular cyclical burns. Permanent recession of the forest took place in some uplands (e.g. southern Pennines, Dartmoor and North York Moors, the treeline being lowered from a maximum of around 700 m in places to around 300 m OD (Simmons, 1979). In this former woodland edge zone there are very dense concentrations of Mesolithic sites (Jacobi et al., 1976).

There is some limited evidence for Mesolithic impact on woodland early in the Flandrian before 8 ka BP. This may represent a response to the increasing forest encroachment on valued grazing areas (Innes and Simmons, 1988). Such activities may have given rise to Mesolithic plant communities that were more diverse than the natural climax, and brought about a mosaic of various stages of succession. That would have been beneficial to the natural carrying capacity of upland areas in the short term, though in the longer term it would have exacerbated the natural tendency of these soils towards leaching, podzolization and peat formation which is the inferred scenario where Mesolithic flints occur in mineral soil below blanket peat.

Simmons (1975) and Mellars (1976) have interpreted evidence for burning as deliberate management by Mesolithic communities to encourage and attract large herbivores, particularly red deer, the most abundant animals in most Mesolithic assemblages. The evidence increases in the later Mesolithic, suggesting that population pressure and a scarcity of resources led to growing environmental manipulation and an increasingly close relationship between people and deer.

An area of lowland with comparable evidence of Mesolithic impact is the sandy heathland of the Weald in south-east England where Mesolithic sites were buried by earthworms and sand blow, the former attesting to a more fertile brown earth soil and the latter to a subsequent loss of structure and heathland development as a result of Mesolithic burning (Dimbleby, 1985). Other sites with evidence for burning and Mesolithic vegetation change are Seamer Carr (Cloutman, 1988), Eskmeals (Bonsall et al., 1986), and Kinloch on the island of Rhum (Hirons, 1990; Hirons and Edwards, 1990). In general, however, lowland and coastal sites, e.g. Star Carr (Cloutman and Smith, 1988), Oronsay/Colonsay (Mellars, 1987) and Westward Ho! (Balaam et al., 1987), show much less evidence of disturbance than the moorlands and heaths.

Some authors have taken a very wide view of the possible effects of Mesolithic people on vegetation history. In pollen records from several sites there is a correspondence between peaks of hazel pollen or a rise in the alder curve, and flints and charcoal, which suggests that people may have brought about the vegetation changes (Smith, 1970). Similar vegetational changes are also apparent, however, at sites with no evidence for human activity and where peaks of hazel-type pollen do not correspond to charcoal peaks (Bennett et al., 1990; Edwards, 1990). Indeed, Rackham (1980) has questioned the assumption that hazel maxima in pollen diagrams necessarily relate to the ability of hazel to regenerate after burning, an inference that he suggests is based on the American rather than British Corylus species. More localized evidence of Mesolithic impact is also in need of critical examination. Some charcoal may derive simply from camp fires or natural forest fires (Edwards, 1989) and some of the open ground indicators like plantains could have been encouraged by a range of ecological factors apart from people. In Holocene records from Scotland, for example, open habitat taxa occur in horizons pre-dating the earliest known Mesolithic sites (Edwards and Ralston, 1984). Indeed by comparison with the English uplands, Scotland has more limited and equivocal evidence of Mesolithic impact. The same is true of Ireland, although here limited impact is more explicable because the impoverished Irish fauna includes fewer ungulates which could have been encouraged by burning, i.e. elk, roe deer and aurochs were absent and red deer were probably few (Woodman, 1985). This faunal impoverishment accounts for the greater focus of Irish Mesolithic communities on aquatic resources.

Overall, however, evidence for the effects of Mesolithic people on vegetation is considerable. Over a hundred instances of apparent Mesolithic clearance are recorded (Simmons, 1979), chiefly, although not exclusively, from the moorlands and heaths. It has been suggested that damp lowland woods would burn less easily. There may also be a problem of resolution of the evidence in these areas in that on more fertile soils minor clearances may have been short lived and thus only be detectable by very close sampling. In contrast on base-poor soils, clearance is likely to have exacerbated an underlying trend towards leaching and peat development. In a similar way, on very exposed upland and coastal sites even small-scale human activity may have disproportionate effects and thus register in a long-term and more easily detectable way (Hirons and Edwards, 1990).

What remains puzzling, in view of all the evidence for Mesolithic impact in Britain, is the relative lack of similar evidence from elsewhere in north-west Europe. In Denmark little has been found to modify Iversen's (1941) view that there was no clearance in the pre-Neolithic forest and people were forced during the Climatic Optimum to give way to the advancing forest without resistance, and consequently concentrated to a greater extent on coastal resources (Iversen, 1973; Jensen, 1982). The likelihood of environmental manipulation in response to stress has been hypothesized (Paludan-Müller, 1978) but little

evidence has been found. Charcoal horizons do occur on a Climatic Optimum site on Zealand but pollen grains of cereals suggest that this may relate to the activities of early farmers (Kolstrup, 1988). Elsewhere in Scandinavia there is a little more evidence for Mesolithic burning from pollen diagrams or charcoal concentrations on four sites in Sweden, others in Norway (Mikkelsen and Høeg, 1979) and one in Finland (Welinder, 1983, 1989). In some cases the vegetation was burnt two or three times and modified for a period, while some burnings correlated with occupancy of adjacent settlements. The scale and number of these effects is, however, much smaller than in many areas of Britain which may point to distinctive insular environmental relationships and economic strategies.

The transition to agriculture in central and north-west Europe

Although significant, hunter–gatherer impacts were geographically limited in their effects, and it was with domestication that the wholesale landscape transformation of Europe began. Dogs, domesticated in the Near East by 10 ka BP, were present in Denmark and Britain during the Early Mesolithic (Simmons and Tooley, 1981). They would have given people greater control over the movements of other animals in hunting and perhaps semi-domestic manipulation, although this remains debatable. By the Upper Palaeolithic, people were following herds of reindeer as they moved from winter grounds on the North European Plain to summer ranges on the German and Swiss uplands. A relationship similar to that between modern Lapps and reindeer is envisaged, in which case the degree of control may have amounted to herding (Jarman et al., 1982). In Britain it has been hypothesized that Mesolithic communities followed cyclical movements of red deer from lowland winter sites to spring and summer upland grazing (Clark, 1972). Upland burning would have created a vegetation mosaic with increased woodland edge grazing. High ivy pollen values in diagrams from some Mesolithic sites have been interpreted as indicating that flowering ivy was collected as winter fodder for deer (Simmons and Dimbleby, 1974). More direct control may be hinted at by the presence of deer on some remote Scottish islands, such as Colonsay and Oronsay, to which they may possibly have been transported by people (Grigson and Mellars, 1987).

The degree of environmental manipulation in Mesolithic Britain is comparable to that among hunter–gatherers in many parts of the world (Harris and Hillman, 1989) and should not necessarily be interpreted as evidence that these people were incipient agriculturalists. There is very little evidence of local domestication in Europe. Among the main Neolithic domesticates, cereals, sheep and goats were not native to Europe and were introduced into south-east Europe from the south-west Asian centre of domestication (p. 115) by 8 ka BP. The Neolithic way of life also involved husbandry of cattle and pigs, both native to Europe, as well as pottery and ground stone technology (Champion et al., 1984). By 6.5 ka farmers using distinctive Linearbandkeramik (LBK) pottery and living in substantial settlements of long rectangular houses had occupied central Europe as far as the west of Germany. The LBK settlement pattern was largely restricted to fertile and light loess soils across which farming spread rapidly, perhaps helped by the particularly favourable conditions of the Holocene Climatic Optimum (p. 70).

Following this rapid initial spread, a delay occurred before farming was adopted in areas peripheral to central Europe, perhaps because in maritime areas ecotonal environments offered a viable alternative, rich enough locally to support semi-sedentary settlement. Zvelebil and Rowley-Conwy (1986) have defined three distinct stages in the transition to farming: an availability phase when it was becoming known through contact with farmers but was not adopted; a substitution phase when farming was becoming part of the ecology of a frontier zone; and a consolidation phase representing the maturation of the social and economic structures associated with an agricultural community. The date and length of these phases vary considerably within Europe depending to a

large extent on local ecology. Denmark is characterized by a particularly lengthy availability phase between 6 ka and 5 ka BP known as the Ertebølle during which settlement was concentrated on large coastal middens (Figure 6.6). People here adopted some of the trappings of a Neolithic way of life such as pottery and ground stone axes but it was more than a millennium before there is clear evidence of agriculture. Its adoption during a brief substitution phase may have resulted from a reduction in littoral resources marked by disappearance of the oyster (Zvelebil and Rowley-Conwy, 1986). In the wetlands of The Netherlands (p. 123) and the alpine lakes, well-preserved environmental evidence from Neolithic settlements shows that hunting, fishing and foraging continued to make major contributions to the economy alongside agriculture of varying extents (Louwe Kooijmans, 1987; Coles and Coles, 1989). In northern Europe the substitution phase was particularly attenuated with little agriculture east of the Baltic and in the European USSR until after 3.5 ka BP (Dolukhanov and Khotinskiy, 1984). Climatically favourable areas in the south of Norway and Sweden adopted agriculture from 5 ka BP but its northerly limits fluctuated with climatic change and hunting remained the economy of the North (Berglund, 1985).

By comparison with other maritime areas of north-west Europe, Britain is usually seen as experiencing an abrupt transition to agriculture around 5 ka BP. There are, however, some earlier radiocarbon dates for Neolithic sites and an increasing number of sites with a few cereal pollen grains in earlier contexts (Edwards, 1988). This makes it difficult to be sure whether clearances in Britain during the late seventh and sixth millennia BP were the work of hunter–gatherers or that of agriculturalists, during a substitution phase which is largely invisible in the archaeological record (Groenman-van Waateringe, 1983; Smith, 1981). The appearance of such Neolithic sites as causewayed camps, flint mines, long barrows and trackways around 5 ka BP could be associated, not with the initial arrival of agriculture, but with the consolidation phase. The advent of these monuments certainly attests to dramatic social

change (Thomas, 1988). The economic base may not, however, have changed so suddenly. Some Neolithic sites were still located in quite dense woodland and there is growing evidence from Neolithic macrofossil assemblages that alongside the new exotic cereals, wild foodstuffs such as nuts and apples continued to be important (Moffet *et al.*, 1989).

The elm decline

A decline in elm pollen occurs between about 5300 and 5000 BP in diagrams from sites throughout northern and north-west Europe. Elm pollen values decline typically to half their former frequency (Figure 6.7). Isopollen maps show that the decline began rather earlier in south-east Europe (Huntley and Birks, 1983), although in north-west Europe it was broadly synchronous within the limits of

radiocarbon dating. Its causes have been much debated and the various hypotheses reflect evolving paradigms in archaeology and environmental studies.

Iversen (1941) used the elm decline, together with a decline in ivy and a rise in ash pollen, to define the boundary between the Atlantic and Sub-Boreal pollen zones. This inferred vegetational change was attributed to increasing continentality as substantiated by the effects of severe winters of 1939–42 on the growth of ivy, mistletoe and holly in Denmark (Iversen, 1944). The existence of the Climatic Optimum with mean temperatures some 2 °C warmer than today is widely evidenced although, as noted in Chapter 3 (p. 71), there are regional differences in the date of its end. A sudden deterioration at 5 ka BP has been suggested (Frenzel, 1966) but current interpretations favour more gradual deterioration during the Neolithic (Smith, 1981). In Britain, for instance, the most climatically sensitive species, ivy, holly and lime, are

Figure 6.7   Schematic Holocene pollen diagram with curves smoothed, from Jutland, Denmark (after Iversen, 1973)

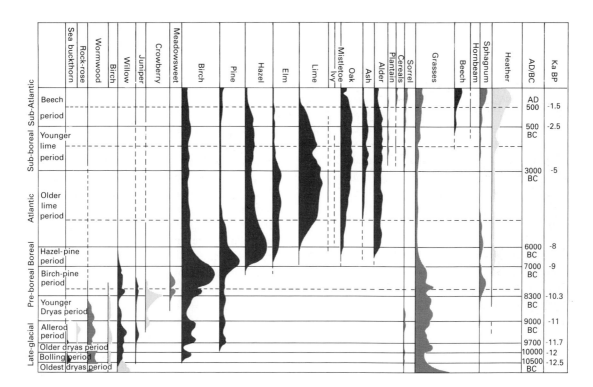

not affected at 5 ka BP. In fact, the overall geographical pattern and chronological spread of the elm decline, does not easily fit any climatic hypothesis such as greater coolness, wetness or continentality (Huntley and Birks, 1983). Indeed the decline is most pronounced in the west.

In many areas of north-west Europe the elm decline coincides broadly with the first signs of human activity. Thus in the record from Åmosen, Denmark, Troels-Smith (1960) found evidence of small-scale activity exactly at the elm decline, although in general Danish woodland clearances are rather later. A number of British sites such as Shippea Hill in the Fens and a group of sites in Northern Ireland (Smith, 1981) show the elm decline as coeval with anthropogenic clearance. Troels-Smith argued that the decline was caused by the advent of the new Neolithic agricultural

economy. Elm, he suggested (together with ivy and mistletoe), declined not for climatic reasons but because the leaves were collected and used to stall-feed cattle. Pollarding (p. 168) every two to three years would have prevented flowering. Elm leaves are nutritious and would have provided an important source of fodder for pioneer animal husbandry in the densely wooded landscape of north-west Europe. There is ethnohistorical evidence for the use of leaf fodder, and elm foliage occurs in the cattle stalls of the Swiss Neolithic lake village at Weier (Troels-Smith, 1960; Robinson and Rasmussen, 1989).

Recent detailed pollen studies of the elm decline horizon show that it sometimes occurs following a series of human impacts, as at Pawlaw Mire in Northumberland (Sturlodottir and Turner, 1985), Soyland Moor on the Pennines (Williams, 1985), on the Isle of Wight (Scaife, 1988) and in a group of sites in Northern Ireland (Hirons and Edwards, 1986). Mitchell (1956) has argued for another anthropogenically based explanation of the decline, suggesting that elms were growing on the best soils and hence were selectively cleared by the first farmers.

All anthropogenic explanations for the elm decline suffer from a fatal flaw, however, in that they fail to take account of the scale of the event. Elms, according to Rackham (1980), made up about one-eighth of the woodland of Britain, covering perhaps in aggregate some 10 million acres (25 million hectares). The pollarding of this area, to the extent that pollen production was halved, would require a human population of something like half a million. Rowley-Conwy (1982) did similar calculations for Denmark which show that the decline implies the pollarding of 50–80 million trees which in turn means the feeding of perhaps 500 000 to 1 million cattle. Such scales of activity rule out any wholly anthropogenic explanation for the decline. Furthermore, an anthropogenic explanation implies synchronous cultural change over a wide area of north-west Europe whereas it was demonstrated in the preceding section that there was considerable variation depending on local ecological conditions.

Some writers have also suggested a link with contemporary soil changes (Sturlodottir and Turner, 1985). In the British Lake District, for example, lake chemistry indicates significant soil leaching and the inwash of minerogenic sediments at the time of the elm decline (Pennington, 1965). Although pedogenic changes may have prevented regeneration, the elm decline itself is too sudden and widespread an event to be accounted for by soil changes alone and demands a more universal explanation.

One such is suggested by the outbreak of Dutch elm disease, the most recent outbreak of which started in Britain around 1965. Within 13 years 60 per cent of non-woodland elms had been killed (Rackham, 1980, 1986). The disease is an ascomycete fungus *Ceratocystis ulmi* which partly blocks water conducting vessels and thus causes death of the tree. It is spread by two species of bark

beetle, *Scolytus scolytus* and *S. multistriatus*. The history of the disease is somewhat obscure, although there is evidence of a French outbreak in 1918 and a British occurrence in the nineteenth century, together with strong literary hints of its presence earlier in the Post-Medieval period (Rackham, 1980).

The idea that a tree pathogen was responsible gains a measure of support from the well-documented effects of disease on North American trees in the twentieth century (Patterson and Backman, 1988). In 1904–50 there was an outbreak of chestnut blight caused by the fungus *Endothia parasitica* which virtually exterminated populations of chestnut (*Castanea dentata*) throughout its range in eastern North America (Davis, 1981). Chestnut pollen dropped from 7 per cent to less than 1 per cent and other species expanded to take its place. The horizon can be traced by pollen analysis wherever chestnut was common and is a valuable marker horizon in the study of recent land-use history in the area, since the progress of the disease is well dated (Anderson, 1974).

About 4850 BP there was a dramatic decline of hemlock (*Tsuga* sp.) which is evident in pollen diagrams from eastern North America and Canada (Davis, 1981). It was sudden and broadly synchronous over an area from New Brunswick to upper Michigan; 1.5 million km^2 were affected. Recovery of the hemlock took up to two millennia, though in some areas it never recovered its former abundance. Again the synchroneity and scale of the phenomenon seems to rule out climatic and human causation, for at that time the region was sparsely populated and without agriculture. One suggestion is that the hemlock decline was caused by looper moths which eat buds and cause defoliation (Davis, 1981). Minor outbreaks occur today every 11–17 years but a catastrophic outbreak 4850 BP might have been caused by the sudden arrival of the loopers or some microevolutionary change. Further support for the disease hypothesis comes from comparative studies of pollen changes in annually laminated sediments at Pout Pond, New Hampshire (Allison *et al.*, 1986). The recent pathogenically caused chestnut decline and the much earlier

Figure 6.8   The chestnut (Castanea) decline (a) map of the
decline in eastern North America (after Anderson, 1974); (b) A
comparison of the timescales of the twentieth century chestnut
decline and the hemlock decline c. 4850 BP (5450 calendar years
BP) in annually laminated sediments at Pout Pond, New
Hampshire, USA (after Allison et al., 1986)

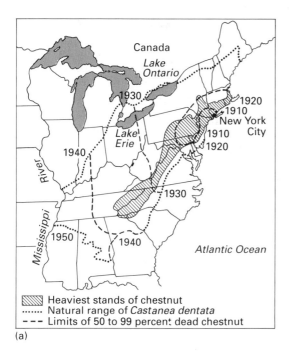

(a)

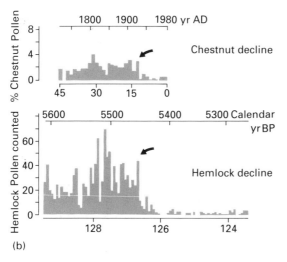

(b)

hemlock decline involved very similar reductions
(around 75 per cent) over timescales as short as 7–8
years (Figure 6.8).

Radiocarbon dates suggest that the elm decline
spread at *c.* 4 km per year which is comparable to
the recent spread of Dutch elm disease. The rapid
diffusion of elm disease throughout the north
European lowland can probably be equated with the
emergence of a more virulent strain (Huntley and
Birks, 1983). The disease hypothesis has also been
strengthened by the discoveries of fragments of the
beetle *Scolytus scolytus* in horizons immediately
below the elm decline in a peat sequence at the
Mesolithic site at Hampstead Heath, London
(Girling, 1988), and wood damaged by another
Scolytus species at Åmosen, Denmark (Kolstrup,
1988).

A recent analogue for the Mid-Holocene elm
decline in north-west Europe is provided by a
pollen diagram from the soils of an elm wood in
Kent, an area which was severely affected by a
1978 outbreak of Dutch elm disease in southern
England (Perry and Moore, 1987). The pollen
evidence shows the effects to have been broadly
similar, with a marked increase in open habitat
plants as the woodland canopy disappeared. This
ties in with the ideas of Groenman-van Waateringe
(1983, 1988) who has argued that the association
between pollen from plants suggestive of
agricultural activity and the elm decline may be
partly illusory. In a totally wooded landscape small-
scale clearances are likely to have been masked in
the pollen record, unless, as Edwards (1982) has
shown, they happened to be very close (i.e. within
c. 30 m) of the site of pollen deposition. Under the
more open conditions following the elm decline
even small-scale clearances would be more likely to
register in pollen diagrams. The effects might be
further amplified, adopting Mitchell's (1956)
hypotheses (see above), if human groups were
specifically attracted to the more open areas on
better soils where the elms had previously been
growing. This opens up the possibility that the
areas most intensively settled by early farming
communities *c.* 5 ka BP may have been partly
determined by those areas where the disease had
created sufficient ready-made openings in the

wildwood. The disease would have been most virulent where hunter–gatherers or early agricultural communities had already been active in vegetation modification. In recent outbreaks of the disease, trees damaged by stock or pollarding were especially susceptible to infection. Hence the disease hypothesis acknowledges the scale of the phenomenon while at the same time explaining its general contemporaneity with the spread of farming; as Rackham (1980) has written: 'civilisation helping the spread of the disease and the disease helping the spread of civilisation'.

Clearance by early farmers

Following the elm decline clearance took place on an ever increasing scale, eventually resulting in the present largely deforested 'cultural landscape' of Europe. Burning is likely to have been one of the main clearance processes and charcoal horizons in pollen sequences and lake deposits are widespread from hunter–gatherer times onwards. Neolithic communities were also capable of clearing forest with polished stone axes as experiments in Draved Forest, Denmark showed (Iversen, 1973). Of greater significance, however, may have been the effect of sustained grazing over a long period, preventing new trees from becoming established so that very gradually, as older trees died, the woodland became more open.

Early clearances often exhibit the characteristic three-fold pattern of vegetational change termed by Iversen (1973) a 'landnam' (Figure 6.9). Stage 1 was the clearance of high forest and a brief peak of herbaceous plants. Stage II was marked by low maxima of willow and poplar, a decline in lime and ash and a birch peak, implying the use of fire in the original clearance. There was also a peak of plantains, herbaceous plants and cereals. Stage III is marked by a hazel maximum and the regeneration of mixed oak forest. During these episodes pollen indicators of grazing activity were evident and there was some smaller scale arable activity. Iversen envisaged animal husbandry and the growing of crops for a few seasons in the ashes

of a subsequently abandoned forest clearing. The whole sequence from burning to regenerated woodland lasted perhaps half a century.

Landnams were seen as part of a shifting slash and burn (swidden) system associated with a frequent settlement mobility. Experiments at Draved and elsewhere (Steensberg, 1979) appeared to support this model by showing that yields fall off in the years after burning. Slash and burn agriculture is well documented in Finland both ethnohistorically and by pollen analysis (Vuorela, 1986), while shifting cultivation was also practised by Indian communities in North American forests (O'Shea, 1989). Rowley-Conwy (1981), however, has questioned the assumption that shifting agriculture in Finland is a remnant of an agricultural system which was widespread in prehistory. He contends that the time expended in clearance would be out of all proportion to the benefits. Moreover, the Draved experiments only lasted three years, were on soils too acid for a good crop and may, therefore, give a misleading impression of the possibility of sustained yields. Indeed the results of radiocarbon dating of 'landnam' episodes, both in Scandinavia (Rowley-Conwy, 1981; Göransson, 1986) and in the British Isles (Smith, 1981), show that some episodes lasted for hundreds of years with open conditions being maintained by grazing and perhaps also by periodic crop growing within managed woodland (Göransson, 1986). Some landnams may also represent an amalgam of several clearance events which may only be detectable in local pollen spectra from small bogs (Ammann, 1988). Although the concept of landnam is fundamental to many theories concerning the Neolithic and the spread of farming, the evidence for the duration of these events remains somewhat contradictory. Nevertheless, the theory of shifting Early Neolithic agriculture has recently received support from the increasing precision provided by dendrochronology, which shows that a number of settlements were indeed only occupied for one or two decades (Coles and Coles, 1989).

Spatial and temporal contrasts in the extent of clearance and early farming activity are evident throughout prehistoric Europe. In Britain many

Figure 6.9   Generalized pollen diagram showing the three stages of a landnam (after Iversen, 1973)

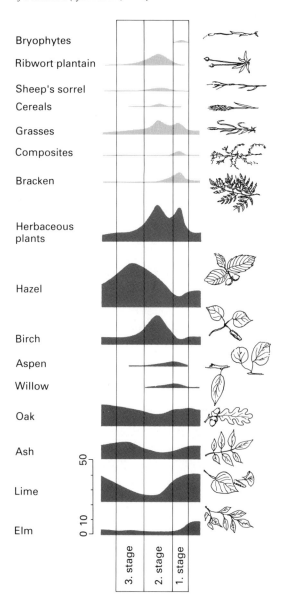

litter, whereas later the 'landnam' type clearance became common.

In Scandinavia, clearance occurs earlier and on a much larger scale in the agriculturally more favourable areas of Denmark and Scania (Figure 6.10) and not until perhaps a millennium later is there extensive and permanent impact in other parts of southern Sweden and Norway (Berglund, 1985). The region as a whole does, however, show at least four phases of broadly synchronous agricultural expansion. In the more northerly areas of Scandinavia impact tends to be weak and late and there are distinct troughs when farming was replaced by a return to hunting and gathering. An example is the Middle Neolithic around 4.4 ka BP when, on present evidence, agriculture retreated into southern Sweden. Further east in the forest zone of the USSR, prehistoric human impact is very limited and the earliest anthropogenic effects are often Post-Medieval (Dolukhanov and Khotinskiy, 1984).

Clearance in Britain

In Britain there are marked regional contrasts in the extent of early clearance which partly reflect environmental factors, and partly also social contrasts in the distribution of power which prehistorians have recently identified (Bradley, 1984). Figure 6.11 summarizes the dates at which clearance occurred in various parts of the British Isles. The main concentration of clearance before 3.5 ka BP was in the chalkland of Wessex, on the East Anglian Breckland, the Cumbrian coastal plain, the magnesian limestone in Durham and in parts of Ireland. Early clearances show a marked concentration in areas of light calcareous soil. Mollusc analyses of palaeosols below Neolithic barrows (from 5 ka BP) and henges (from 4.5 ka BP) in Wessex show these were built in areas which had been grassland for some time (Evans, 1972). Similarly, the Neolithic and Beaker tombs in the Boyne Valley in Ireland were laid out within grassland (Groenman-van Waateringe, 1983). The implication is that these areas were particular foci of human activity in advance of the construction of ritual monuments.

The South Downs are geologically and

landnam episodes begin at the time of the elm decline whereas in Denmark they appear slightly later during the Middle Neolithic around 4.5 ka BP. This led Troels-Smith (1960) to argue for two distinct types of Neolithic agricultural system; from the elm decline to 4.5 ka BP he envisaged the economy as based on stall feeding of cattle with leaf

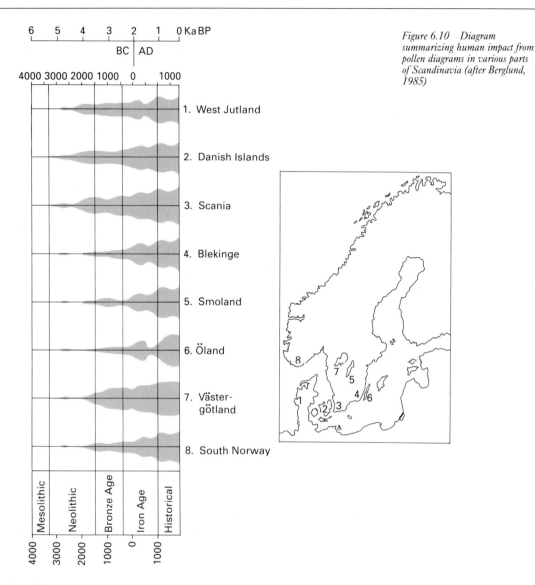

Figure 6.10   Diagram summarizing human impact from pollen diagrams in various parts of Scandinavia (after Berglund, 1985)

topographically very similar to Wessex but they do not have the same concentration of Neolithic ritual monuments and they were not cleared so early. Here causewayed camps and flint mines were within small clearings in the forest (Thomas, 1982) and both pollen (Thorley, 1981) and molluscan evidence from valley sediments (Bell, 1983) indicates that these areas were cleared between 4 ka and 3.5 ka BP in the Early to Middle Bronze Age. It needs to be emphasized that not all early clearances were permanent and even in parts of Wessex pollen evidence shows that there were a

number of episodes of regeneration and further clearance (Waton, 1982; Scaife, 1987).

In much of southern Britain, and in the moorland areas considered below (p. 177), there was considerable woodland clearance during the Bronze Age (4–2.5 ka BP). However, in parts of Wales and most of northern England and southern Scotland clearances were small scale and limited with a major episode of woodland destruction during the Iron Age (2.5–2 ka BP) (Turner, 1979, 1981). During the Late Iron Age and especially during the Roman period, extensive clearances

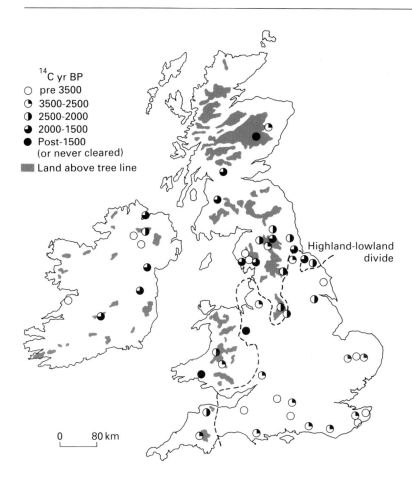

^{14}C yr BP
○ pre 3500
◑ 3500-2500
◐ 2500-2000
◑ 2000-1500
● Post-1500
(or never cleared)
▓ Land above tree line

Highland-lowland
divide

0   80 km

Figure 6.11   The dates at which clearance occurred in various parts of the British Isles from pollen and mollusc evidence (after Roberts, 1989)

occurred throughout Europe even in areas beyond the Roman frontier (Behre, 1988). This was not, however, the case in every area: parts of the Lake District and Scotland were not cleared until post-Roman times and in Ireland there was extensive clearance and agricultural activity during the early Christian period (Mitchell, 1986).

When Hoskins (1955) wrote his classic text, *The Making of the English Landscape*, he envisaged vast tracts of forest remaining as late as Domesday (AD 1086) and argued that the direct prehistoric contribution to the landscape had been small. Clearly that view now needs to be revised and Jones (1986) has shown in detail the extent to which people modified the pre-Domesday landscape. England may have lost 50 per cent of its woodland cover by the Iron Age and only 15 per cent

remained by Domesday, with much of that disappearing during the agricultural expansion of the Middle Ages (Rackham, 1986). In many areas of mainland Europe, by comparison, woodland destruction was less severe with large areas of forest still remaining until at least the eighth century AD (Behre, 1988).

Woodland management

Despite the extent of prehistoric and historic woodland clearance it is important to avoid the impression that human activity has been associated only with woodland destruction. Even in the almost entirely cultural landscape of modern Britain, often extensive stands of 'ancient' woodland survive to the present day. This has been demonstrated by an eclectic combination of historical studies, fieldwork on wood bounds and botanical composition (Rackham, 1980, 1986, 1988). An example of a survivor is the small-leaved lime (*Tilia cordata*) which was the predominant wildwood tree in much of lowland England (Figure 6.5) and is today considered to be an indicator of ancient woodland. Similarly the existence of rare lichens (Rose, 1974), molluscs (Kerney and Stubbs, 1980) or beetles may be used to show that certain woodlands are of considerable antiquity. Such woodland stands have survived not as unwanted waste but often as a carefully conserved and husbanded resource. They provided timber, by coppicing (cutting close to the ground to produce a crop of long straight poles) and pollarding (cutting above the height of browsing cattle), as well as a wide range of plant and animal resources including wood pasture and, particularly on the continent, leaf fodder. In Britain the techniques of woodland management can be traced as far back as the Neolithic period in the Somerset Levels where the Neolithic Sweet Track incorporated poles which were grown specifically for the purpose (Rackham, 1986). Later Neolithic hurdle tracks, such as Walton Heath Track (2300 BC) were made of hazel rods from coppice (Figure 6.12). In view of the very early origins of woodland management it seems probable that some woods may have survived, albeit in a much modified form, since the time of the wildwood. One example is Epping Forest where pollen analysis confirms the hypothesis of continuous woodland cover since 4 ka BP, though with changes in species composition, including the loss of lime, which resulted from Anglo-Saxon management (Baker *et al.*, 1978). Another apparently ancient wood is Wychwood, Oxfordshire, but this contains many prehistoric sites and Roman villas (Schumer, 1984) and must, therefore, be at least partly the result of post-Roman regeneration.

On the continent of Europe there is widespread historical evidence for the feeding of leaf fodder to stalled animals (Austad, 1988) and, as already noted (p. 162), this is attested archaeologically as early as the Neolithic. In Westphalia, Germany, Pott (1986) has found pollen evidence that beech coppice was established by 2 ka BP. In general, however, early woodland management practices are not as well documented in continental Europe as in England (Behre, 1988).

Clearance in North America

The extensive impact of prehistoric farmers in most areas of Europe is in dramatic contrast to the situation in North America where in the eastern states European colonists encountered a landscape which was still largely deciduous woodland (Figure 6.13). According to McAndrews (1988) prehistoric agricultural impacts register in less than 7 per cent of Holocene pollen diagrams. This is much more limited than might be expected in view of the historical evidence for Indian burning for both hunter–gatherer and agricultural purposes (p. 154). The main reason for the limited anthropogenic impact in America is that the North American Indians did not keep domestic stock (Davis, 1984). They did cultivate maize, legumes and squash beginning 7 ka BP in Mexico (p. 118), although the first clearance phases associated with maize pollen are not found in pollen diagrams until four millennia later (McAndrews, 1988). This reflects the much slower transition to a reliance on agriculture in America compared to Europe (p. 158). Maize was introduced into the southern United States by 3 ka BP but only reaching New England by the first millennium BP (Mulholland, 1988), but in many areas agriculture supplemented rather than replaced hunting and gathering.

Evidence of human activity in the United States is more marked along the major river corridors such as the Mississippi and Tennessee and their tributaries, where agricultural systems, such as those of Mississippian communities, based on maize and floodplain agriculture developed between 1 ka and 0.5 ka BP (Delcourt, 1987; Butzer, 1990). In eastern North America there is only limited evidence of Indian impact (Davis, 1984), for instance on Rhode Island (Thorbahn and Cox, 1988) and at Crawford Lake, Ontario where an episode of maize cultivation began AD 1360, as dated by annual sediment laminations (McAndrews, 1988). A coeval decline in cedar pollen may relate to the use of its timber and bark in construction of

Figure 6.12   Late Neolithic hurdle, Walton Heath trackway, 2300 radiocarbon years BC, Somerset Levels, England (photo Somerset Levels Project)

a nearby village which was probably occupied for 10–20 years. Archaic sites in Kentucky have evidence of some human manipulation, particularly to encourage acorn production, from 4 ka BP (Crawford, 1987) and the Kumeyaay of southern California planted oak groves (Shipek, 1989). Even where farming eventually came to constitute a major element in the economy, wild resources continued to play a key seasonal role and an important part in times of particular shortage (O'Shea, 1989). This was a strategy requiring a degree of balance between the needs of agriculture and those of nature conservation. That is likely to have engendered a different attitude to nature from that of fully agricultural communities. When Europeans settled in North America they introduced alien crops and agricultural practices which had evolved over millennia in quite different environmental conditions. The forests of the eastern United States were cleared on a vast scale

(Figure 6.13) particularly in the second half of the nineteenth century (Conzen, 1990; McCracken Peck, 1990). Hence the most profound impact on the Holocene woodland of North America came not, as in north-west Europe, from the activities of prehistoric farmers but from the depradations of technologically sophisticated immigrants within the timeframe of the last 150 years.

Biological consequences of clearance and farming

Clearance and the advent of farming had profound consequences on the ecosystems concerned. Species requiring closed woodland habitats declined and those preferring open conditions, disturbed habitats and the woodland edge were able to expand dramatically beyond the limited refugia they occupied in wildwood times. Domesticates were introduced from other geographical areas,

Figure 6.13   The United States showing the estimated extent of virgin forest in 1620, 1850 and 1920. This diagram does not include secondary regenerated woodland and thus underestimates the extent of twentieth century tree cover (after Greeley, 1925; Williams, 1990)

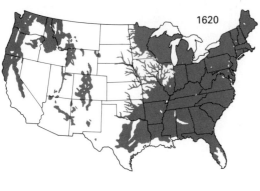

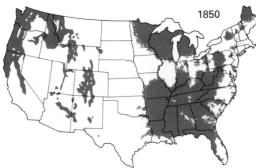

Europe were carried to colonies all round the world (Crosby, 1986), so much so that 80 per cent of New Zealand weeds are of European origin (Salisbury, 1964). Common plantains introduced to America by Europeans were called 'whiteman's foot' because they followed the colonists everywhere (Iversen, 1941; McCracken Peck, 1990). Today human activity in many parts of the world has created landscapes dominated by monocultures of introduced species, be it cereals or genetically identical conifer plantations.

In Europe people were the major cause of biological change after 5 ka BP, obscuring most climatic effects on plant and animal distributions. The gradual disappearance of the thermophilous lime (*Tilia*), for example, once the dominant tree in the wildwood of much of northern Europe (Greig, 1982), may partly relate to a fall in temperatures following the Climatic Optimum (Chapter 3, p. 71). However, the decline in lime pollen is a markedly non-synchronous event occurring at various dates in the Neolithic and Bronze Age records, depending on the extent of human activity and particularly on the timing of woodland clearance of the best soils where this species was growing (Turner, 1962). The Holocene expansion of beech (*Fagus*) into central and northern Europe followed the spread of farmers and the success of both beech and hornbeam (*Carpinus*) is partly due to their ability to colonize abandoned agricultural land (Bottema, 1988). A particularly pronounced phase of beech expansion, for example, is associated with the post-Roman migration period (Behre, 1988). In North America there are instances where Indian activity involved transplantation of medicinal and food plants (Shipek, 1989) and others where it may have caused changes in the dominance of particular tree species or introduced ruderal plants to new areas with crops, thus paralleling, although to a far more limited extent, the situation in Europe (Delcourt, 1987).

The later Holocene saw reductions in the ranges of many mammals, islands suffering particularly early and severe losses. Aurochsen, or wild cattle (Figure 6.14) were common in the European wildwood but on the Danish island of Zealand

bringing with them their diseases and parasites and the weed floras of crop plants (di Castri *et al.*, 1990). Near Eastern and Mediterranean weed species colonized north-west Europe with the first farmers (Sykora, 1990). Much later, weeds from

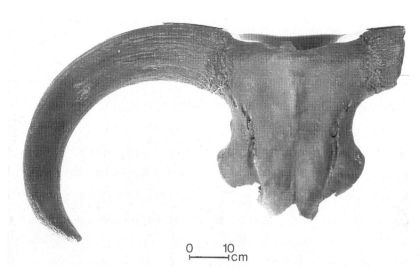

Figure 6.14   Aurochs skull from submerged forest at Whitesands Bay, Wales (photo T Harris)

became extinct, along with five other mammals, early in the Climatic Optimum (Aaris-Sørensen, 1980). These extinctions have been seen as a result of isolation by rising Holocene sea levels but it is likely that hunting pressure by Mesolithic communities was also important. Aurochsen became extinct in Britain during the Bronze Age but survived on the continent until total extinction in Poland in AD 1627. Beaver, bear, wolf and boar all suffered habitat destruction and human predation resulting in their extinction in Britain in early historical times (Rackham, 1986) and in drastic reductions in their mainland European ranges.

The effects of anthropogenic activity on invertebrates such as beetles (Buckland and Coope, 1991) are also very marked. Human activity has led to the introduction of species to new areas. There is the now well-documented role of Norse communities in introducing beetles, particularly synanthropic species (those dependent on people), to the north Atlantic islands with the trappings of a European farming system. Introductions now make up around half the beetle fauna of Iceland and Greenland (Buckland *et al.*, 1991; Sadler, 1991). Equally dramatic are the effects of woodland destruction in reducing the range of wildwood species; even where woods survive they became increasingly managed, leading to the removal of

dead and decaying trees which provide important invertebrate habitats. The majority of insect species which have become extinct in Britain are those of old woodland (Girling, 1982). The beetle *Ernoperus caucasicus* is an exclusive lime feeder and was widespread during the Climatic Optimum; today it survives on only two British sites. *Rhysodes sulcatus* is an inhabitant of rotting wood known from three Neolithic and Bronze Age contexts in Britain (Buckland, 1979; Speight, 1991). Its European range still continues to contract and populations recorded in the nineteenth century in south Sweden and Germany have since died out (Figure 6.15). The increasingly open country assemblages of Bronze Age and later times have been described by Girling (1982) as derived from 'culture steppe' because of the similarities to faunas from the natural steppes of eastern Europe and to assemblages from temperate interstadial treeless episodes.

Some changes of species range can, however, be related to climatic factors. Associated with the Neolithic Sweet Track, England, there are certain non-woodland beetle species whose present-day ecology, range and thermophilous preferences suggest that the Early Neolithic climate of the area was more continental than today, rather like southern Denmark with mean temperatures slightly lower in January and higher in July (Girling, 1984).

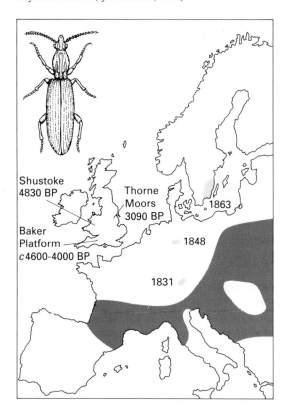

Figure 6.15   *The past and present distribution of the beetle*
Rhysodes sulcatus *(after Buckland, 1979)*

woodland of lowland zone England but is now
restricted to relict patches in Snowdonia and the
Lake District where it survives largely in stone walls
(Figure 6.16). The removal of woodland in
calcareous areas from Early Neolithic times led to a
dramatic increase in open country species (Kerney,
1966; Evans, 1972). As with the insects, some of
these had been characteristic of the open
environments of the Lateglacial (p. 55). These have
been joined by other species which have been
introduced to Britain by people. Members of the
Helicellid family, for example, have been
accidentally introduced from the Mediterranean,
and have flourished in Britain and western France
not only in dry coastal dunes but in the seasonally
hot and dry habitat of arable land and in many
other habitats. Most changes in European
molluscan distributions over the last 5 millennia are
clearly the result of human habitat destruction,
although a few may reflect the Late Holocene
climatic deterioration (Chapter 3, p. 71). *Pomatias
elegans*, for example, has retreated from the

Figure 6.16   *The past and present distribution of the mollusc*
Vertigo alpestris *(after Kerney, 1976)*

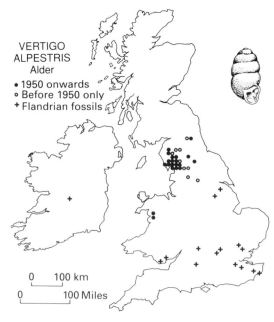

It also seems that the Post-Medieval Little Ice Age
(Chapter 3, p. 73) caused a southward reduction
in the range of the death watch beetle (*Xestobium
rufovillosum*) which was once present in the
wildwood of northern Britain where today there are
only disjunct populations in the artificial habitat of
old buildings (Buckland, 1975).

Land molluscs present a comparable picture.
The wildwood contained a group of species which,
though not totally extinct in Britain, are now
restricted to small relict distributions, due to the
removal of woodland and the drying up of habitats
brought about by clearance and drainage.
Spermodea lamellata was widespread in the
wildwood but in southern England it is now
reduced to a few sites with similar relict patches in
Holland and West Germany. *Vertigo alpestris* was
likewise widespread in the Climatic Optimum

northern part of its European range (Figure 6.17a) and has also undergone a marked size reduction (Figure 6.17b) which may have a climatic cause (Burleigh and Kerney, 1982).

Figure 6.17   The mollusc Pomatias elegans. *(a) Comparison of past and present European distributions (after Kerney, 1968) (b) Scattergram showing post-glacial size reduction at Brook, Kent (after Burleigh & Kerney, 1982)*

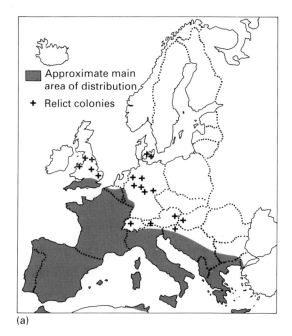

(a)

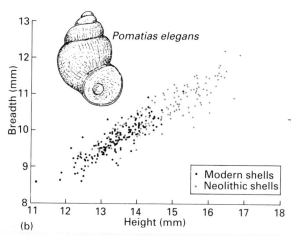

(b)

Holocene pedogenesis

Pedogenic processes are reviewed by Duchaufour (1982) and a historical dimension is provided by palaeosols which provide evidence of Early Holocene pedogenesis and the effects of subsequent natural and anthropogenically induced changes (Limbrey, 1975; Catt, 1986; Macphail, 1986). During the protocratic phase (Chapter 4, p. 95) the soils of north-west Europe developed, often on glacially or periglacially worked material, and were frequently base-rich. Under climax woodland **brown earths (cambisols)** characterized the mesocratic phase. Trees, being deep rooted, draw nutrients from the lower part of the soil profile but shed leaves, etc., on the soil surface; this recycling helps to counteract the natural tendency of soils to leach (Figure 6.18). In the same way evapotranspiration by trees draws water from the soil. The tree canopy also intercepts rainfall and facilitates its gradual infiltration into the soil. Base-rich soils in deciduous woodland support an active

Figure 6.18   Diagram showing the cycling of calcium and water through the soil in a wooded environment (after Evans, 1975)

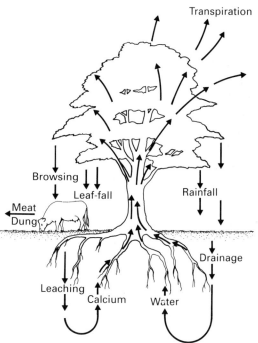

soil fauna (e.g. earthworms) which mix the organic and mineral fractions to create a good crumb structure of mull humus.

In the boreal coniferous woodland of northern Europe the vegetation produces a more acid litter and low temperatures lead to limited microbiological activity. Under these conditions bases are leached from the upper (eluvial) horizons which eventually become bleached as clay is lost and iron, aluminium and organic matter are moved down the soil profile and deposited as illuvial horizons. This characteristic horizonation, known as a **podzol** in European terminology or as **spodosols** in North America, represents the natural climax soil in the boreal vegetation zone. To its south in Denmark and eastern England there is evidence of localized natural (i.e. pre-human impact) podzol formation (Iversen, 1969; Valentine and Dalrymple, 1975). Even so, widespread podzol occurrence in Europe south of the boreal zone largely follows the loss of woodland on base-poor strata. The processes involved can be summarized as follows. Deforestation meant that nutrients were no longer recycled from depth in the profile, there was a loss of soil structural stability and gradually nutrient-demanding plants were replaced by those tolerant of low nutrient levels. The litter produced was poor in bases and rich in tannins which inhibit decay. Consequently mull gave way to a mor humus with low levels of faunal mixing in which organic plant material accumulates on the surface, a trend exacerbated by associated hydrological changes. Modern studies show that woodland removal leads to a reduction in evapotranspiration and other hydrological changes which may have the effect of increasing stream discharge by between 10 and 40 per cent (P. D. Moore, 1985, 1986) as well as leading to increased waterlogging, the development of gleyed profiles and erosion. Increased groundwater also reduces microbial activity and soil temperatures, further reducing decomposition rates. Drainage is often impeded by the development of a semi-permeable podzol ironpan. The trend towards increased leaching and hydromorphism (impeded drainage) would have been further exacerbated by the effects of the Holocene climatic deterioration (Chapter 3, p. 71).

Blanket bog formation

Blanket bogs or mires are a form of ombrogenous (rain-fed) peat development (p. 30) which is restricted to areas where precipitation exceeds 1200 mm yr^{-1}. They are particularly well developed on the west coast of Ireland, Scotland and Norway. Mires are also extensive in more continental areas of the boreal region and make up about one-third of Finland. Almost all areas covered by blanket bog once carried woodland, as shown by pollen analysis and by the presence of waterlogged tree stumps at the junction between the underlying mineral soil and peat. A dramatic landscape change has occurred but opinion is divided as to what brought this about. Scholars variously emphasize the role of climate, natural pedogenesis and anthropogenic factors in what is essentially a classic illustration of the equifinality problem (p. 47).

Climatic factors have been regarded by some as the prime cause of increased wetness leading to peat initiation (Godwin, 1975). Supporters of this view could point to some sites in Britain, for example, where peats started to form $c.$ 7 ka BP (Early Atlantic) and $c.$ 3 ka BP (Early Sub-Atlantic), both episodes of apparently increased wetness. However, as the number of sites with radiocarbon dates for peat initiation increases, it is becoming clear that it occurred over a protracted time period without clear clustering in wetter phases (Smith, 1981).

Natural pedogenic processes have also been seen as a cause of blanket peat formation in high rainfall areas (Smith and Taylor, 1969). In west Wales, for instance, acid brown soils supporting woodland gradually became more base deficient and leached, leading to podzol development and the accumulation of mor humus; in some areas this was underway before significant human impact in the Bronze Age (Taylor, 1980). Some Scottish and Irish sites also show evidence of blanket peat initiation before extensive human impact (Pennington et al., 1972). Evidence for retrogressive pedological changes has been detected in previous interglacials (p. 95) but blanket peat inception

(a)

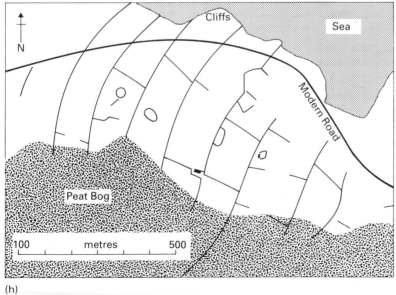

(b)

Figure 6.19 Neolithic field walls at Behy, Co. Mayo, Ireland buried by blanket peat (photo (a) M Bell, Plan (b) after Caulfield, 1983)

Figure 6.20   Blanket peat formation on the Momyr drumlins of central Norway showing the dates of peat initiation and the direction of its spread (after Solem, 1986)

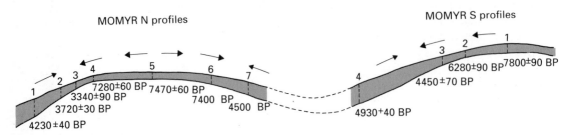

cannot be seen purely as the result of pedogenic process because some blanket peats show no evidence of earlier podzolization and a number of studies show close links with anthropogenic environmental changes (Moore, 1975; Smith, 1981).

Reference has already been made (p. 156) to sites on the Pennines and other uplands where blanket peat started to form during periods of Mesolithic activity. Elsewhere, peat initiation occurred at about the time of the elm decline, coeval with evidence for small-scale human activity in the form of, for instance, plantain and bracken pollen. This has led to the suggestion that even small-scale clearance or limited grazing within a wooded landscape could have been sufficient to cross critical thresholds and trigger peat formation (Wiltshire and Moore, 1983).

More extensive archaeological evidence for human activity at the mineral soil/peat interface has been revealed by peat cutting on the north coast of County Mayo, Ireland. There a complete buried landscape consisting of field walls (Figure 6.19), settlements and tombs was laid out in the Neolithic (Caulfield, 1983). Traces of similar systems are widely reported elsewhere in the west of Ireland (Herity, 1971; F. Mitchell, 1989). Localized nuclei of blanket peat formation were established here from 6 ka BP, before evidence of major human impact. Peat cover did not, however, become widespread until the Bronze Age from 4 ka BP in the north of Ireland, and in Iron Age and later times between 3 ka and 1 ka BP in south-west Ireland (Lynch, 1981; O'Connell, 1990). On many sites peat inception follows burning, and grazing

(Smith, 1981) and on a few sites, such as Goodland, surface waterlogging can be directly attributed to the effects of soil changes (see above) brought about by Neolithic agriculture (Dimbleby, 1985).

Some of the west Norwegian blanket mires also started to form following woodland clearance, well-developed blanket bog being restricted to areas of early clearance (Kaland, 1986, 1988). On the island of Haramsøy (Solem, 1989) blanket mire began to form at 3 ka BP on an unwooded upland plateau following intensification of land use indicated by charcoal evidence for regular burning. The plateau was largely used for grazing and some crop growing. However, there are other sites where human activity was minimal and peat initiation is believed to have been caused by increased waterlogging as a result primarily of climatic factors (Solem, 1986). In such a situation, on drumlins in central Norway a complete blanket peat cover took 4 ka to form with initial growth on plateaux and hollows eventually coalescing to mantle the intervening slopes (Figure 6.20). Blanket bogs around the alpine forest limit in Norway also have histories unrelated to that of human activity and appear to reflect heavy precipitation levels.

The present state of the blanket peat debate may be summmarized by noting general agreement that in some very high rainfall areas, such as the west of Scotland, nuclei in the west of Ireland and upland Norway, peat formation was underway before human activity on any scale and may, therefore, be seen as a result of natural Holocene climax conditions. Further south there is considerable evidence for human activity, albeit often on a small scale, at the critical soil/peat interface and it seems that people played a key role in tipping the ecological balance towards peat inception. The result may be that in some areas a process which would have happened much later in the present interglacial was greatly accelerated by human activity, while elsewhere the area covered by blanket peat may have been extended by the dramatic ecological changes which people brought about.

The development of moorland and heathland

Nearly all heathland and moorland areas were tree covered at the Climatic Optimum. Today, trees are absent or few and the vegetation consists of a restricted range of species tolerant of poor soils.

Moorland is best developed in the Highland Zone on the west of the British Isles where rainfall is high, and the soils are gley-podzols, often with a peaty top forming blanket peat in the highest and wettest areas. The vegetation consists of such species as heather (*Calluna*), bilberry (*Vaccinium*), grasses and *Sphagnum* moss (Pearsall, 1950). The very clear contrast we see today between moorland and the surrounding agricultural landscape (Figure 6.21) is largely a human artefact. In some areas a distinctive moorland character had developed as early as the Neolithic but in many instances even in the Early Bronze Age those areas which are today moorland had a similar woodland vegetation and resource potential to their surroundings. It was particularly Bronze Age clearance and grazing which led to the expansion of moorland vegetation.

Bronze Age activity has been studied in particular detail on Dartmoor, England (Fleming, 1988; Balaam *et al.*, 1982). The Early Bronze Age

Figure 6.21   The moorland edge near Widecombe-in-the-Moor, Dartmoor, England (photo M.J.C. Walker)

Figure 6.22   Bronze Age landscape of reaves, houses and cairns on Holne Moore, Dartmoor, England (after Fleming, 1988)

(4–3.5 ka BP) saw the construction of large numbers of burial and ritual monuments such as cairns, stone rows, and standing stones in a landscape which was becoming increasingly open as a result of grazing and regular burning. Around 3.5 ka BP there was a dramatic change with the construction of stone boundary walls or reaves which delimit territory on the periphery of the moor (Figures 1.1 and 6.22), and which were associated with enclosures and hut circles. Collectively they represent perhaps the largest area of preserved prehistoric landscape in Europe. The reaves are all the more remarkable because those which have been radiocarbon dated appear to have been constructed over a short period between 3300 and 3100 BP and the concept governing their layout was so powerful that it sometimes completely ignored major topographic features such as river valleys. This major human impact on the landscape occurred at a time of accelerated clearance which gave rise to scrubby grassland maintained by

grazing. Land use was predominantly pastoral with only small-scale crop growing. At the time of reave construction some areas had already begun to develop podzolic soils and it has been suggested that the onset of soil deterioration led to increased concern over land ownership and access. Such concerns may also have been prompted by the local occurrence of tin (used in bronzes) which may explain the greater Bronze Age impact on Dartmoor than on areas such as Exmoor (Merryfield and Moore, 1974) and the North York Moors (Simmons et al., 1982), where extensive clearance did not occur until Iron Age times. Dartmoor, by contrast, seems to have been largely abandoned in the Iron Age and Romano-British periods but, in common with other moorlands, saw extensive agricultural reclamation of its margins during the Medieval period (p. 144).

The fluctuating use of moorland may partly relate to climatic factors, both the Bronze Age and Medieval episodes being warmer and drier than the intervening and succeeding periods. Land-use pressure was also a factor and social considerations would also have played a part. It may be that the degree of consensus or coercion which created the Dartmoor reaves could not long be sustained. It is equally the case that, principally as a consequence of human activity, these areas became increasingly marginal and unable to support long-term intensive settlement and exploitation. They were in a sense set aside, not as waste, but for very specific forms of exploitation, principally grazing, based on settlements beyond the moor. The present-day vegetation is to a large extent a reflection of these regimes.

Heathland has much in common with moorlands botanically and in terms of its origin. Heaths occur at lower elevations and in areas of lower rainfall, often on sandy podzolic soils. Trees are similarly sparse and the vegetation is dominated by evergreen dwarf shrubs, particularly members of the Ericaceae such as *Calluna* (heather) (Gimingham, 1972; Webb, 1986). Heathland extends in a coastal belt along the Atlantic seaboard from north Portugal to just beyond the Arctic Circle in Norway (Figure 6.23). Its most extensive development is in the British Isles, The Netherlands, north Germany and Denmark. Plant communities with many of the same characteristics occur in the circumpolar region and in mountains above the treeline. It is clear that climatic factors play an important part in governing the extent of the north-west European lowland heaths as mapped in Figure 6.23. The traditional view was that heathland represents a natural climax vegetation type, and this does seem to be the case in extreme oceanic situations such as the Faeroes, St Kilda, parts of Orkney, Shetland, the Hebrides and Caithness, where pollen evidence shows virtually treeless conditions with grass and heathland communities throughout the Holocene (Walker, 1984; Birks, 1986). The exposure factor did not, however, preclude tree growth in all coastal areas (Wilkins, 1984). Submerged forests in some parts of Britain and Ireland indicate that wildwood extended to the shore (Figure 5.8) while pollen analysis shows that the Scilly Isles, today devoid of native woods and with a grass and heath flora, had

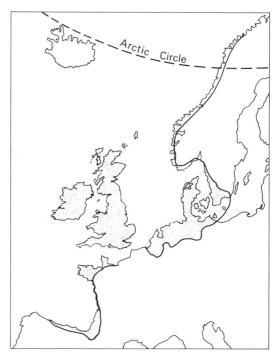

Figure 6.23 The distribution of heathland vegetation in Europe (after Kaland, 1986)

been wooded until the Bronze and Iron Ages (Dimbleby *et al.*, 1981; Thomas, 1985).

Inland heaths were also formerly wooded and there is evidence from both Denmark and southern England that the beginnings of heathland formation corresponded to phases of woodland clearance and occurred at a wide variety of dates (Dimbleby, 1962; Iversen, 1973; Scaife and Macphail, 1983). Some heaths in south-east England, for example, originate with Mesolithic human activity (p. 157), while in the Neolithic, woodland was replaced by heath on the Breckland of East Anglia and Hampstead Heath, London. Some Bronze Age barrows in present-day heathlands were constructed on brown earth soils but the majority sealed soils which had already begun to podzolize. Heathland barrows often have a core of clearly recognizable turves picked out by the bleached horizon and overlying dark humus horizon which characterizes a podzol (Figure 6.24). These changes follow anthropogenic impact on woodland and an expansion in heathland plants. Sometimes

Figure 6.24   Section of a Bronze Age barrow at Moor Green, Hampshire, England, showing old land surface with evidence of podzolization and podzol turves making up the core of the barrow (photo P Ashbee)

Figure 6.25   Heathland at Højstrup, north-west Jutland, with round burial mounds and in the background standing stones in the shape of ships, these form a Viking cemetery of the ninth century AD (photo M Bell)

the soil changes pre-date an increase in *Calluna* pollen which reflects the full development of heathland. It is clear that the Bronze Age barrows were constructed at a time of profound vegetation and soil change. In the northern Netherlands some Neolithic barrows and many more Bronze Age barrows buried podzol profiles with heath pollen spectra whereas in the central and southern Netherlands widespread heath formation mostly occurred from the Late Bronze Age/Early Iron Age onwards (Casparie and Groenman-van Waateringe, 1982).

The heaths of western Jutland (Figure 6.25; Odgaard, 1988; Odgaard and Rostholm, 1987) always carried more open forest than was present in eastern Denmark and some *Calluna* was present throughout the Flandrian, although there is a dramatic rise in *Calluna* pollen soon after the elm decline and the first widespread evidence for human impact on vegetation. Soil profiles beneath barrows dating around 4.5 ka BP show evidence of clearance by grazing and burning with podzolization already underway, and an increase in heath species towards the top of the old land surface. By the fourth millennium BP large areas of Jutland were heath, with further expansion in the second millennium BP. Some Danish heaths, however, formed during the Viking period and some of those in southern Sweden as late as the sixteenth century AD. At one time the heaths of Jutland were seen as forming on abandoned and exhausted agricultural land, but the more recent view is that this habitat was deliberately created by burning as a grazing and fodder resource (Andersen *et al.*, 1983).

From early Medieval times onwards an important factor in heathland expansion in The Netherlands, Flanders, North Germany and parts of Denmark was the creation of 'man-made' plaggen soils (Westeringh, 1988; Gimingham and de Smidt, 1983). These are artificially raised soils characterized by a dark humic topsoil more than 50 cm thick. They were created by the addition to the soil of organic-rich material often mucked out from animal byres. This material included turves, grass sods, forest litter and straw. A major source of turves was heathland and their removal is believed to have been an important factor in the extension of heathland. This is in addition to the exploitation of these areas for litter, fuel and the grazing of animals. The creation of plaggen soils continued from Medieval times until the nineteenth century AD. In human terms it played an important part in sustaining long-term agriculture on poor soils of Pleistocene coversands, but it also extended heathlands and sharpened the contrasts between heaths and their surroundings.

A particularly revealing picture of the cultural context of heath formation comes from near Bergen, Norway where heaths form a 25 km wide coastal belt (Kaland, 1986). Figure 6.26 shows the dates at which heathland developed on 26 farms on a west to east transect. The heaths are of widely differing ages, and the dates of development correlate in many cases with archaeological and place-name evidence for farm foundations. The earliest dates are from two of the most exposed islands in the west where heaths formed in the fifth millennium BP and the later dates tend to come from more easterly sites where there was a rapid development of heathland from 2 ka to 1.5 ka BP. High charcoal values and pollen evidence, even from sites with the earliest evidence of heath formation, show that anthropogenic activity was a key factor causing the replacement of woods by heath. Furthermore, it is clear that many Norwegian heaths have been maintained by grazing and burning as part of a traditional agricultural system which has only declined in the present century leading to woodland regeneration on some heaths.

While certain coastal heaths do seem to represent climax communities, the Norwegian evidence suggests that this may not apply to all such areas. The key role of people emerges clearly in relation to heathlands and moorlands and it is generally accepted that no Holocene climatic change was on its own of sufficient magnitude to bring about the demise of woodland in most of these areas. Heathland formation becomes extensive in the Bronze Age apparently as a result of deforestation, grazing and burning rather than arable activity. In Britain and Denmark Bronze Age barrows are particularly concentrated on the heaths which had perhaps by this time become traditional grazing lands partly maintained by burning.

Figure 6.26   The dates of heathland formation along a west-east transect in Nordhorland, Norway (after Kaland, 1986)

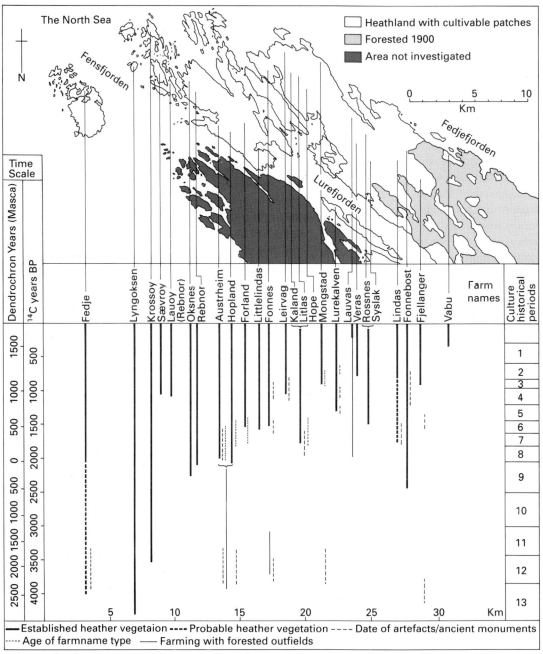

1 Late medieval period 2 High medieval period 3 Early medieval period 4 Viking period 5 Merovingian period
6 Migration period 7 Late Roman period 8 Early Roman period 9 Pre Roman period 10 Late bronze age
11 Early bronze age 12 Late neolithic period 13 Middle neolithic period

The origins of grasslands

In areas climatically too dry or cold for the development of climax woodland, extensive tracts of natural grassland exist such as the African savanna, the American prairies and the Asiatic steppe; natural grasslands also exist above the treeline in mountainous areas. Some grasslands such as the African savanna have existed for millions of years (p. 112); today their ecology is often influenced by natural fire regimes and they have been extended geographically by human activity (Harris, 1980).

The first Europeans in North America who ventured beyond the forests of the eastern states encountered vast tracts of prairie extending from beyond the Mississippi to the Rockies. Increasing numbers of well-dated Holocene pollen sequences show that the position of the prairie forest boundary had advanced eastward between 9 ka and 7 ka BP and then retreated to the west between 7 ka and 0.5 ka BP (Figure 6.27). The large-scale fluctuations in north American plant distributions correspond closely to the changes predicted by general circulation models of Holocene climatic change with prairie expansion during phases of

Figure 6.27   The north American midwest showing changes in the position of the prairie forest boundary at 9, 8, 7, 6, 3 and 0.5 ka BP as indicated by pollen sites with 20% of selected prairie plant types. The higher percentages are to the west and the locations of pollen sites are marked by dots (after Webb et al., 1984)

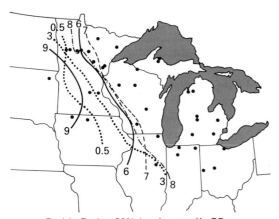

Prairie Forbs 20% Isochrones Ka BP

decreased precipitation (Webb *et al.*, 1984, 1987). Locally, however, there is evidence that the position of the prairie boundary was determined by topographic obstacles to the spread of fires, many of which were started by Indians. It is also argued that expansions and contractions of the prairie/forest interface may have been locally exacerbated and retarded by Indian burning (Whitford, 1983). This is supported by the fact that the near extinction of bison by Europeans and a much more recent reduction in cattle herding has led to woodland invasion of some prairies (Rackham, 1986). The extent to which burning and natural grazing played a part in the formation and maintenance of prairies themselves remains controversial (McCracken Peck, 1990; Hudson, 1990).

In parts of eastern Europe, such as the Ukraine, natural steppe existed in pre-Neolithic times but this subsequently expanded south-eastwards into parts of central Europe where pollen analysis shows that even some very dry areas were formerly wooded. The Hungarian steppe, for instance, is essentially a cultural artefact, some parts of which came into existence as a result of human activity as late as the seventeenth century (Behre, 1988).

The lowland temperate grasslands, which characterize some coastal areas and are typically found on calcareous strata such as chalk and limestone in Britain, are frequently species-rich, and contain relict species which occurred in the Lateglacial steppe. This led to the belief that they were natural climax communities. Ecological change on the chalk disproved this, however, when the introduction of rabbit myxomatosis in 1954 and a post-war reduction in sheep grazing, led to invasion of grassland by scrub and woodland (Smith, 1980). The grassland was, therefore, a plagioclimax – a succession prevented by grazing and human pressure from progressing to full climax. A wide range of palaeoenvironmental evidence supports and amplifies this conclusion, with molluscs (Evans, 1972) and pollen (Waton, 1982) in particular indicating that the chalkland formerly carried extensive woodland. Buried soils frequently show bowl-shaped fossil tree holes containing mollusc assemblages indicative of closed

Figure 6.28 A fossil tree-hole at Itford Bottom, Sussex, England, overlain by colluvial deposits of Bronze Age to Romano–British date. The present short-turf flora has been maintained by grazing since then i.e. over the last 1.5 ka, scale 2 m (photo Brenda Westley)

Figure 6.29 Landsnail diagram from the buried soil below the bank of Avebury Neolithic Henge, this shows the decreasing abundance of shade-loving species and their replacement by taxa indicative of open conditions such as grassland (after Evans, 1972)

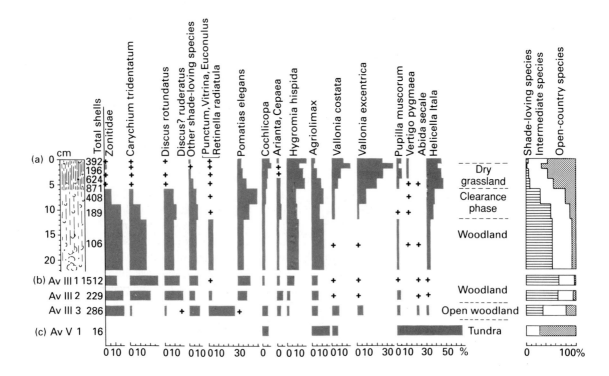

woodland (Figure 6.28). The woodland soils were truncated by prehistoric agriculture which gave rise to the present thin rendzina soils supporting grassland. Short turf communities were established in parts of Wessex prior to the later Neolithic construction of Silbury Hill and Avebury (Figure 6.29). Around Stonehenge, Bronze Age barrows seal typical rendzina grassland soil profiles and contain mollusc assemblages indicative of dry short grassland. This picture of the Stonehenge landscape is amplified by the preservation, by waterlogging, of a wide range of biological evidence in the base of Wilsford Shaft (Figure 6.30) dated around 3150 BP (Ashbee *et al.*, 1989). By that time trees were absent and grassland well established. It was maintained by grazing animals, particularly sheep, and herbivore dung was attested by the beetle remains.

Plant refugia in the wildwood

Refugia for grassland, heathland and moorland plants and other biota during the period when most of Europe was covered by woodland are difficult to establish. It is generally supposed that the main areas were coastal or upland but other disturbed areas and openings would have been created by the activities of fauna (Buckland and Edwards, 1984) and the effects of floods and storms. Pollen evidence from the Great Wold Valley, Yorkshire shows evidence for the survival of chalk grassland through the Early Holocene and it seems possible that here Mesolithic activity or ungulate grazing pressure round a wetland area, in an otherwise dry landscape, enabled some grassland to survive throughout the wildwood episode (Bush, 1988).

Figure 6.30   Wilsford Shaft, Wiltshire, England, a 30 m deep shaft dating to the middle Bronze Age around 3150 BP. The waterlogged base contained a wide range of biological evidence indicative of open grassland, grazing animals and some arable. Scale in feet (photo E Proudfoot)

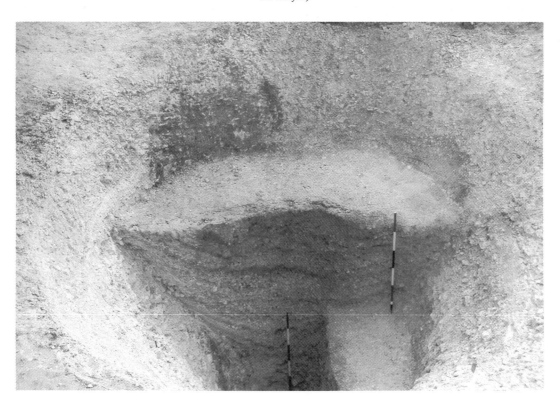

Geographical areas which were particularly prone to episodic environmental change or attractive to herbivores would have supported more open vegetation and this mosaic would in turn have supported a greater animal biomass and have attracted larger numbers of people. This may go some way towards explaining the patchiness of prehistoric settlement distributions during wildwood times.

7

People, Climate and Erosion

Introduction

Erosion is a universal geomorphological process. It occurs most rapidly and effectively on unvegetated, or sparsely vegetated, ground where soil is exposed to subaerial weathering. During the earlier Holocene, natural erosional processes were the dominant influences on the landscape, their effectiveness of operation being governed by such factors as climatic regime (especially precipitation levels and distribution as well as temperature ranges), slope angle and aspect, and bedrock structure. From the Mid-Holocene onwards, however, with increasing human impact on the landscape, the anthropogenic factor became increasingly important in determining rates and patterns of soil erosion in the mid-latitude temperate zone of the Northern Hemisphere. Identifying the extent of human influence in the erosional history of a region is far from straightforward, however, for the proxy records often contain both anthropogenic and climatic

signals, the relative importance of which is difficult to disentangle.

The purpose of this chapter is to examine the phenomenon of erosion in the later Holocene record in order to establish the extent to which this process was naturally (mainly climatically) or anthropogenically induced. Case studies are taken from North America and various parts of Europe. First, however, some of the processes and products of soil erosion are considered.

Processes and products of soil erosion

According to Morgan (1986), the main types of erosion are as follows:

1. *Rainsplash*. Rain impact causes soil aggregates to shatter and produces a hard surface crust which reduces infiltration. Soil particles are flung into the air by impact and on a slope move downwards.

2. *Overland flow*. The soil's infiltration capacity is exceeded and water flows across the surface.

3. *Rill erosion*. Overland flow becomes channelled and forms ephemeral gully-like features.

4. *Gully erosion*. Relatively permanent channels formed by running water.

5. *Mass movement*. Mud flows, landslides and debris avalanches.

6. *Subsurface flow*. Movement of fine particles through voids in the soil and movement of minerals in ionic solution.

7. *Wind erosion*. Movement of silt particles in suspension (e.g. loess) and coarser sand grade particles by saltation and surface creep (e.g. dunes).

In addition to these processes any activity such as frost heave, cultivation or grazing which disturbs soil on a slope will involve a component of downslope movement due to gravity.

The products of erosion accumulate in valley bottoms and other sediment traps where the

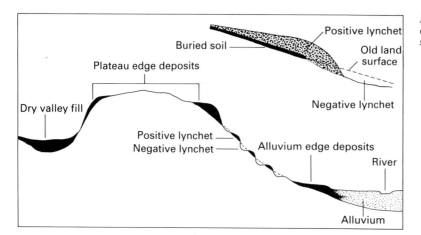

Figure 7.1   Schematic diagram of the situations in which eroded soils occur

deposits are of two basic types:

Colluvium, poorly sorted or unsorted sediment laid down by slope processes including slopewash and downslope creep, generally occurring under cultivation.

Alluvium, laid down by running water and consequently sorted. Coarser sands and gravels reflect high energy conditions, such as a river's bed load, whereas silts and clays comprise the suspended sediment laid down, for instance, during overbank flooding.

Under conditions of limited mineral deposition, peats and tufas also form, the first as a result of impeded drainage which inhibits biological decay, the second where springs rich in calcium carbonate discharge in a damp well-vegetated situation. Figure 7.1 is a schematic representation of typical situations in which these sediment types are found. Colluvial deposits accumulate in dry valleys and as lynchets (ancient field banks, Figure 7.2) on slopes. More amorphous colluvial accumulations are found at the perennial boundaries of cultivation, and at breaks of slope such as plateau edges or the alluvium edge in river valleys.

Riverine sediments are of more complex derivation on account of much larger catchment size. Sediment deposition implies erosion upstream, whether from channel downcutting, bank erosion or colluvial inputs. Conversely, the occurrence of erosion does not necessarily lead to the local deposition of sediment which may be transported elsewhere. Key factors are the relationship between sediment supply and the competence of a river to remove that material; both are affected by climate and land use. The principal climatic factors affecting erosion are rainfall (its intensity and duration), the amount of runoff and indirect effects on vegetation cover, especially in semi-arid zones.

Some erosion may occur on a regular and gradual basis, for instance during rain of a certain frequently achieved intensity. Of greater importance geomorphologically, however, may be events of medium frequency and magnitude. Recent studies in England and central Europe have shown that up to 80 per cent of erosion occurs in major storms which take place two to five times a year (Morgan, 1986; Richter, 1986). The overall contribution of such events may often be greater than that of, for instance, catastrophic storms which recur every hundred or thousand years. Infrequent events can, however, be particularly devastating in human terms as was the 1952 Exmoor flood which carried 100 000 tonnes of boulders, soil and uprooted trees and devastated the small town of Lynmouth, killing 34 people (Kidson, 1953).

Increased erosion and deposition will not necessarily occur during major secular episodes of higher rainfall. Indeed storm intensity and thus erosion may be greater in periods of lower mean rainfall. However, periods when extreme storms

Figure 7.2   A prehistoric and Romano–British lynchet at Bishopstone, East Sussex, England, scale 2 m (photo B Westley)

and floods are more frequent may lead to the transgression of geomorphological thresholds, triggering a change such as stream incision. A particular difficulty is that of distinguishing the effects of individual climatic events from those of major secular episodes of climatic change in recent sediment accumulations, such as landslips (Brunsden and Thornes, 1979).

Although erosion is a naturally occurring process, by removing the vegetation cover people accelerate the rate at which it takes place. Innumerable studies demonstrate greater runoff and erosion with decreasing vegetation cover

(Lockwood, 1983), for vegetation, especially trees, dissipates the kinetic energy of rainfall and facilitates gradual infiltration. Data from America suggest that sediment load doubles for every 20 per cent loss of forest cover (Goudie, 1981). On sandy soils in Bedfordshire, for example, sediment yield reached 17.7 tonnes per ha per year on bare ground as compared to 2.4 under grass and zero under woodland (Morgan, 1986). Thus human activity can lead to dramatically increased erosion, particularly from bare cultivated ground. Cultivation also results in soil changes including loss of nutrients with crops, reduction of organic matter and soil fauna. Increased silt movement down the profile may also occur producing 'agric horizons', the formation of which is exacerbated by ploughs, which turn a furrow and cause smearing at the base of the plough zone. Each of these processes reduces infiltration capacity and soil structural stability, and hence increases erosion. Micromorphological studies of both buried soils and soils in experimental fields cultivated using prehistoric methods have produced evidence for soil change and erosion following clearance and cultivation (Courty et al., 1989; Macphail et al., 1990).

The occurrence of erosion only constitutes a problem in human terms if its rate is greater than that of soil formation. Present-day erosion rates may be measured in the field using various instruments (Morgan, 1986). The short-term information which they provide can be complemented by estimates of longer term erosion rates from the archaeological and geological records (Costa, 1978). The archaeological record helps to establish the effect of human land use on erosion trends while the geological record provides evidence of the long-term natural erosion rate, the so-called 'geologic norm'.

Valley sediments: regional studies

North America

For most of the Holocene, erosional processes throughout North America appear to have been governed by climate as opposed to human agency. In the river systems different geographical areas show the same pattern of major alluvial discontinuities and some areas show broadly coeval periods of alluviation and incision (Figure 4.16). Variations in the magnitude and frequency of floods are seen as an important climatic control (Knox, 1984), their occurrence may have disrupted equilibria (Figure 1.3) and initiated new erosion cycles. The extent to which climatic change registers in alluvial sequences depends to a large extent on vegetation cover, the effects being most

pronounced in semi-arid areas and prairies in the central and south-west United States. In the south-west alluvial changes correlate with pollen evidence for changing effective moisture levels and with dendroclimatic evidence for changing climatic variability (Chapter 5, p. 137). In this area a subject of particular debate among geomorphologists has been the origins of deep erosion gullies known as arroyos (Figure 5.21; Graf, 1983). A major phase of arroyo formation coincided with the onset of cattle grazing in the 1880s. However, earlier cycles appear to be unrelated to land use and are interpreted in terms of climatic factors (Leopold, 1963; Gumerman, 1988). In the eastern United States alluvial changes correlate with pollen evidence for fluctuations in the prairie/forest boundary in Wisconsin (Knox, 1985) whereas in the Eastern Woodlands the inferred climatic changes do not register as strongly (Knox, 1984).

The sequence of colluvial loess-derived deposits at the Koster site, Illinois, which separates Indian occupations into 26 horizons spanning 9450–750 BP, is also thought to be largely a reflection of changes in tree cover in response to climatic influences. Only during the Archaic period of settlement, c. 5.5 ka BP, is it suggested that accelerated colluviation related to human activity. It is notable that even when crop growing became the basis of the Koster economy, around 1 ka BP, erosion rates did not increase markedly (Butzer, 1977b). Dramatic changes did, however, occur with the introduction of alien European agriculture in the nineteenth century and subsequent geomorphological changes here, and widely across America, are clearly related to land-use factors (p. 199).

Mediterranean

Debate about the causes of Holocene valley sedimentation is particularly sharply focused in the Mediterranean. Two main phases of alluviation have been identified: an Older Fill of last glacial date and a Younger Fill which overlies classical buildings, contains classical pottery and is dated to the Late- or Post-Roman period (Vita-Finzi, 1969; Bintliff, 1977). Apparent synchroneity of the

Figure 7.3   Collapsing agricultural terraces around a deserted Medieval settlement in the Alpes-Maritimes, France (photo M Bell)

Younger Fill throughout the Mediterranean suggested a climatic cause and Vita-Finzi favoured a southward shift of European depression fronts causing a change from seasonal to perennial river flow. It was argued that human activity could not have been responsible for the accelerated erosion, since the areas concerned were initially cleared much earlier and, moreover, intensive exploitation and unwise husbandry today tends to lead to stream incision rather than aggradation.

The climatic hypothesis is weakened, however, by the absence of independent evidence for climatic change coeval with Younger Fill deposition. It is also now clear that the deposits themselves are by no means as synchronous as once believed. Increasingly, prehistoric and Hellenistic aggradation episodes are being recognized (Davidson, 1980) and, in areas such as Greece, correspond closely with land-use histories (van Andel *et al.*, 1990).

Also contentious is the source of the sediments comprising the Younger Fill. It was initially thought to consist of reworked Older Fill sediments although the extent of erosion on the interfluves was far from clear (Vita Finzi, 1975). Subsequently, attention has been directed towards the possible contribution of colluviation (Wagstaff, 1981), particularly following the collapse of the Roman agricultural system (Butzer, 1976, 1980). Cultivation terraces (Figure 7.3) may have retained material on the slopes until abandonment, after which accelerated colluviation and increased stream loading led to aggradation.

Although many Mediterranean soils are highly susceptible to erosion (Morgan, 1986), its extent remains controversial. Parts of arid south-west Spain with considerable evidence of gullying have none the less experienced only limited soil erosion since the Bronze Age (Gilman and Thornes, 1985). The situation clearly varies regionally; there is a

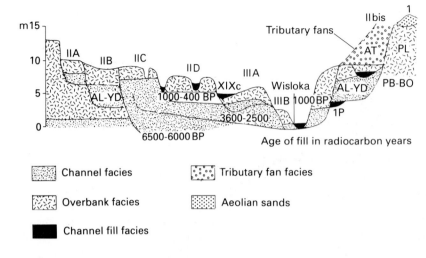

Figure 7.4 The Wisloka Valley, Poland showing terrace levels and alluvial fills of various ages (after Starkel et al., 1981)

need for detailed and quantified case studies of the ecological context of erosion (Thornes, 1988). A monocausal climatic hypothesis appears unsatisfactory and the contribution of anthropogenic factors and soil erosion deserve more detailed consideration in this area with its long history of intensive land use.

Central and north-west continental Europe

Valley sediments in northern Europe show a more complex sequence of changes than that generally recognized in the Mediterranean, and the role of climatic and anthropogenic factors has been more clearly defined. Climatic influences are most apparent in continental areas where many German, Polish and Soviet rivers show evidence of downcutting and aggradation which are broadly coeval from one basin to another and are, in some cases, dendrochronologically dated (Starkel, 1983a, b, 1985; Schirmer, 1988). Figure 7.4 illustrates the sequence from the Wisloka Valley, Poland where rhythmical variations in river discharge and sediment load of 2–2.5 ka were identified. Changes in river regime are seen as possibly triggered by 100–300-year episodes of higher precipitation and increased flood frequency. River regime changes correspond in some cases to other proxy climatic indicators such as alpine glaciers and lacustrine sediment sequences. Over the last two millennia

the pattern has been made more complex by increased erosion arising from human activity which resulted in extensive Late Roman and Medieval alluviation in the river valleys of Poland and Germany (Starkel, 1988; Schirmer, 1983). In Germany, comparative studies of past and present erosion by Richter (1980, 1986) identified cycles of erosion and stability correlated not with climatic cycles, but with periods of population expansion, which led to the extension of cultivated land, and also with periods when new crops were introduced. Detailed survey of a 20 km section of the Leine Valley near Gottingen showed that 25 million m^3 of alluvial loam had been deposited in the last 2.5 ka. More than half of this derived from two comparatively brief episodes between AD 600 and 900 and AD 1500 and 1700. In the Luxembourg Ardennes the history of erosion is seen to correlate with palynologically attested agricultural expansions (Kwaad, 1977; Imeson et al., 1980).

British Isles: chalk and limestone

These geologies were once regarded as too permeable for significant runoff and erosion and they were thought always to have had thin rendzina soils (Godwin, 1967). Recently it has been shown that on the chalk extensive colluvial deposits are in storage (due to the lack of stream activity) in dry valley fills and as lynchet deposits on the slopes

Figure 7.5   Colluvial sequence at Kiln Combe, East Sussex, England. Main layers (1) Beaker and (2) Iron Age, both predominantly silt (3) Romano–British, flint nodules probaby representing colluvium in which the fine particles have been washed away (4) Post Romano–British truncated soil (5) Medieval chalky colluvium (6) Topsoil, scale 2 m (photo B Westley)

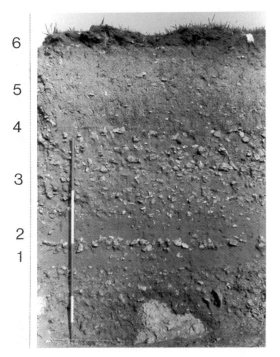

(Bell, 1983, 1986; Allen, 1988). In subsoil horizons, hollows representing fossil tree holes with woodland mollusc faunas are present (Figure 6.28). In one or two cases there is evidence of some colluviation in wooded landscapes (Evans *et al.*, 1978; Burleigh and Kerney, 1982) but in the vast majority of instances erosion follows clearance and agriculture, as shown by Mollusca and micromorphology (Macphail *et al.*, 1990). Early colluvial increments are largely non-calcareous and contain redeposited Pleistocene loess (Catt, 1986). Later colluvium, commonly of Iron Age or later date, contains a higher proportion of stones and chalk granules which reflect progressive thinning of soils (Figure 7.5).

The archaeological evidence is complemented by a growing body of data on current erosion processes (Boardman, 1990a). Heavy rainfall creates rills on the valley sides (Figure 7.6) which discharge fans of chalk granules on to the floor, where a very stony deposit is sometimes left by temporary streams. A comparable Romano-British deposit (Figure 7.5) can be associated with a land-use regime similar to today's with much of the valley under arable as the field archaeological evidence confirms (Drewett, 1982). Recent erosion

Figure 7.6   Severe erosion on fields recently drilled with winter cereals at Rottingdean, England, October 1987. The rills and gullies have cut through thin topsoil to the underlying chalk and fans of coarse material including chalk have been deposited on the valley floor (photo J Boardman)

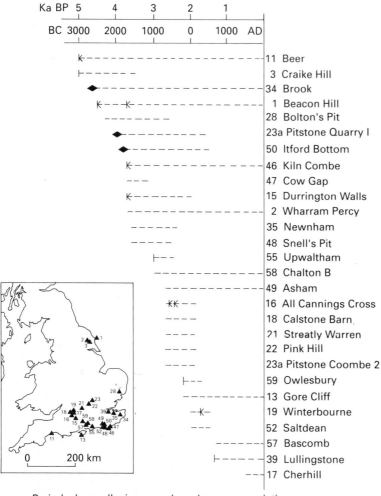

Figure 7.7   The dates at which colluviation was initiated and was occurring in Britain (after Bell, 1982)

--- Period when colluvium may have been accumulating

ᴋ- Occupation horizon below colluvium

◆ Radiocarbon-dated basal horizon

is almost all on arable land and given these conditions the frequency with which erosion occurs is determined largely by rainfall factors (Boardman, 1990a). Erosion is particularly associated with the practice of autumn sowing and it may be significant that there is plant macrofossil evidence from weed floras that this was also practised during the Bronze and Iron Ages (Jones, 1981) when many colluvial deposits started to form. If the dates of known colluviation are plotted out they do not show evidence of clear temporal clustering such as would

be expected if the process was associated with particular secular climatic episodes (Figure 7.7).

Erosion rates can be quantified on the basis of present-day events, such as that at Bevendean, which had a recurrence interval of *c.* 25 years and involved soil losses greatly in excess of acceptable levels (Boardman and Robinson, 1985). Assuming continuance of the same land-use regime this could completely remove the topsoil in 300–500 years. Logically it might be argued that present erosion rates may be higher than those in the past because many of the factors which exacerbate erosion today did not obtain in prehistory (e.g. large fields, hedge removal, vehicle tracks, etc.). 'Celtic fields', which formed between 3.5 ka and 1.5 ka BP, were small (*c.* 1–0.6 ha) by modern standards, and separated by grassy lynchet banks; both factors would have led to reduced erosion, as would the better developed weed flora and prehistoric scratch ploughs which, by comparison with contemporary methods of tillage, caused less soil disturbance, oxidation of organic matter and loss of crumb structure. This accords with pottery evidence that prehistoric colluvium built up by gradual increments over a long period. Even so, there are areas where the amount of colluvium in storage implies erosion rates greater than those of today, perhaps because their soils were more susceptible before erosion of surface loess horizons (Evans, 1990).

Calculation of long-term erosion rates may be based on sediment in storage within specific catchments. Studies of 74 mapped areas (on a variety of bedrocks) indicate that surviving colluvium accounts for the erosion of between 9 and 204 mm of soil, not including that washed out to sea or deposited as coastal alluvium (Evans, 1990). These figures are of a similar order of magnitude to those obtained from archaeological investigations on the chalk; this indicated that before the Bronze Age the soils of these areas were about twice their present depth (Bell, 1986). That figure does not, however, take account of material lost and reworked in river valleys. We do, for instance, know from molluscan and micromorphological evidence and the eroded remains of fossil tree holes that the former woodland soils in many of these areas had been totally eroded away by about the Bronze Age.

There is also evidence of significant erosion in a number of other limestone areas. Cultivation of a steep slope behind the Middle Bronze Age settlement at Brean Down, Somerset (Figure 5.16; Bell, 1990) led to the complete removal of soil on the upper part of the slope down to an underlying Pleistocene breccia. While that was a localized occurrence, on the karst upland of the Burren in the west of Ireland (Drew, 1982) the soil from a

Figure 7.8   A megalithic wedge tomb on the Burren, Ireland, surrounded by karst pavement (photo D Drew)

formerly wooded landscape was stripped away, probably in the third and second millennia BP, revealing vast areas of limestone pavement (Figure 7.8). Traces of the former soil are preserved below megalithic tombs and stone walls and in eroded forms in rock fissures, depressions and caves.

Palaeoenvironmental evidence and modern analogues show that in many of these areas erosion is largely restricted to phases of intensive arable farming, although today most of the erosion occurs during periodic high rainfall events. In the course of millennia this combination of anthropogenic factors and climatic vagary has changed the soils and land-use potential of many of Britain's calcareous regions.

British Isles: river valleys

Extensive changes in river regime occurred as a result of Late Devensian and Early Holocene climatic changes (Chapter 4, p. 101), but many river valleys contain the following greatly simplified sequence of deposits (Bell, 1982). At the base are gravels representing high energy fluvial conditions and physical weathering on valley sides during the Late Devensian period. These are overlain by deposits that accumulated more slowly, including peats or tufas which often contain biological evidence of deposition during an episode of climax woodland. Finally, there is a layer of mineral-rich sediment which biological evidence suggests is largely derived from an open agricultural landscape. In particular valleys there is frequently interdigitation of these sediment types reflecting several cycles of erosion and alluviation (Burrin and Scaife, 1988).

Tufas highlight the problem of establishing the cause of sedimentary change. Most date from the Holocene Climatic Optimum (Evans, 1972) and may be thought to reflect higher rainfall and humidity, because such deposits are not forming on a comparable scale today. However, some tufas began to form before the Climatic Optimum suggesting that these deposits are not so much the product of a particular climatic regime, as indicative of swampy calcareous woodland environments which disappeared in the Mid- to Late Holocene.

Although the transition to mineral-rich alluvium in river sediment sequences during the Late Holocene has been interpreted by some authors as reflecting climatic change, and indeed links have been suggested with the Mediterranean Younger Fill (p. 190; Potter, 1976), the balance of evidence would seem to favour an anthropogenic explanation for these minerogenic alluvial deposits. In the Severn/Avon Valleys alluviation has been dated to the third millennium BP and related to the development of a more continuously open agricultural landscape and perhaps also to the introduction of autumn-sown crops (Shotton, 1978). Subsequent more detailed study of the entire Severn catchment confirmed greatly accelerated alluvial sedimentation, perhaps five times the present rate, between 2 ka and 3 ka BP (Brown, 1987). Pollen sequences show that the change correlates broadly with deforestation episodes, though a time lag in some cases suggests that subsequent soil deterioration may also have been important.

In the Thames Valley the riverine sediments are well dated by their relationship to a series of archaeological sites which show that flooding and alluviation have been largely restricted to the last three millennia (Robinson and Lambrick, 1984; Limbrey and Robinson, 1988). During the period of climax woodland the Thames was incised in the present floodplain which was free from flooding. This situation continued throughout the Neolithic and Bronze Ages when settlement became established on the floodplain. There was some localized Neolithic alluviation in the lower Thames Valley, and in the upper Thames significant infilling of low-lying areas had begun by the Late Bronze Age. In the third millennium BP, however, a major environmental change occurred leading to increased flooding, alluviation and a higher water-table and this correlates with accelerated forest clearance and agricultural intensification in the catchment. A similar, more extensive alluviation episode in the ninth to thirteenth centuries AD has been correlated with Medieval agricultural expansion and runoff from ridge-and-furrow cultivation. Comparable evidence comes from a small chalk stream at Avebury, Wiltshire (Evans *et al.*, 1988), where sediment deposition in Early–Mid-Holocene swamp woodland was followed by the development of a soil containing molluscan evidence of dry grassland and later Mesolithic and Neolithic artefacts. Much of the surrounding area is known to have been cleared in the Neolithic but flooding and alluviation were limited until the third millennium BP when an extensive silt loam deposit began to accumulate, near the base of which were buried field boundary walls.

Recent work on river sedimentation in the Weald in south-east England has established a similar relationship between anthropogenic activity and erosional episodes, but the chronology contrasts with areas discussed above. Major alluviation began in the Ouse Valley as early as the Mesolithic and was largely complete by the Iron Age (Scaife and Burrin, 1983); elsewhere alluviation began in the Neolithic and Bronze Ages (Burrin and Scaife, 1988). That there should be such early anthropogenically generated alluviation in this area is surprising in view of the limited evidence for prehistoric activity in the Weald (Drewett *et al.*, 1988).

The dates of major alluviation in British river valleys, therefore, show considerable variation and this seems largely to reflect land-use contrasts. A different view is, however, expressed by Macklin and Lewin (1991) who see river alluviation in Britain as climatically driven but culturally blurred. They see increasing chronological precision as pointing to particular phases of deposition comparable to, and broadly contemporary with, the previously noted alluvial discontinuities in North America (p. 190). Even so, alluviation occurred widely in south-east England from 5 ka BP onwards and in the north from 2.2 ka BP and this does correspond to pollen evidence for the dates of major clearance in these different parts of Britain (p. 165). However, in a number of areas major phases of alluviation occur significantly later than clearance episodes. In some cases this seems just as likely to relate to the impact of Iron Age societies practising more intensive land use as to climatic change in the first millennium BC.

Aeolian sediments

The effects of aeolian erosion are most spectacularly demonstrated by the American dust bowl on the Great Plains in the 1930s which resulted from a series of hot dry years and depleted vegetation cover, exacerbated by years of overgrazing and the rapid extension of cereal growing facilitated by mechanization (Goudie, 1981). In 1934 one four-day storm transported 300 million tonnes of sediment!

In Europe there is evidence for prehistoric and early historic aeolian erosion in specific areas which were susceptible because of the particle size of their soils. For instance, the Pleistocene coversands of The Netherlands were originally of aeolian origin and were subject to reworking in the later Holocene. There was some aeolian deposition on archaeological sites on the Drenthe Plateau during the Neolithic and Bronze Ages but a particular concentration of erosion occurred in the first millennium BC, relating to the maximum extent of 'Celtic fields' on the Plateau (van Gijn and Waterbolk, 1984). Extensive agriculture and soil exhaustion triggered sand blow and the morphology of 'Celtic fields' in The Netherlands and north-west Germany may relate to attempts to counteract excessive erosion (Groenman-van Waateringe, 1979). Further south the Veluwe Plateau remained wooded until the eighth century AD when heathland began to expand with increasing settlement and charcoal production. Extensive aeolian deposition began in the tenth century AD and buried the Early Medieval settlement of Kootwijk (Heidinga, 1987; Groenman-van Waateringe and Wijngaarden-Bakker, 1987). The effects of deforestation and agriculture were exacerbated by a dry episode which first dried up the settlement's pond and then lowered water-tables, as reflected in the levels of successive wells (Figure 7.9). That dry episode may correlate with the onset of Younger Dune formation on the Dutch coast (Chapter 5, p. 126). In north Germany the sandy moraines are susceptible to wind erosion today when vegetation is removed. Earlier cycles of aeolian deposition have also been identified, including one in the first century AD, a particularly

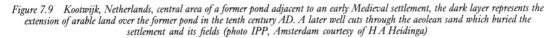

Figure 7.9   Kootwijk, Netherlands, central area of a former pond adjacent to an early Medieval settlement, the dark layer represents the extension of arable land over the former pond in the tenth century AD. A later well cuts through the aeolean sand which buried the settlement and its fields (photo IPP, Amsterdam courtesy of H A Heidinga)

destructive cycle in the late eighteenth century which possibly relates to overpopulation, and others following cultivation after both world wars (Richter, 1980).

In Britain, despite the maritime nature of the climate, wind erosion occurs, particularly in eastern parts of the country where lower rainfall is recorded, and on cultivated fluvioglacial deposits during dry spells (Radley and Sims, 1967; Fullen, 1985). Coversands originally deposited by wind during the Devensian cold stage became mobile once again following Mid- and Late Holocene forest clearance. At Heslerton, Yorkshire, for example, various aeolian deposits bury and preserve a long sequence of occupations of Neolithic and later date, while sand blow on the Brecklands of East Anglia buries Iron Age and Anglo-Saxon occupations at West Stow, Suffolk (Macphail, 1987). On the chalk of southern England, where

wind erosion does not seem to occur even under the intensive arable regime of today, there is some evidence of localized aeolian deposition in ditches during the Bronze Age (Cornwall, 1953; Evans and Jones, 1979), which may relate to a drier (Sub-Boreal) climatic regime. In coastal dune systems increasing encroachment and mobility followed removal of woodland (Spencer, 1975; Evans, 1979) but alternating episodes of sand blow and stabilization in Britain, The Netherlands and Denmark correlate largely with marine transgressions and regressions rather than land use (Chapter 5, p. 118). Conversely, inland areas of western Europe would not have experienced later Holocene aeolian erosion were it not for open conditions, cultivation and grazing which people created. Within this cultural landscape a combination of both climatic and land-use factors appear to have triggered individual erosion events.

Lakes

Lakes are key contexts for the study of human impact (Berglund, 1986; Lang and Schlüchter, 1988). They preserve long sediment sequences frequently spanning the Lateglacial and Holocene and occasionally extending even further back in time. It is a record datable by radiocarbon, sometimes by archaeomagnetism (Thompson and Oldfield, 1986) and dendrochronology and occasionally by the presence of annual varves. Where lake carbonates are present, isotope curves can provide Holocene palaeoclimatic records (Figure 2.30), e.g. on the island of Gotland, Sweden (Mörner and Wallin, 1977) and some Polish lakes (Rozanski et al., 1988). Lake shores are ecotones providing human communities with the opportunity for an economic strategy combining the resources of adjoining habitat types. Their favourability for settlement is emphasized by the wealth of waterlogged sites on lake margins, particularly in the Swiss Alps (Coles and Lawson, 1987) and the crannogs (artificial islands) of Ireland and Scotland (Morrison, 1985), despite the vulnerability of such sites to changes in lake level. Furthermore, a wide range of palaeoenvironmental evidence is often preserved (Lowe and Walker, 1984), particularly pollen, diatoms, sediment chemistry and magnetic properties, facilitating comparative studies of a range of data sets (Chapter 3, p. 30).

Changes in properties of American lake sediments following European contact have been intensively researched (McAndrews, 1988) and provide a useful analogue for possible earlier effects in Europe. Before contact sedimentation rates and lake sediment characteristics reflect the operation of natural processes (Haworth, 1972; Cwynar, 1978). Dramatic changes occur, however, with the arrival of Europeans which is typically marked by an increase in pollen of the agricultural indicator *Ambrosia* (ragweed). A further useful marker horizon in recent lake sediments is the chestnut decline (p. 162) which occurred in eastern North America during the 1930s. European agriculture is also reflected in changes in lake sediment chemistry, in diatom assemblages (Davis and

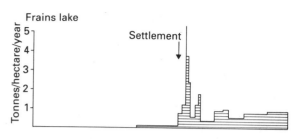

Norton, 1978) and in a change from organic to minerogenic sediment with increased ash from clearance burning. In Frains Lake, Michigan, for example, the establishment of settlement and associated clearance in 1830 was accompanied by an initial dramatic increase in sedimentation to 30 times its previous level, erosion then stabilized at 10 times previous levels (Davis, 1976b; Figure 7.10).

The margins of Alpine lakes have revealed many waterlogged prehistoric settlements which flourished at times of low lake levels (Joos, 1982). Artefact dating and more particularly dendrochronological work on the building timbers show that some of these low lake levels were contemporaneous at the widely separated lakes of Constance, Zurich and the Neuchatel area (Figure 7.11). These widespread episodes of lower lake level are evidently due to climatic factors, apparently warmer or drier episodes which seem to have been relatively brief while more localized changes in lake level can be ascribed to anthropogenic clearance and the damming effects of landslides.

In north-west Europe present evidence suggests anthropogenic factors were the main influence on lake sedimentation rates. For example, Lake Bussjösjön in southern Sweden (Figure 7.12) received minimal sediment input between the Early Holocene and the onset of anthropogenic clearance c. 2.7 ka BP. This was followed by a steady increase, paralleled by more intensive land use (Dearing et al., 1990). Even so, peak rates of topsoil erosion between 0.5 ka and 0.3 ka BP may be partly caused by the Little Ice Age since the most intensive erosion today follows periods of frozen

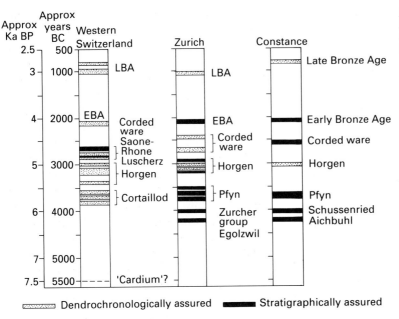

Figure 7.11   The chronology of Neolithic and Bronze Age lake shore settlements in Switzerland (after Joos, 1982)

ground. Braeroddach Loch, Scotland shows similarly low Early to Mid-Holocene inputs (Edwards and Rowntree, 1980), until the onset of pastoral farming *c.* 4.6 ka BP (Figure 7.13) when sediment yield increased by a factor of three. A further modest increase was associated with the first growing of cereals between 3.4 ka and 2.1 ka BP. Within the last 400 years a dramatic

Figure 7.12   Holocene sedimentation in Lake Bussjösjön, southern Sweden (after Dearing et al., 1990)

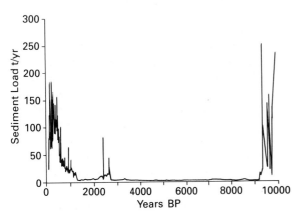

acceleration in deposition has occurred, during which a quarter of the lake basin sediment accumulated. Elsewhere, clearance of the catchment of Llangorse Lake in Wales around 1.8 ka BP resulted in the replacement of nekron muds by silty clays (Jones *et al.*, 1985), while in many lakes of the English Lake District, a lithological change from brown organic mud to pink clay mud occurred more or less contemporaneously with the elm decline and the first evidence of cereals and weeds of cultivation around 5 ka BP (Pennington, 1978). Further increases in sediment inwash occurred in the third millennium BP, the Late Roman period and around AD 1000, all periods which the pollen evidence shows were associated with increased clearance. Particularly dramatic changes took place after AD 1820 when increased sediment and sewage input led to eutrophication (nutrient enrichment) reflecting the tourist popularity of the area. Many European lakes show a different form of human impact over the course of the last two centuries, with evidence of increasing acidification linked to the problem of acid rain (Chapter 9).

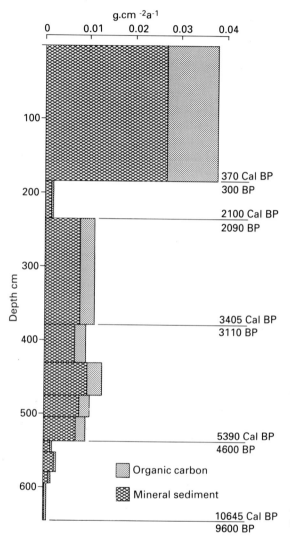

Figure 7.13   Sediment deposition rate at Braeroddach Loch, Scotland (after Edwards & Rowntree, 1980)

north-western Europe anthropogenic effects occur much earlier and appear to mask and distort climatic patterns in many areas. Modern analogues show that erosion and deposition occur during major storms and floods and, in periods when these were more frequent, the crossing of geomorphological thresholds may have instigated river incision or deposition. The effects of climatic changes and stochastic events are, however, likely to have been much more pronounced under certain land-use regimes, particularly arable, which increase sediment supply. In many parts of western Europe land-use changes have obscured underlying secular climatic signals from 5 ka BP, although these signals do register more clearly at sites in Alpine and continental Europe. Interestingly, as chronological precision improves, both studies which favour climatic causation (Starkel, 1985; Macklin and Lewin, 1991) and those which emphasize the role of human activity (Richter, 1986; van Andel *et al.*, 1990) are identifying brief phases of particular erosive activity, and/or deposition, separated by longer periods of relative stability. Ultimately the debate between those who advocate the primacy of anthropogenic or climatic factors may be partly resolved by the extent to which alluvial discontinuities prove to be coeval from one area to another.

Erosion perception

Most erosion does not take place imperceptibly slowly but is concentrated during specific observable events which, depending on the expectation of their recurrence, may have elicited a human response. It is clear that Iron Age and Medieval communities had to abandon particular areas of coversand in The Netherlands because of aeolian erosion and that the Thames floodplain in England became unsuitable for settlement because of increased flooding in the later Iron Age (Robinson and Lambrick, 1984). There are occasional instances of settlements having been destroyed by catastrophic erosion, such as the Ozette Indian settlement in Washington, USA

Spatial and temporal contrasts

Holocene erosion in the mid-latitude region of the Northern Hemisphere shows clear regional contrasts. In North America there is evidence for climatically induced changes of river regime but minimal anthropogenic impact before the dramatic changes wrought by the arrival of European agriculture. In the Mediterranean and

which was buried by a mud slide in *c.* AD 1750 prior to European contact; the memory of this disaster was preserved by Indian communities and the site, with magnificent decorated wooden artefacts, has now been excavated (Gleeson and Grosso, 1976; Coles and Coles, 1989). It should not, however, be assumed that past erosion events necessarily had widespread social repercussions. At Brean Down, England, Bronze Age communities continued intermittent settlement for some 600 years after their immediately adjacent arable resource had been totally eroded away. Their attachment to this site, despite the additional threat of sand inundation (p. 129), probably reflected its pivotal location for the exploitation of coastal grazing and marine resources (Bell, 1990).

Much erosion, though occurring in the form of perceptible events, may only represent an insidious long-term threat to the agriculture of an area. Even where current high erosion rates have been demonstrated, this has so far prompted little change in land use. Furthermore, there is little hard evidence regarding the perception of erosion, even by present-day agricultural communities in Britain (Boardman, 1990b). It is little wonder, therefore, that archaeologists have not really begun to investigate whether the erosion of arable land in prehistory elicited a human response. It could be argued that the 'Celtic fields' of north-west Europe represent deliberate attempts to retard soil erosion. Some formed against boulder walls or very stony deposits which may represent a combination of field clearance and attempts to reduce erosion just as the stone-faced terraces of the Mediterranean lands (Figure 7.3) must, in part, be seen in terms of human response to perceived soil erosion when steeper slopes were used for agriculture. Lynchets and terraces are relatively neglected areas of archaeological enquiry which should have much to offer in improving our knowledge of both former agricultural techniques and past human responses to erosion.

8

Conserving the Cultural Landscape

Introduction

The last three chapters on people/environment relationships provide a time perspective for consideration of conservation issues. Cultural landscapes (Birks *et al.*, 1988) contain relics of past human ecological relationships and their effective conservation demands a detailed knowledge of that habitat's history. Human impact on the environment has increased through time, albeit with periods of dramatic change punctuating episodes of relative equilibrium. At the same time people have practised conservationist strategies to varying extents. This may have the effect of keeping environmental change within predictable bounds, thus reducing the need for immediate response to random fluctuations when thresholds are transgressed (Simmons, 1989). An emphasis on conservation is likely to have been a feature of societies making significant use of wild resources, either because they did not practise agriculture or because wild resources played a key role in buffering a partly agricultural economy against natural environmental changes. A characteristic of Australian Aboriginal and many American Indian communities is a respect for, and community with, nature. This is very different from attitudes prevalent until recently in most industrialized societies. Now, however, views are changing with people becoming more conservation-minded in response to ever diminishing resources and natural habitats. The conservation of living organisms and their habitats has, understandably, attracted most widespread attention but the role of archaeological and palaeoenvironmental evidence deserves greater emphasis in conservation debates (Greeves, 1989) and thus forms the subject of this chapter.

Palaeoenvironmental studies and nature conservation

Many of the most important contexts for investigation of past people/environment relationships are areas of relict landscape, marginal areas such as heathland, moorland, ancient woodland and pasture which have escaped major recent disturbance. In addition to rare plant and animal communities these localities frequently contain environments of deposition which preserve palaeoenvironmental sequences. Paradoxically, just as people have come to appreciate the value of relict landscapes, so the economic pressure for the destruction of many of them has mounted. In Britain, for example, half the area of ancient woodland which existed in 1930 had been grubbed up or replanted by 1983 (Rackham, 1986) while in former East Germany only 1 per cent of peatland remained in a natural or semi-natural state, some 85 per cent having been given over for agriculture (Succow and Lange, 1984). Similar dramatic reductions have been estimated for heathland cover in Scandinavia (Figure 8.1).

Effective conservation requires both a detailed knowledge of the ecology of the biota concerned and, what is less often appreciated, the past trajectory of their habitat, including the history of human land use (Lambrick, 1985). Early attempts at conservation assumed that many habitats

Figure 8.1   The decline in heathland in southern Sweden between 1909 and 1957 (after Gimingham, 1972)

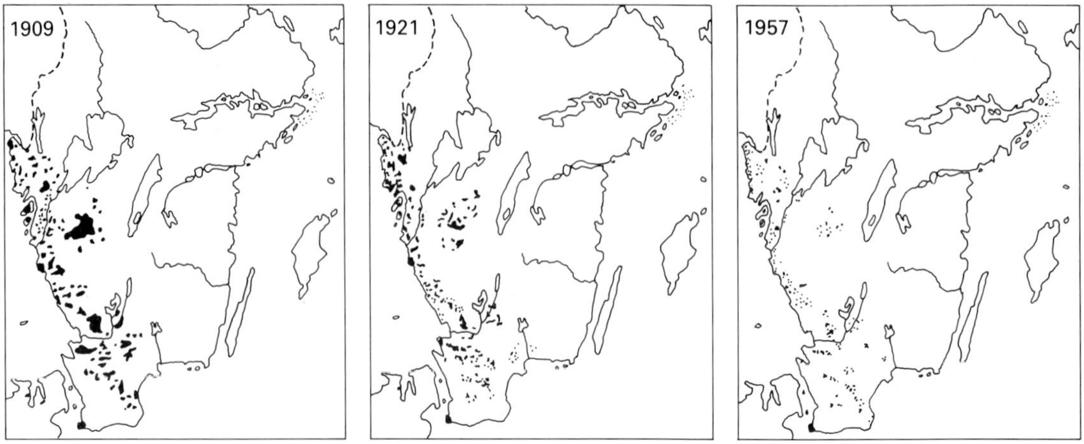

containing rare and interesting species were remnants of natural ecosystems which could be preserved by putting a fence round them to prevent disturbance. However, because the character of these reserves had evolved from particular patterns of human use, reduction in grazing and burning (p. 183) led to major vegetational changes and many grasslands and heaths, for example, became invaded by shrubs and trees. It is now appreciated that the more populous parts of the world have very few surviving 'natural' habitats, i.e. those totally unaffected by human activity. Many biologically important habitats are very much cultural artefacts, on a par with prehistoric settlements, pots and metalwork in the degree of human agency in their making but unlike them in that they are living and subject to continuing change and evolution.

Positive management is needed to retain the special features of a habitat and this may involve grazing, coppicing, or the cutting of reeds and peat. Occasionally there may be a difference of opinion as to the most appropriate management strategy. Conservationists in The Netherlands have created ecologically diverse clearings in woodland nature reserves on the assumption that this replicates conditions in the wildwood resulting from the grazing activities of wild herbivores. Palaeoenvironmental evidence from this area suggests, however, that the wildwood was less diverse than this management strategy implies

(Bottema, 1988). Comparable problems attend the management of old grassland in the Thames Valley, England, where nature conservation practice has been to try to eradicate thistles as inappropriate members of the community. Palaeoenvironmental evidence shows, however, that they were equally prevalent in these communities during the Iron Age (Lambrick and McDonald, 1985). Such examples raise important issues about the status and origins of what we are aiming to preserve.

Palaeoenvironmental evidence is equally important in establishing whether individual rarities are native or otherwise. Small leaved lime (*Tilia cordata*) is a classic case; once regarded as a minor, or even non-native, species in many areas, it is now recognized as having been abundant in the Climatic Optimum wildwood of much of Britain, Belgium, Germany and southern Scandinavia (Greig, 1982). Many of its contemporary British sites are arguably wildwood remnants and are, therefore, deserving of conservation. Conversely, some rarities can be shown to be more recent human introductions; in Britain the corncockle (*Agrostemma githago*) and the cornflower (*Centaurea cyanus*) were introduced in the last two millennia but are now rare due to the abandonment of the agricultural regimes on which they depended (Robinson, 1985). The fact that they are introductions should not of course limit their value in conservation terms. A clear understanding is, however, necessary as to why a particular species is being protected, whether as a relict of former native habitats or a vestige of a former agricultural regime. This information forms an essential aspect of an effective management strategy, for both habitats and individual biota.

Cultural resource management

Both archaeological sites and recent sediment sequences (e.g. bogs, lakes, dunes and estuaries, etc.) are valuable repositories of information about past people/environment relationships. They constitute a non-renewable resource which, unlike living organisms, cannot reproduce and spread. Of critical importance is the relationship between the evidence, whether artefacts, sediments or biological remains, and the context in which this occurs. Disturbance by building or agricultural activity destroys or damages that contextual relationship and results in a loss of meaning. Even scientific archaeological excavation is an essentially destructive process unless a perfect record is obtained of the contextual relationships of every find. Needless to say this ideal is never completely achieved. Only in recent years has due emphasis been given to the finite nature of these resources. Archaeological excavation is now seen as a last resort with preservation as the favoured option.

An emphasis on preservation is encapsulated in the label 'Cultural Resource Management' first used by the National Park Service in the USA in 1971. A broad definition of the cultural resource is 'physical features both natural and man-made associated with human activity' (Fowler, 1982). The International Union for Conservation of Nature and Nature Reserves (1980) has stated that archaeological sites are a resource which is, and will continue to be, useful and needs to be maintained. According to Lipe (1984) 'all cultural material including cultural landscapes that have survived from the past are potentially cultural resources, that is, have some potential value or use in the present or the future'. Such statements, of course, raise interesting issues about the meaning of the words 'value' and 'useful' which are capable of

interpretation in either spiritual or monetary terms.

The various ways in which people have made use of, and been affected by, the past have been explored at a philosophical level by Lowenthal (1985). Sometimes politicians have rejected the past as a stifling influence, as was the case in post-revolutionary America (Lowenthal, 1976). Elsewhere it has helped to consolidate national or cultural identity, just as the site of Zimbabwe became such a powerful symbol of cultural achievement before European influence that it has given its name to an African state. Sometimes the archaeological record has been monstrously distorted for political ends as it was in legitimating the racist theories of Nazi Germany (Arnold, 1990). Conversely, the archaeological record frequently shows the recent and ephemeral nature of present-day nations and ethnic groups (Lipe, 1984). The many publications arising from the 1986 World Archaeological Congress (e.g. Cleere, 1989) have emphasized the concept of a global heritage as opposed to national and regional heritages, while the designation of World Heritage sites is another acknowledgement of the same philosophy. UNESCO has galvanized international action to record and conserve the archaeological heritage in threatened areas such as the Aswan Dam in the Nile Valley, Carthage and Sri Lanka. Despite such initiatives, recent reviews of Cultural Resource Management in a range of countries reveal marked disparities in the degree and type of protection offered to archaeological sites (Cleere, 1984, 1989). National arrangements are seen to be largely a function of administrative tradition. The high level of protection in Denmark, for example, contrasts with the widespread destruction of sites in neighbouring Germany where only 5–8 per cent of the monumental heritage recorded since 1830 survives (Reichstein, 1984).

Problems of Cultural Resource Management vary geopolitically. In Russia state ownership of land should theoretically make protection straightforward, but sites have been destroyed by overzealous attempts to meet state economic targets (Masson, 1989). In America preservation is more difficult on the east coast where most land is privately owned, as opposed to the area west of the

Mississippi where there are substantial holdings of state and federal land (McGimsey and Davis, 1984). In both America and Britain there is a conflict, which is not always fully acknowledged, between the need for greater protection of sites and the enshrined property rights of individuals. The current British approach is to make owners more aware of the importance of sites and to encourage management by agreement (Darvill, 1987).

Archaeological sites in their landscape

In most countries legal protection concerns individual archaeological sites but that raises issues of how a site and its boundaries are defined. It is often simply regarded as the area containing buildings, but it might equally be thought to extend to the cemetery, trackways, fields and the neighbouring bogs or ponds containing biological evidence essential to an understanding of its environmental history. The protection of small individual elements within the landscape is typified by the image of a Bronze Age round barrow as a protected grassy mound in a vast expanse of arable land (Figure 8.2). Frequently, cultivation is insidiously encroaching on the monument, reducing its size and the information it contains.

Only recently have archaeologists and other historical scientists begun to give real emphasis to the need to conserve sites within their landscape. Marginal areas, which have only been subject to human exploitation during occasional episodes of economic pressure or favourable climate, preserve relict landscapes. Figure 8.3 shows a classic example of British chalk grassland with sites of many periods, in this case protected by National Trust ownership. Relict landscapes have suffered serious agricultural encroachment to the extent that many are now mere fragments of what survived until fairly recently. However, even within more intensively used farmland, walls and trackways etc., may fossilize boundaries established at a much earlier date as they do in parts of West Penwith, Cornwall, where fields integral with prehistoric

settlements are still in use. In East Anglia, coaxial field systems are cut across by Roman roads and clearly reflect land boundaries established more than two millennia ago (Fleming, 1988).

In order to examine the land use and economy of past communities it is necessary to reconstruct the relationship between settlements, fields and other landscape elements. These geographical relationships are also central to any interpretation of environmental evidence preserved in particular landscape segments. They establish its relationship to the pattern of human activity during successive phases. The term palimpsest (borrowed from the study of parchment documents showing several partly erased and superimposed texts) has been used to describe superimposed phases of landscape activity. This metaphor is especially apt because some aspects of the cultural landscape can be seen as analogous to a written text, in that particular conjunctions of banks, tombs, natural vegetation and other features may be designed to communicate meaning to an observer. Such aspects are likely to have been particularly important to pre-literate societies. The 'Dreamtime' of Australian Aborigines is a tradition making symbolic use of the material world. The landscape is 'charged with awe, every rock, spring and waterhole represents a concrete trace of a sacred drama carried out in mythical times' (Eliade, 1973). Young initiates learn of the deeds of their ancestors from inscriptions on the landscape. Australian legislation originally protected only those sites with physical Aboriginal remains but subsequently this has been extended to cover natural features with significance for Aboriginal communities (Flood, 1989). Similarly, in the United States there is growing awareness of the need to take cognizance of the significance of landscape features for native Indian communities (Laidlaw, 1989). The role of particular landscape facets in the non-material lives of Iron Age communities in north-western Europe is demonstrated by the ritual deposition within peat bogs, in some cases over several centuries, of human sacrifices (Figure 8.4) and quantities of metalwork.

Within a landscape the significance of individual features may relate not so much to their specific

Figure 8.2   A preserved Bronze Age barrow at Beech Pike, Gloucestershire, England, surviving as an island within agricultural land (photo T Darvill)

properties but to their relationship to one another and to the total landscape, both natural and cultural. As Hodder (1990) notes, the placing of Neolithic monuments in the landscape seems to have been designed to represent a monumental intervention in nature. According to Lipe (1984) 'because humans generally modify the landscape in which they live and because they attach names, myths and effective value to features of the territory they inhabit, the landscape of the past may also qualify as a cultural resource'. Unfortunately such concepts remain innovative and rare, and the vast majority of archaeological site preservation is on an individual site basis with the result that potentially significant relationships are all too readily obscured and lost.

A case for integrated conservation

The foregoing discussion has highlighted some distinctions between the conservation of living plant and animal communities and that of both

Figure 8.3   Aerial photograph of Whitesheet Hill, Wiltshire a relict landscape on chalk grassland with archaeological sites of many periods including a Neolithic causewayed camp and later barrow (middle right), a cross-ridge dyke (centre) and an Iron Age hillfort (top left) (photo Cambridge University Collection, Copyright Reserved)

palaeoenvironmental resources and archaeological sites. Such distinctions go against the integrated view of people and nature which lies behind this book, but they do reflect much current administrative practice. The significance of living cultural landscapes as resources for investigation of the continuum of people/environment relationships is not sufficiently acknowledged in the present conservation machinery of many countries. In general, there is a tendency towards

compartmentalization so that the case for conservation is weakened by a lack of coordination, with one interest group accepting compromises which weaken the case of others (Rackham, 1986). In most countries separate government bodies deal with archaeological and natural history conservation and it is noticeable that in recent surveys of Cultural Resource Management (Cleere, 1984, 1989) some contain no reference to the links between these two strands, e.g. the pre-1990 state

Figure 8.4   Tollund Mose, Denmark, the wooden post marks the spot where Tollund Man was found in 1950 (photo M Bell)

of West Germany (Reichstein, 1984) and the USSR (Masson, 1989).

England provides an example of a country in which the conservation agencies are quite separate, despite being funded from the same ministry (Department of the Environment). Sites of natural historical interest are the province of the Nature Conservancy Council which designates Sites of Special Scientific Interest (SSSIs). Some of these contain relict plant and animal communities which owe their present character to the past history of human activity, while others contain often unique records of natural environmental change in the form, for example, of geological sections. Archaeological sites, by contrast, are the

responsibility of English Heritage which manages 'guardianship monuments' in state care, maintains a 'schedule' of protected sites in private, or public, ownership and funds rescue excavations on threatened sites. The schedule mostly concerns individual sites in isolation from their landscape context although there are areas with several sites, such as the uninhabited island of Samson, one of the Isles of Scilly. Relict landscapes receive some protection in a variety of other ways, particularly through local and central government planning mechanisms in which the emphasis of policy has recently shifted to some extent in favour of preservation (Department of the Environment, 1990). Many sites are also protected as a result of ownership by non-governmental organizations, such as the National Trust. The recent designation of Environmentally Sensitive Areas, in which it is proposed to harmonize agricultural practice with conservation objectives, may also help to protect some sites (Fowler, 1987; Darvill, 1987). Many sites are also protected incidentally within nature reserves and SSSIs although archaeological considerations paradoxically form no part of the criteria for designating SSSIs.

The common interests of archaeology and natural history are increasingly being recognized (Lambrick, 1985) and are particularly well illustrated by the case of the Neolithic Sweet Track in Somerset, now protected within an SSSI, into which water has to be pumped in order to maintain both the botanical community and the wooden trackway (Figure 8.5). Environmental Assessment (EA) of the effects of major engineering projects on archaeology, natural history and a range of other factors has recently been introduced in response to a European Economic Community Directive. The EA concept was pioneered in the USA from the late 1960s. It required the assessment of cultural resources affected by federal projects, reservoir construction and the programmes of various agencies such as the Forestry Service. In the USA archaeological sites within National Parks are well preserved in a landscape context but those beyond the park boundaries enjoy much less protection (Kristiansen, 1989).

The world leader in terms of an integrated conservation strategy is clearly Denmark where ancient monuments, wildlife and landscape are protected by a single law: the Conservation of Nature Act (Kristiansen, 1989). That strategy is said to reflect the perceived importance of archaeological sites within the entirely

Figure 8.5   Sweet Track, National Nature Reserve, England into which water is pumped to preserve a buried Neolithic trackway and to maintain the living biological community (photo M Bell)

anthropogenic landscape of Denmark. It is a country with a very high level of public awareness of archaeology, natural history and past environmental conditions. In all, 5 per cent of the country has been designated for conservation purposes and areas have been identified based on their value in terms of biology, cultural history and recreational aspects. Furthermore, archaeological sites are surrounded by a 100 m protected zone in order to help preserve the relationship between monuments and their landscape. This Danish approach reflects an almost unique national appreciation of the relationship between culture and landscape in both the past and the present.

Presentation and experiment

Progressive attitudes to conservation in Denmark have come about because of widespread public interest in both natural history and archaeology. In other countries archaeologists and palaeoenvironmentalists have not yet capitalized fully on growing public concern for nature conservation which could lead to wider interest in the past dimension of present-day landscape types. One way of achieving this is to integrate natural historical and archaeological field presentations in display boards and interpretation centres. That is straightforward where one agency, such as a national park, is concerned, but elsewhere it is often hampered by the involvement of several different public bodies. In relation to archaeological aspects, traditional museum displays, with their emphasis on artefacts, need to be supplemented by presentations concerned with past ecosystems. This has, for example, been achieved with great success at the Søllerød Museum (1985), Denmark, where material from a complex of Mesolithic sites at Vedbaek is shown alongside dioramas depicting various contemporary habitats. Another approach is the physical reconstruction of past sites and landscapes by means of experimental archaeology. Examples of such projects are the creation of an Iron Age village at Lejre, Denmark (Figure 8.6; Hansen, 1982), the Iron Age farmstead at Butser,

England (Reynolds, 1987), a reconstructed Medieval village at Melrand, Brittany (Chalavoux, 1988) or the recreation of the town of Colonial Williamsburg in the USA. More ambitious still are current proposals by Dutch archaeologists to create Archeon, an archaeological theme park which will reconstruct settlements *and environments* throughout the Holocene (Ijzereef, forthcoming). Such projects may have much to offer by increasing the accessibility to a wider public of results from archaeological and palaeoenvironmental studies. It must, at the same time, be acknowledged that reconstructions present fundamental problems, for they may give the impression of presenting one particular model of the past as if it were the only valid interpretation. One answer may be presentation within an experimental framework in which various alternative hypotheses can be evaluated, thus promoting a methodology based on more explicit hypothesis testing (Chapter 2, p. 11). In the past there has often been difficulty in reconciling educational and experimental objectives, particularly because funds for experiment have had to be generated by presentational aspects which accordingly take precedence.

Experimental reconstructions have, none the less, contributed in a very significant way to our understanding of past environments. Experimental clearance and crop growing at Draved Forest, Denmark, though relatively small scale and short term, still underpin many interpretations of the changes consequent upon early woodland clearance (p. 164; Steensberg, 1986). Work at Butser Farm has helped to identify agricultural techniques and crops which might be better adapted than modern varieties to certain soil conditions and may produce better yields without recourse to the application of chemical fertilizers (Reynolds, 1981). Experimental results also impinge on the related management policies of archaeologists and nature conservationists. Botanical study of the British Association experimental earthwork constructed at Overton, Wiltshire in 1960 shows that within 30 years it acquired a classic old grassland flora richer in species, as a result of disturbance by the earthwork, than the surrounding National Nature

Figure 8.6   Reconstructed Iron Age Village at Lejre, Denmark (photo M Bell)

Reserve (Fowler, 1989). Such examples highlight the relevence of a dialogue between past and present and serve to emphasize that much of what is increasingly recognized as worthy of conservation is not a purely natural but a cultural landscape.

9

The Impact of People on Climate

Introduction

In previous chapters, the interactive relationships of people and the environment have been explored against a background of environmental change. A central theme throughout has been the historical scientist's adherence to the uniformitarian principle (Chapter 2), with environmental reconstructions based on geological and biological proxy records relying heavily on contemporary analogues. Our understanding of human environmental perception and response to change is also heavily dependent on the recent ethnohistorical record. Thus the present provides a methodological key to unlock the mysteries of the past. More recently, however, with growing concern about the future of the planet, research in the historical earth sciences has taken on an additional dimension with the search for clues as to future climatic trends and their likely impact on the global environment. Analogues from the past are now being sought to provide climatic scenarios for the future, while historical proxy data

are being widely employed in increasingly sophisticated simulation models which seek to predict the course of climatic change (Houghton *et al.*, 1990). Hence, while the present may well be regarded as the key to the past, it might equally be said that the past is the key to the future. The point, of course, is that environmental scientists and archaeologists are working along a temporal continuum. Consequently, any discussion of the historical interaction between people, climate and landscape would be incomplete without some reference to the future. It is this prospective view, as opposed to the hitherto retrospective analysis, which constitutes the basis for this final chapter.

The greenhouse effect

Although natural forcing factors (solar output variations, volcanic aerosols, etc.) may be responsible for some of the global climatic changes that are apparent in the recent meteorological records (see below), there is now a substantial body of evidence pointing to the increasing influence of human activity on the world's climate. Well over a century ago it was suggested that atmospheric gases could retard heat output and hence raise surface temperatures of the planet, and by the turn of the century a quantitative relationship had been established between atmospheric carbon dioxide and global temperatures (Revelle, 1985). It was not until the 1950s, however, that the full significance of what has now become known as the greenhouse effect, namely the warming of the atmosphere arising from the accumulation of anthropogenically produced gases, began to be fully appreciated (Kellogg, 1987). This issue is now central to the current environmental debate and is high on the political agenda of most industrialized nations. So great is the global concern that pressure is growing for an international law of the atmosphere to control the emission of carbon dioxide and other trace gases (Goreau, 1990).

Atmospheric carbon dioxide

The major element contributing to the greenhouse effect is carbon dioxide (CO_2) which absorbs thermal longwave infrared radiation. This leads to atmospheric warming and hence an increase in global surface temperatures. Measurements of atmospheric concentrations of CO_2 in polar ice cores show that the mean concentration during the Holocene was around 270 ppmv (parts per million by volume), and comparable values have been observed for the mid-eighteenth century, i.e. the immediate pre-industrial period (Neftel *et al.*, 1982). Thereafter, the ice core data reflect a progressive increase in CO_2 concentrations with values of 315–320 ppmv being recorded for the 1950s (Neftel *et al.*, 1985). Continuous monitoring of atmospheric CO_2 began in 1958 and these observations show an increase from around 315 to *c.* 343 ppmv in the period up to 1984 (Bolin *et al.*, 1986). The estimate for 1990 was 353 ppmv, and atmospheric CO_2 is currently rising at about 1.8 ppmv each year due to anthropogenic emissions (Watson *et al.*, 1990). These data indicate an increase in atmospheric CO_2 of 20–30 per cent in less than 200 years, attributable principally to an increase in fossil fuel combustion and, to a lesser extent perhaps, to extensive clearance of forests and the conversion of biomass to CO_2 (Kellogg, 1987).

Past and predicted future changes in CO_2 concentration are shown in Figure 9.1. The future trend is uncertain, however, because awareness of the effects of increasing atmospheric CO_2 may influence the amounts of CO_2 released into the atmosphere from industrial and related sources (Keepin *et al.*, 1986). Analysis by the Intergovernmental Panel on Climate Change (IPCC) suggests four possible scenarios for future CO_2 emissions (Houghton *et al.*, 1990). The first of these assumes that few or no steps are taken to limit greenhouse gas emissions (the 'business as usual' scenario) in which case atmospheric CO_2 concentrations will have doubled by around 2070 (Figure 9.1). The other three scenarios assume that progressively increasing levels of controls will reduce the growth of emissions (scenarios B, C and D). Under scenario D, for example, stringent controls in industrialized countries, involving a major shift to renewable and nuclear forms of energy production, combined with a moderate growth of emissions in developing countries, could

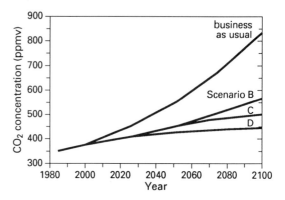

Figure 9.1   Atmospheric concentrations of CO_2 resulting from the four IPCC Emissions Scenarios (after Houghton et al., *1990)*

effectively stabilize atmospheric CO_2 concentrations.

Other atmospheric trace gases

Carbon dioxide is not the only constituent of the atmosphere that is important in determining global heat budget, however, for it is now recognized that other less abundant trace gases are also increasing and are likely to have an effect on climate comparable with that of CO_2 (Dickinson and Cicerone, 1986). These gases which, like CO_2, act to block the escape from the atmosphere of thermal infrared radiation, include methane (CH_4), nitrous oxide (N_2O), tropospheric ozone (O_3) and the chlorofluorocarbons or 'Freons' (CFCs) – CFC-11 ($CFCl_3$) and CFC-12 (CF_2Cl_2).

Methane

Of the trace gases listed above, methane is perhaps the most significant, its estimated future contribution to global climatic warming being about 25 per cent of that due to CO_2 (Raynaud *et al.*, 1988). CH_4 is produced by microbial activities during the mineralization of carbon under anaerobic conditions, e.g. in waterlogged soils or in the intestines of animals. It is also released by anthropogenic activities such as exploitation of natural gas, biomass burning and coal mining (Bolle *et al.*, 1986). Ice core studies show that atmospheric CH_4 concentrations have risen steadily since the middle of the eighteenth century (Figure 9.2) from

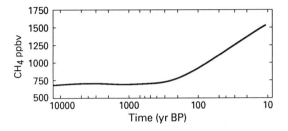

Figure 9.2   Holocene variations in methane (CH₄) based on measurements of air samples from ice cores in Greenland and Antarctica. The timescale is logarithmic to emphasize the marked increase in atmospheric CH₄ concentrations around 200 years ago (after Bolle et al., 1986)

a pre-industrial level of about 650 ppbv (parts per billion volume) to a value of 1650 ppbv in the 1980s (Stauffer *et al.*, 1988). The current estimate for CH₄ concentratons is around 1720 ppbv (Watson *et al.*, 1990). Since 1951 an increase of over 500 ppbv has been observed (Rinsland *et al.*, 1985), which suggests that atmospheric CH₄ concentrations are presently increasing at a rate of 1–2 per cent yr^{-1}.

Nitrous oxide

Although N₂O is also increasing in concentration (Figure 9.3), the rate is considerably less than that of CH₄ (Jones and Henderson-Sellars, 1990). The flux of N₂O into the atmosphere is due primarily to microbial processes in soil and water and is part of the nitrogen cycle. Principal sources of atmospheric N₂O include fossil fuel burning, biomass burning, and mineral fertilizers. Estimates of N₂O emissions

Figure 9.3   Trend of estimated nitrous oxide (N₂O) emission rates between 1880 and 1980 (after Bolle et al., 1986) 1Tg = 1 gm × 10¹²

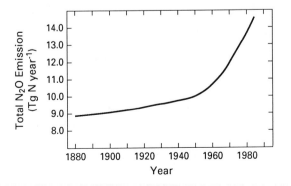

suggest an increase from around 280 ppbv during anthropogenically undisturbed conditions to 310 ppbv in 1990 (Watson *et al.*, 1990). The emission rate has increased dramatically from around 0.1 per cent yr^{-1} at the turn of the century to 1.3 per cent yr^{-1} over the past 10 years (Bolle *et al.*, 1986). By the year 2000, atmospheric concentrations of N₂O of 320–330 ppbv are likely, an increase of around 15 per cent over pre-industrial levels (Dickinson and Cicerone, 1986).

Ozone

Despite the problems of measurement and a relative paucity of data, there is evidence to suggest that concentrations of tropospheric O₃ are increasing as a result of photochemical processes, principally related to the enhanced emission of CH₄, carbon monoxide (CO) and nitrogen oxides (NOₓ) (Bolle *et al.*, 1986). Over the last 100 years or so, surface ozone levels over parts of Europe and North America appear to have risen by a factor of 2–3 on average (Watson *et al.*, 1990), while between the late, 1960s and the early 1980s an increase in near surface ozone of 1–3 per cent yr^{-1} and in middle tropospheric ozone of 1–2 per cent yr^{-1} has been detected (Logan, 1985). Tropospheric O₃ concentrations in the Northern Hemisphere around 50 per cent higher than in the Southern Hemisphere have also been observed (Ramanathan *et al.*, 1985). It must be emphasized that these increases are recorded only in the lower atmosphere, for in some parts of the world, a significant decrease in ozone content has been recorded in the overlying stratosphere. As will be shown below, this decline in ozone levels in the middle atmosphere, which appears to be due largely to anthropogenic activity, poses a different set of problems to life on earth.

Chlorofluorocarbons

First introduced some 60 years ago, manufactured chlorofluorocarbons (CFCs) have a variety of uses including solvents, refrigerator coolant fluids and propellants for aerosol sprays. As the name implies, they are complex compounds of chlorine, fluorine and carbon, and while they are ideal industrial chemicals in that they are highly stable, unreactive

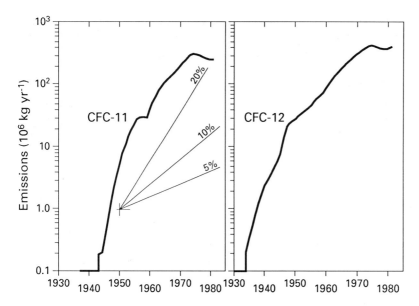

Figure 9.4   Emission of CFC-11 (CFCl₃) and CFC-12 (CF₂Cl₂) over the past 40 years. The curves in the left-hand figure correspond to a 5%, 10% and 20% annual increase (after Bolle et al., 1986)

and non-toxic, they do not degrade readily in the atmosphere. Hence, by 1980 atmospheric concentrations of these compounds were increasing by around 6 per cent yr^{-1} (Figure 9.4), the highest values being recorded for the northern middle latitudes where most emissions occur (Bolle et al., 1986). In the period, 1975–85, increases of over 100 per cent were observed in both CFC-11 and CFC-12 (Jones and Henderson-Sellars, 1990). In the period 1980–2000, worldwide emissions for CFC-11 and CFC-12 are projected to increase by 2.8–5.1 per cent yr^{-1} and 2.5–3.5 per cent yr^{-1} respectively. Unless steps are taken to restrict global output of chlorofluorocarbons, atmospheric concentrations of both CFC-11 and CFC-12 could still be increasing at a rate of 4.5–5.0 per cent yr^{-1} by the year 2025 (Dickinson and Cicerone, 1986).

Consequences of the greenhouse effect

Global temperature changes

There is a considerable body of empirical evidence for recent climatic warming. Global surface air temperature data for the present century (Figure 9.5) show an initial episode of maximum warmth around 1940, but a significant increase in temperature in the, 1980s (Jones et al., 1986a, b). Indeed, the three warmest years of the present century occurred after 1980 with 1987 being the warmest. In the Southern Hemisphere, the recent warming trend is even more apparent with seven of the eight warmest years this century occurring during the 1980s. This global increase in surface warmth is reflected in a similar rise in temperatures in the lower troposphere (Jones et al., 1988). A general rise in world temperature is also evident in sea-surface temperature measurements derived from satellite data, for these show that in the period

Figure 9.5   Global mean surface air temperatures 1901–1987 (after Jones et al., 1988)

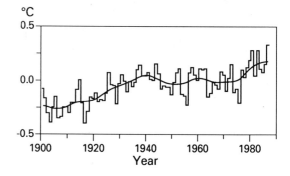

1982–88, the global oceans were undergoing a gradual but significant warming at a rate of *c.* 0.1 °C yr^{-1} (Strong, 1989).

There is no doubt that this apparent warming trend is consistent with model predictions for greenhouse-gas-induced climatic change. Detecting the greenhouse effect in empirical climatic data is another matter, however, for the records display a variability that *could* be explained entirely in terms of natural fluctuations within the climatic system (Mitchell, 1989). Hence, although a cause-and-effect relationship *appears* to exist, the observational data are not sufficiently consistent to *confirm* the hypothesis that trace-gas forcing has been the sole cause of recent global warming (Bolin *et al.*, 1986; Wigley and Barnett, 1990). Indeed, it has been suggested in some quarters that the magnitude of future global warming has been overestimated (Newell and Doplick, 1979; Idso, 1980), and that, under certain conditions, increased atmospheric CO_2 could even lead to a decrease in global temperatures (Idso, 1984)! Although this is undoubtedly now a minority view, it does reflect the uncertainty that still exists in some quarters of the scientific community, principally because of the present inability to detect unequivocally the greenhouse signal in empirical climatic data.

Future climatic trends are derived from Atmospheric General Circulation Models (AGCMs) which simulate global climatic patterns and hence enable predictions to be made about the influence of atmospheric trace gases (e.g. Cubasch and Cess, 1990). Under the IPCC 'business as usual' scenario (see above), a rate of increase of global mean temperature during the next century of about 0.3 °C per decade is predicted, which could result in a rise in temperature of about 1 °C above the present value by 2025 and 3 °C before the end of the next century (Houghton *et al.*, 1990). Rates of increase in global mean temperature of about 0.2 °C per decade, just above 0.1 °C per decade and about 0.1 °C per decade are predicted under scenarios B, C and D respectively (Figure 9.6). The current 'best estimate' of the likely rise in global temperatures resulting from a doubling of CO_2 is around 2.5 °C.

Global temperature increase is likely to be most

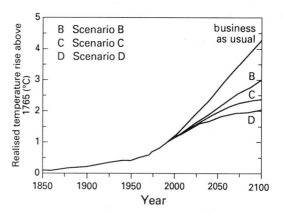

Figure 9.6   Estimates of the increase in global mean temperature from 1850–1990 resulting from increases in greenhouse gases, and predictions of the rise between 1990 and 2100 under IPCC Scenarios B, C and D, with the business as usual case for comparison (after Houghton et al., 1990)

pronounced in the high latitudes, with winter warming in the polar regions by the middle of the next century projected to be around 2 to 2.4 times above the global average, accompanied by a summer warming 0.5–0.7 times that average (Maxwell and Barrie, 1989). Models predict that by 2030, annual temperatures will have risen by 1–2 °C in southern Asia and by 1–3 °C in the Sahel. In southern Europe and central North America, summer temperatures may have risen by 2–3 °C, with winter temperatures in the latter region increasing by up to 4 °C (Mitchell *et al.*, 1990).

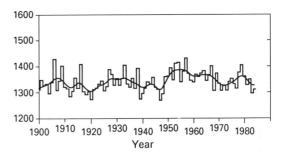

Figure 9.7   Annual precipitation (measured in mm) for the Northern Hemisphere landmasses 1900–1984 (after Jones, 1988)

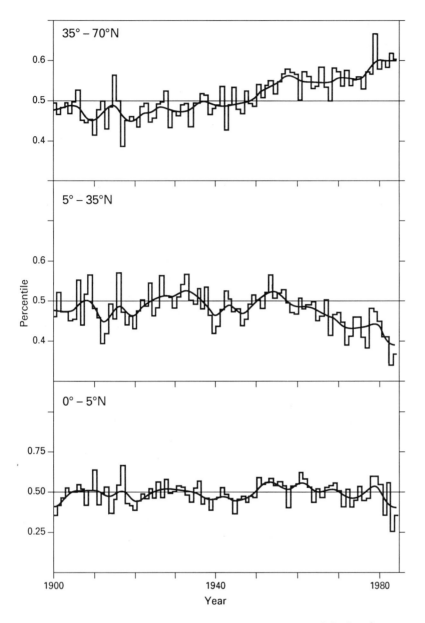

Figure 9.8   Precipitation indices for zones 35°–70°N, 5°–35°N and 0°–5°N. Precipitation data from over 1400 sites have been converted to probability estimates and then averaged to produce these indices of Northern Hemisphere regional precipitation variability since 1900 (after Bradley et al., 1987)

Global precipitation changes

Annual precipitation for the Northern Hemisphere land masses since 1900 is shown in Figure 9.7. These data show how precipitation values have fluctuated in a quasi-cyclical manner, with distinct maxima during the 1930s and 1950s (Jones, 1988). However, this general curve masks distinctive regional trends. When Northern Hemisphere precipitation data are considered on a latitudinal basis, for example, precipitation in the higher latitude zone (35°–70°N) is seen to have increased markedly over the last 30–40 years, whereas in the lower latitudes (5°–35°N) a pronounced fall in precipitation values occurred from around 1950 onwards (Figure 9.8).

The recent trend in the north equatorial region (0°–5°N) is also downward. Although these trends in zonally averaged precipitation over the past 30–40 years may simply reflect large-scale natural climatic variability, they are consistent with changes predicted by AGCM models associated with a doubling of atmospheric CO_2 levels (Bradley *et al.*, 1987). Changes in the global pattern of precipitation as a consequence of the build-up of atmospheric greenhouse gases, therefore, remains a real possibility.

Sea-level changes

Empirical data show that global sea level has risen by 10–15 cm over the course of the last 100 years (Robin, 1986; Titus, 1987), and a significant correlation is apparent between this trend and that for global warming (Figure 9.9). Estimates suggest that of the observed rise in sea level since 1880, thermal expansion of ocean waters could have contributed 2–5 cm, melting of small glaciers 2–7 cm, long-term isostatic adjustments a further 2 cm, with the remainder coming from the melting of the polar ice sheets (Wigley and Raper, 1987). The IPCC estimates that global sea level will have risen by 18 cm by the year 2030, 44 cm by 2070 and 66 cm by 2100, that is to say an acceleration of 3–6 times the current rate of sea-level rise and a rise of about 6 cm per decade over the course of

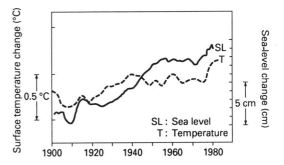

Figure 9.9   Comparison of the global mean sea-level trends during the twentieth century with global mean surface temperatures (after Bolin et al., *1986)*

the next century (Houghton *et al.*, 1990). The considerable range within these estimates (Figure 9.10), however, reflects the degree of uncertainty within the scientific community not only with regard the future course of global warming, but also over the terrestrial, hydrological and especially glaciological response to increased surface temperatures. For example, while there has been speculation that greenhouse warming could lead to melting of the polar ice caps and a rise in sea level, increased precipitation in the high latitude regions (Figure 9.8) *could* counter this effect, with the Antarctic ice sheet in particular possibly gaining in

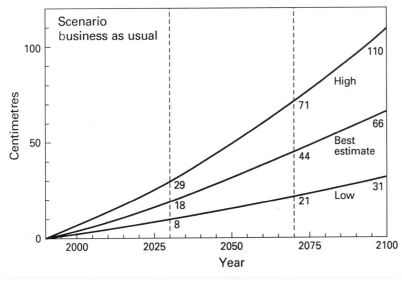

Figure 9.10   Predicted rise in global sea level 1990–2100 under the IPCC 'business as usual' scenario (after Warrick & Oerlemans, 1990)

mass and thereby contributing to a lowering of global sea level (Oerlemans, 1982). Similarly, fears have been expressed over the stability of the marine-based west Antarctic ice sheet, for it has been suggested that the rapid destruction of this ice mass as a result of CO_2-induced warming during the course of the next century could raise global sea levels by over 6 metres (Mercer, 1978). However, current glaciological data appear to indicate that such an ice-sheet collapse is not imminent (Robin, 1986) and, moreover, even if such a process was initiated, it would take several centuries for the ice sheet to disintegrate completely (Thomas et al., 1979; Hughes, 1983). Hence, it seems likely that the principal mechanisms by which greenhouse warming will raise global sea level during the course of the next century are the melting of low latitude glaciers (with the Greenland and Antarctic ice sheets probably playing minor roles), and the thermal expansion of ocean water (Warrick and Oerlemans, 1990).

The effects of a gradual rise in sea level will be accelerated coastal erosion and shoreline retreat, salt intrusion into estuaries and freshwater aquifers, flooding of new areas and increased storm damage, and the progressive dislocation of human activity in coastal regions (Barth and Titus, 1984; Viles, 1989). Areas particularly at risk are the thousands of hectares of coastal wetlands of western Europe and North America where marsh development has kept pace with the course of Holocene sea-level changes. Crustal subsidence around the North Sea coasts of southern England and along the Atlantic seaboard of the United States will tend to increase the rate of greenhouse-induced sea-level rise in these and other similarly affected localities (Kana et al., 1984).

Hydrological changes

The impact of global warming on hydrological systems is likely to be considerable, although on present data the effects are difficult to predict. Experimental results suggest that an increase in CO_2 leads to a reduction in the rate of transpiration from plants and an increase in their water use efficiency. Hence, decreased rates of transpiration resulting from higher atmospheric CO_2 levels could increase surface runoff (Wigley and Jones, 1985). This would tend to offset the effects of CO_2-induced reductions in precipitation in the low latitude areas of the world, or enhance the effects of precipitation increases in the mid- and high-latitude regions. Changes in river regimes and in flood frequency would have a profound effect not only on river systems, but also on ecosystems adjoining the rivers, and on a range of human activities that depend on reliable quantity and quality of water (Warrick et al., 1986a). In addition, numerical simulation models predict major changes in soil water balance as global temperatures rise. In the mid- and high latitudes, for example, increased summer dryness could lead to a general reduction in soil moisture content (Manabe et al., 1981), and although the magnitude of summer drying is likely to vary with physical attributes of the soil (Mitchell and Warrilow, 1987), the consequences for agriculture of these possible changes in soil water balance are potentially serious.

Effects on agriculture

The IPCC has concluded that while food production at the global level can, in the face of estimated changes of climate, be maintained at essentially the same level as would have occurred without climatic change, the costs of achieving this are unclear and could be very large (Parry, 1990). The effects are likely to be uneven, however, and certain regions could actually see increases in food production. Evidence from laboratory experiments on cultivated plants, for example, suggests that a doubling of atmospheric CO_2 concentration could cause a 0–10 per cent increase in growth and yield of such crops as maize, sorghum and sugar cane, and a 10–50 per cent increase for crops such as wheat, soya bean and rice (Warrick et al., 1986b). However, when the climatic consequences of increased atmospheric CO_2 are taken into acount, the likely impact on agriculture appears more variable. For example, in the mid-latitude regions of North America and western Europe, crop impact analyses suggest that a warming of 2 °C could reduce average yields of cereals by as much as 17 per cent (Warrick *et al.*, 1986b). While these reductions in grain yield might be partly offset by changes in precipitation, decreases in precipitation would accentuate them, particularly in the 'humid' grain regions of western Europe. Hence potential shifts in the boundaries of cereal regions of the order of several hundred kilometres per °C change might be anticipated in the mid- and high latitudes over the course of the next century (Warrick, 1988). In the mid- and low-latitude regions of the world, projected patterns of climatic change suggest that reduced crop water availability could lead to a marked diminution of the agricultural resource base over large areas of the semi-arid tropics and subtropics (such as western Arabia, the Mahgreb, the Horn of Africa, southern Africa and eastern Brazil). Parts of the humid tropical and equatorial regions including, possibly, South East Asia and Central America, are likely to experience similar effects (Parry, 1990).

Effects on forest ecosystems

On current evidence it is difficult to predict the impact of enhanced atmospheric CO_2 levels on the world's forests, although simulation modelling suggests that the projected temperature increases arising from the greenhouse effect are likely to produce substantial intermediate and long-term responses in the composition, size and location of the forests of the world (Shugart *et al.*, 1986). Modelling results indicate that, in higher latitudes, the increase in temperature associated with a doubling of atmospheric CO_2 could reduce both the boreal forest and tundra by over one-third, and cause a substantial poleward shift of the treeline (Boer *et al.*, 1990). For low-latitude areas, modelling predicts a decrease in area of subtropical forest and an expansion of subtropical thorn woodland and deserts (Emanuel *et al.*, 1985). Overall the models suggest that, in the high-latitude regions, higher temperatures resulting from the increased concentration of greenhouse gases are likely to exert the dominant control on natural vegetation patterns, whereas, in the low latitudes, the forests will be more sensitive to future

variations in precipitation rather than temperature (Shugart *et al.*, 1986).

The ozone layer

During the 1970s it was established that not only did the increase in atmospheric CFC concentrations contribute to the greenhouse effect (Ramanathan, 1975), but also continued emission of CFCs posed a possible threat to the ozone layer (Molina and Rowland, 1974). Stable CFC compounds rise through the troposphere into the stratosphere where they are exposed to the intense UV radiation that is absorbed by O_3 at lower altitudes. Exposure to radiation leads to a breakdown of the normally stable CFCs into more reactive forms such as chlorine (Cl), a chemical which is known from laboratory studies to destroy ozone (Stolarski, 1988). In 1985, a significant decline (by 40–50 per cent) in springtime atmospheric O_3 levels was reported over Antarctica (Farman *et al.*, 1985). Subsequent measurements of atmospheric trace gases over Antarctica showed that periods of O_3 depletion coincided with marked increases in concentration of halogenated hydrocarbons, especially Cl (Farmer *et al.*, 1987). More recently, a similar reduction in atmospheric O_3 has been detected in the winter months over the Arctic (Hofmann *et al.*, 1989; Proffitt *et al.*, 1990). Although it is possible that such levels of O_3 depletion could reflect natural atmospheric variations, there is a growing body of evidence to suggest that the seasonal thinning of the ozone layer is directly attributable to man-made chlorine pollution.

Progressive depletion of stratospheric ozone levels could have far-reaching implications for life on this planet. Although O_3 constitutes less than 1 ppm of the atmospheric gases, it aborbs much of the potentially harmful incoming UV radiation and prevents it from reaching the earth's surface. An increase in UV radiation could lead to a higher incidence of skin cancer, cataracts and immune deficiencies, as well as affecting crops and aquatic ecosystems (Stolarski, 1988). Particularly vulnerable to increased fluxes of UV radiation are polar regions, where components of both terrestrial and marine ecosystems are very susceptible to increased UV exposure. Indeed, preliminary experiments in Antarctica are already beginning to show increased UV stress levels in marine phytoplankton (Voytek, 1990). The recent ratification of the Montreal Protocol, an agreement signed by more than 60 nations to protect the global atmosphere from stratospheric ozone depletion, is an indication of the seriousness with which the international community views the problem (Koehler and Hajost, 1990).

Acid rain

The term 'acid rain' has been used to describe a form of atmospheric pollution which is currently affecting many of the industrialized regions of the world. In a sense the term is misleading for the pollution occurs in the form of snow, hail, gas clouds, fog, mist and dry dust; moreover, most rainfall is naturally acidic (i.e. pH < 7). These inputs are, therefore, more accurately referred to as acid deposition (Elsworth, 1984). The basic elements involved in acid deposition are sulphur dioxide (SO_2) and the two nitrogen oxides (NO_x), nitric oxide (NO) and nitrogen dioxide (NO_2). These chemical pollutants are released into the atmosphere through the burning of fossil fuels and petroleum products, from the smelting of metallic ores, from petrochemical and related industries and from vehicle exhausts. Ice core data from Greenland show that, prior to 1900, atmospheric sulphur and nitrate levels were generally low, but since the turn of the century concentrations of nitrates have doubled and sulphates have trebled (Wolff and Peel, 1985). This dramatic increase in atmospheric acidity is reflected in twentieth-century precipitation measurements (Brimblecombe and Stedman, 1982). It is also apparent in the changing composition of diatom assemblages (Chapter 2) which show increasing acidity of lake waters in upland areas of western Europe over the course of the past two centuries (Battarbee, 1984). Within the

Figure 9.11   Recent trends in atmospheric pollution: (a) Total emission of SO_2 over the British Isles since 1900 (after Park, 1987) (b) Total emissions of SO_2 and NO_x over the United States since 1870 (after Galloway, 1989)

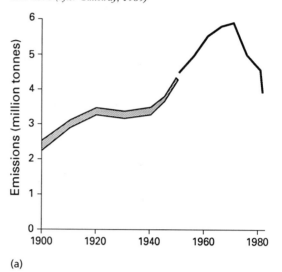

(a)

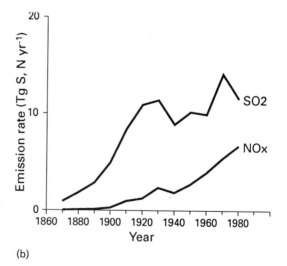

(b)

last ten years, emissions of SO_2 and NO_x from the major industrialized nations have fallen (e.g. Heden et al., 1987) due to increasing pollution control, reductions in fossil fuel use and a lower energy demand during economic recession. While it seems unlikely that the recent trend in Europe and North America (Figure 9.11) will be reversed over the coming decades, future emissions of SO_2 and NO_x

from the newly industrialized nations of the Third World could considerably exceed current emissions to the global atmosphere (Galloway, 1989). Hence, acid deposition seems likely to continue to pose a major threat to the global environment well into the next century (Rodhe, 1989).

The effects of acid deposition are considerable. Increased acidification of lake ecosystems has resulted in a decline in fish stocks in lakes and rivers throughout north-west Europe and North America (Muniz, 1984; Magnuson *et al.*, 1984). This appears to be due partly to the increased acidity of lake waters, and partly to higher concentrations of dissolved metals (such as aluminium) which are often found in waters of low pH. These metals are not only toxic to many forms of aquatic life, but may also constitute a threat to human health in more remote areas where water

treatment is rudimentary (Moghissi, 1986). Enhanced acidification of soils leads to nutrient depletion, which is exerting increasing environmental stress on forested regions of North America and Europe (Cowling, 1989). Data from southern Sweden, for example, indicate that over the past 35 years, soil pH has fallen by up to 1.5 units and a long-term decline in tree growth seems inevitable (Falkengren-Grerup, 1989). Building damage, crop damage and human health problems are further consequences of acid deposition (Park, 1987). In the atmosphere, higher concentrations of SO_2 and NO_x will affect the global radiation balance, influence atmospheric chemical reactions, and contribute to the scattering of solar radiation as the aerosol load of the troposphere increases (Likens, 1989).

10

Postscript

Realization of the extent to which human activity has affected atmospheric gas concentrations, and the far-reaching implications that this is likely to have in terms of global warming (Chapter 9), has placed Late Quaternary environmental change firmly on the political agenda throughout the world. Although these atmospheric changes have taken place largely within the last 200 years of intensive industrialization, a proper evaluation of their significance can only be made if they are set in the context of natural climatic rhythms, many of which operate over much longer timescales (Chapter 3). Other environmental changes, such as deforestation, erosion and the extinction of biota, occurred as a result of human activities, in many cases with histories extending back many thousands of years before the present. Contemporary interpretations of those histories both affects, and is affected by, our prognosis for the future.

Simmons (1989) has identified three very different trajectories for future people/environment relationships: (1) a high technology path with sustained economic growth; (2) an environmentalist future based on conservation which aims to stabilize population growth and consumption; and (3) a decoupled future in which the natural world is, as far as possible, separated and insulated from the effects of human activity by living, for example, under domes and recycling waste. Which, if any, of these scenarios we favour depends on how much faith we place in human adaptability and technological solutions to change induced by both natural and anthropogenic processes. The remarkable adaptive capacity of people is undeniable, and is reflected in the range of responses to environmental change on many different timescales (Chapter 5). This has led some to conclude that environmental problems can be solved by human ingenuity and an ever-increasing mastery over nature. As Evans (1975) has observed: 'By attempting to maintain the environmental *status quo*, are we not denying ourselves and our progeny the opportunity and the ability to exploit challenging new environments both created by our own individual and agricultural needs and by natural climatic shifts?' However, in the present over-populated world, maintenance of the *status quo* may simply not be an option; change and adaptation are inevitable. The question is to what degree?

Consideration of the extent of anthropogenically induced environmental change during the course of the Holocene (Chapters 6 and 7) has prompted many scientists to adopt a conservationist perspective, and to view past and future people/environment relationships in pessimistic terms. Zohary (1983), for example, has noted the longstanding uneasy relationship between human communities and landscape, observing that 'from his very beginning man was an enemy of the rest of nature and a crucial obstacle in the more or less balanced interactions between populations with the biosphere'. Whether the spiral of deleterious impact implied by Zohary is inevitable, however, is a moot point, although there is evidence in both the archaeological and anthropological record to lend weight to his argument. For example, although Australian Aborigines and American Indian communities possess a deep respect for nature, a concept of balance that has been described as a kind of mutualism (Ingold, 1990), these groups have changed their environments by burning to the extent that the stability of many habitats was reduced. This, together with evidence of other major environmental impacts, such as widespread forest clearance (on a scale comparable with that currently taking place in the tropical rainforests) and the extinction of the Late Pleistocene megafauna, challenges the image of 'the noble savage' in harmony with nature. Even where human activity has sought to enhance the environment by the creation of more fertile man-made soils (p. 180), by the establishment of irrigation projects in areas of low rainfall, or by the creation of gardens, there have sometimes been undesirable ecological repercussions. Heathland expansion in The Netherlands and north Germany was brought about by the removal of turves for the creation of 'man-made soils', upland peat formation may have followed the removal of the natural forest cover, and the first attempts at intensive cultivation may have led to soil deterioration and erosion (Chapter 7), while many arid areas have suffered salinization as a consequence of early irrigation. It is yet to be established whether human nature allows people to enjoy an absolutely balanced, that is to say, non-destructive relationship with their environment. Future experiments in ways of living may ultimately address this question, but unless these experiments are to be complete leaps in the dark, they will need to be informed by the past dimension, i.e. the palaeoenvironmental and archaeological record.

Careful scrutiny of that record, however, will show that some human communities have experienced a more balanced environmental relationship and have been more stable than others. Some have evolved over a long period in harmony both with natural (i.e. climatic) and cultural changes. Palaeoenvironmental investigations may help to identify those forms of land use that have proved more damaging in the past, just as others may be identified which, although they may have fallen into disuse, should perhaps be considered for revival. One particularly compelling reason for the conservation of past landscapes and their associated human activities (Chapter 8) is that they represent field laboratories for the investigation of past environmental problems, the results of which may have implications for both the present and the future.

It is not being claimed that past perspectives alone will provide a panacea for our contemporary environmental difficulties. The levels of uncertainty relating to future climatic trends discussed in the previous chapter, coupled with the ramifications of chaos theory (Chapter 2), make prediction hazardous. Moreover, constantly changing social parameters serve to complicate further the nature of human response. Despite the many insights that the uniformitarian perspective has brought to historical reconstructions, there can of course be no *absolutely* precise analogues in the past for the future. Where the palaeoenvironmental, archaeological and ethnohistorical perspectives are important, however, are in the *contributions* that they make to interpretations and predictions, by broadening our awareness of possible alternative scenarios and thereby helping to identify those ways of life that are most likely to be relatively stable over protracted timescales.

That the remarkable human capacity for change and adaptation will be tested to the limits in the coming decades and centuries appears undeniable. It is not clear at present, however, the precise direction that our adaptive response will take. What is equally uncertain is the extent to which we will make use of the palaeoenvironmental record to inform that response.

References

Aaby, B (1976) Cyclic climatic variations over the past 5500 years reflected in raised bogs. *Nature*, **263**, 281–284.

Aaby, B (1986) Palaeoecological studies of mires. In *Handbook of Holocene Palaeoecology and Palaeohydrology* (edited by B E Berglund). John Wiley, Chichester and New York, 145–164.

Aaris-Sørensen, K (1980) Depauperation of the mammalian fauna of the island of Zealand during the Atlantic period. *Videnskabelige Meddelelser Dansk Naturh Forem*, **142**, 131–138.

Abbink, A A (1986) Structured allocation and cultural strategies. In *Gedacht over Assendelft Working Paper 6* (edited by R W Brandt, S E van der Leeuw and M J A N Kooijman). University of Amsterdam, Amsterdam, 23–32.

Aharon, P (1984) Implications of the coral-reef record from New Guinea concerning the astronomical theory of climatic change. In *Milankovitch and Climate* (edited by A Berger, J Imbrie, J Hays, G Kukla and B Saltzman). Reidel, Dordrecht, 379–390.

Aikens, C M (1984) Environmental archaeology in the western United States. In *Late Quaternary Environments of the United States* Volume 2 (edited by H E Wright Jr). Longman, London, 239–251.

Aitken, M J, Michael, H N, Betancourt, P P and Warren, P M (1988) The Thera eruption: continuing discussion of the dating. *Archaeometry*, **30**, 165–182.

Albritton, C C (1975) *Philosophy of Geohistory 1785–1970*. Hutchinson and Ross, Dowden.

Allen, J R L (1987) Late Flandrian shoreline oscillations in the Severn Estuary: The Rumney Formation at its typesite (Cardiff area). *Philosophical Transactions of the Royal Society, London*, **B315**, 157–174.

Allen, J R L and Fulford, M G (1986) The Wentlooge Level: a Romano-British saltmarsh reclamation in south-east Wales. *Britannia*, **17**, 91–117.

Allen, J R L and Fulford, M G (1987) Romano-British settlement and industry on the wetlands of the Severn Estuary. *Antiquaries Journal*, **67**, 237–289.

Allen, J R L and Rae, J E (1987) Late Flandrian shoreline oscillations in the Severn Estuary: a geomorphological and stratigraphical reconnaissance. *Philosophical Transactions of the Royal Society of London*, **B315**, 185–230.

Allen, M J (1988) Archaeological aspects of colluviation in south-east England. In *Man-made Soils* (edited by W Groenman-van Waateringe and M Robinson). British Archaeological Reports, Oxford, IS 410, 67–92.

Allison, T D, Moeller, R E and Davis, M B (1986) Pollen in laminated sediments provides evidence for a mid-Holocene pathogen outbreak. *Ecology*, **67**, 1101–1105.

Alvarez, L W, Alvarez, A, Asaro, F and Michel, H V (1980) Extraterrestrial cause for the Cretaceous–Tertiary extinction. *Science*, **208**, 1095–1108.

Ammann, B (1988) Palynological evidence of prehistoric anthropogenic forest changes on the Swiss Plateau. In *The Cultural Landscape – Past, Present and Future* (edited by H H Birks, H J B Birks, P E Kaland and D Moe). Cambridge University Press, Cambridge, 289–299.

Ammann, B, Chaix, L, Eicher, U, Elias, S A, Gaillard, M-J, Hofmann W, Sigenthaler U, Tobolski, K and Wilkinson, B (1984) Flora, fauna and stable isotopes in Late-Würm deposits at Lobigensee (Swiss Plateau). In *Climatic Changes on a Yearly to Millennial Basis* (edited by N-A Mörner and W Karlén). Reidel, Dordrecht, 69–74.

Andel van, T H, Zangger, E and Demitrack, A (1990) Landuse and soil erosion in Prehistoric Greece. *Journal of Field Archaeology*, **17**, 379–396.

Andersen, B G (1981) Late Weichselian ice sheets in Eurasia and Greenland. In *The Last Great Ice Sheets* (edited by G H Denton and T G Hughes). John Wiley, Chichester and New York, 3–65.

Andersen, S T, Aaby, B and Odgaard, B V (1983) Current studies in vegetational History at the Geological Survey of Denmark. *Journal of Danish Archaeology*, **2**, 184–196.

Andersen, S Th (1986) Palaeoecological studies of terrestrial soils. In *Handbook of Holocene Palaeoecology and Palaeohydrology* (edited by B E Berglund). John

Wiley, Chichester and New York, 165–177.

Anderson, A J (1988) Coastal subsidence economies in prehistoric southern New Zealand. In *The Archaeology of Prehistoric Coastlines* (edited by G Bailey and J Parkington). Cambridge University Press, Cambridge, 93–101.

Anderson, A J (1989) *Prodigious Birds: Moas and moa-hunting in prehistoric New Zealand*. Cambridge University Press, Cambridge.

Anderson, H A, Berrow, M L, Farmer, V C, Hepburn, A, Russel, J D and Walker, A D (1982) A reassessment of podzol formation processes. *Journal of Soil Science*, **33**, 125–136.

Anderson, I W (1974) The chestnut pollen decline as a time horizon in lake sediments in eastern North America. *Canadian Journal of Earth Sciences*, **11**, 678–685.

Andrews, J T (1979) The present ice age: Cenozoic. In *The Winters of the World* (edited by B S John). David and Charles, London, 173–218.

Andrews, J T (1982) On the reconstruction of Pleistocene ice sheets: a review. *Quaternary Science Reviews*, **1**, 1–30.

Andrews, J T (1987) The Late Wisconsin glaciation and deglaciation of the Laurentide Ice Sheet. In *The Geology of North America, Volume K-3. North America and adjacent oceans during the last deglaciation* (edited by W F Ruddiman and H E Wright). Geological Society of America, Boulder, Colorado, 13–37.

Andrews, J T and Barry, R G (1978) Glacial inception and disintegration during the last glaciation. *Annual Reviews of Earth and Planetary Science*, **6**, 205–228.

Angell, J K and Korshover, J (1976) Global analysis of recent total ozone fluctuations. *Monthly Weather Review*, **104**, 63–75.

Arnold, B (1990) The past as propaganda: totalitarian archaeology in Nazi Germany. *Antiquity*, **64**, 464–478.

Ashbee, P, Bell, M and Proudfoot, E (1989) *Wilsford Shaft: Excavations 1960–62*. English Heritage Archaeological Report 11, London.

Atkinson, T C, Lawson, T J, Smart, P L, Harmon, R S and Hess, J W (1986) New data on speleothem deposition and palaeoclimate in Britain over the last forty thousand years. *Journal of Quaternary Science*, **1**, 67–72.

Atkinson. T C, Briffa. K R and Coope. G R (1987) Seasonal temperatures in Britain during the past 22,000 years, reconstructed using beetle remains. *Nature*, **325**, 587–592.

Austad, I (1988) Tree pollarding in western Norway. In *The Cultural Landscape: Past, Present and Future* (edited by H H Birks, H J B Birks, P E Kaland and D Moe). Cambridge University Press, 11–29.

Austin, D (1985) Dartmoor and the upland village of the South-West of England. In *Medieval Villages: A review of current work* (edited by D Hooke). Oxford University Committee for Archaeology Monograph 5, 71–79.

Bacon, C R (1983) Eruptive history of Mount Mazama and Crater Lake Caldera, Cascade Range, USA. *Journal of Volcanology and Geothermal Research*, **18**, 57–115.

Bailey, G and Parkington J (1988) *The Archaeology of Prehistoric Coastlines*. Cambridge University Press, Cambridge.

Baillie, M (1989) Do Irish bog oaks date the Shang Dynasty? *Current Archaeology*, **117**, 310–313.

Baillie, M G L and Munro, M A R (1988) Irish tree–rings, Santorini and volcanic dust veils. *Nature*, **332**, 344–346.

Baker, C A, Moxey, P A and Oxford, P M (1978) Woodland continuity and change in Epping Forest. *Field Studies*, **4**, 645–669.

Baker, D G, Watson, B F and Skaggs, R H (1985) The Minnesota long-term temperature record. *Climatic Change*, **7**, 225–236.

Baker, N V and Payne, G K (1978) G.K. Gilbert and modern geomorphology. *American Journal of Science*, **278**, 97–123.

Baker, R G (1984) Holocene vegetational history of the western United States. In *Late Quaternary Environments of the United States Volume 2 The Holocene* (edited by H E Wright Jr). Longman, London, 109–127.

Baker, V R (1981) *Catastrophic Flooding: The Origin of the Channeled Scabland*. Dowden, Hutchinson & Ross, Stroudsberg, Pennsylvania.

Baker, V R and Bunker, R C (1985) Cataclysmic Late Pleistocene flooding from Glacial Lake Missoula: a review. *Quaternary Science Reviews*, **4**, 1–41.

Balaam, N D, Smith, K and Wainwright, G J (1982) The Shaugh Moor Project: fourth report. *Proceedings of the Prehistoric Society*, **48**, 203–278.

Balaam, N D, Bell, M G, David, A E U, Levitan, B, Macphail, R I, Robinson, M and Scaife, R G (1987) Prehistoric and Romano–British sites at Westward Ho!, Devon. Archaeological and palaeoenvironmental surveys 1983–84. In *Studies in Palaeoeconomy and Environment in South West England* (edited by N D Balaam, B Levitan and V Straker). British Archaeological Reports, Oxford, BS 181, 163–264.

Ball, I R (1976) Nature and formulation of biogeographical hypotheses. *Systematic Zoology*, **24**, 407–430.

Ball, T F and Kingsley, R A (1984) Instrumental temperature records at two sites in central Canada: 1768–1910. *Climatic Change*, **6**, 39–56.

Ballantyne, C K (1984) The Late Devensian periglaciation of upland Scotland. *Quaternary Science Reviews*, **3**, 311–343.

Ballantyne, C K (1987) The present-day periglaciation of upland Britain. In *Periglacial Landforms and Processes* (edited by J Boardman). Cambridge University Press, London, 113–126.

Balsam, W L (1981) Late Quaternary sedimentation in the

western North Atlantic: stratigraphy and paleoceanography. *Palaeogeography, Palaeoclimatology, Palaeoecology*, **35**, 215–240.

Barber, K E (1981) *Peat Stratigraphy and Climatic Change*. Balkema, Rotterdam.

Barber, K E (1985) Peat stratigraphy and climatic change: some speculations. In *The Climatic Scene* (edited by M J Tooley and G M Sheail). George Allen & Unwin, London, 175–185.

Barber, K E and Coope, G R (1987) Climatic history of the Severn Valley during the last 18,000 years. In *Palaeohydrology in Practice* (edited by K J Gregory). John Wiley, Chichester and New York, 201–216.

Bard, E, Arnold, M, Maurice, P, Duprat, J, Moyes, J and Duplessy, J-C (1987) Retreat velocity of the North Atlantic polar front during the last deglaciation determined by ^{14}C accelerator mass spectrometry. *Nature*, **328**, 791–794.

Bard, E, Fairbanks, R G, Arnold, M, Maurice, P, Duprat, J, Moyes, J and Duplessy, J-C (1989) Sea–level estimates during the last deglaciation based on δ^{18}O and accelerator mass spectrometry ^{14}C ages measured in *Globigerina bulloides*. *Quaternary Research*, **31**, 381–391.

Bard, E, Hamelin, B, Fairbanks, R J and Zindler, A (1990a) Calibration of the ^{14}C timescale over the past 30,000 years using mass–spectrometric U–Th ages from Barbados corals. *Nature*, **345**, 405–410.

Bard, E, Hamelin, B and Fairbanks, R G (1990b) U–Th ages obtained by mass spectrometry in corals from Barbados: sea level during the past 130,000 years. *Nature*, **346**, 456–458.

Barnett, T P (1984) The estimation of 'global' sea-level change, a problem of uniqueness. *Journal of Geophysical Research*, **89**, 7980–7988.

Barnola, J M, Raynaud, D, Korotkevich, Y S and Lorius, C (1987) Vostok ice core provides 160,000 year record of atmospheric CO_2. *Nature*, **329**, 408–413.

Barnosky, A D (1986) Big game extinctions caused by late Pleistocene climatic change: Irish Elk (*Megaloceros giganteus*) in Ireland. *Quaternary Research*, **25**, 128–135.

Barnosky, C W, Anderson, P M and Bartlein, P J (1987) The northwestern US during deglaciation: vegetational history and palaeoclimatic implications. In *The Geology of North America. Volume K–3: North America and adjacent oceans during the last deglaciation* (edited by W F Ruddiman and H E Wright Jr). Geological Society of America, Boulder, Colorado, 289–321.

Baron, W R and Gordon, G A (1985) A reconstruction of New England climate using historical materials, 1620–1980. *Syllogeus*, **55**, 229–245.

Barry, R G (1983) Late Pleistocene climatology. In *Late-Quaternary Environments of the United States, Volume 1. The Late Pleistocene* (edited by S C Porter). Longman, London, 390–407.

Barth, M C and Tutus, J G (eds) (1984) *Greenhouse Effect and Sea Level Rise*. Van Nostrand Reinhold, New York.

Bartholin, T S (1984) Dendrochronology in Sweden. In *Climatic Change on a Yearly to Millennial Basis* (edited by N-A Mörner and W Karlén). Reidel, Dordrecht, 261–262.

Bar–Yosef, O and Kislev, M E (1989) Early farming communities in the Jordan Valley. In *Foraging and Farming* (edited by D R Harris and G C Hillman). Unwin Hyman, London, 632–642.

Battarbee, R W (1984) Diatom analysis and the acidification of lakes. *Philosophical Transactions of the Royal Society, London*, **B305**, 451–477.

Battarbee, R W (1988) The application of diatom analysis in archaeology: a review. *Journal of Archaeological Science*, **15**, 621–644.

Battarbee, R W and Charles, D F (1987) The use of diatom assemblages in lake sediments as a means of assessing the timing, trends, and causes of lake acidification. *Progress in Physical Geography*, **11**, 552–580.

Battarbee, R W, Flower, R J, Stevenson, J and Rippey, B (1985) Lake acidification in Galloway: a palaeoecological test of competing hypotheses. *Nature*, **314**, 350–352.

Battistini, R and Verin, P (1967) Ecological changes in protohistoric Madagascar. In *Pleistocene Extinctions* (edited by P S Martin and H E Wright). Yale University Press, New Haven, 407–424.

Begin, Y and Payette, S (1988) Dendrochronological evidence of lake-level changes during the last three centuries in subarctic Quebec. *Quaternary Research*, **30**, 210–220.

Behre, K-E (1988) The role of man in European vegetation history. In *Vegetation History* (edited by B Huntley and T Webb). Kluwer, Dordrecht, 633–672.

Behre, K-E (1989) Biostratigraphy of the last glacial period in Europe. *Quaternary Science Reviews*, **8**, 25–44.

Behre, K-E, Dorjes, I and Irion, G (1985) A dated Holocene core from the bottom of the southern North Sea. *Eiszeitalter und Gegenwart*, **35**, 9–13.

Belknap, D F and Kraft, J C (1977) Holocene relative sea-level changes and coastal stratigraphic units on the northwest flank of the Baltimore canyon trough geosyncline. *Journal of Sedimentary Petrology*, **47**, 610–629.

Bell, M G (1982) The effects of land-use and climate on valley sedimentation. In *Climatic Change in Later Prehistory* (edited by A F Harding). Edinburgh University Press, 127–142.

Bell, M G (1983) Valley sediments as evidence of prehistoric land-use on the South Downs. *Proceedings of the Prehistoric Society*, **49**, 119–150.

Bell, M G (1986) Archaeological evidence for the date, cause and extent of soil erosion on the chalk. In *Soil Erosion* (edited by C P Burnham and J I Pitman). SEESOIL, **3**, 72–83.

Bell, M G (1987) Recent molluscan studies in the south

west. In *Studies in the Palaeoeconomy and Environment in South West England* (edited by N D Balaam, B Levitan and V Straker). British Archaeological Reports, Oxford, BS 181, 1–7.

Bell, M G (1990) *Brean Down Excavations 1983–1987*. English Heritage, Archaeological Report 15.

Bell, W T and Ogilvie, A E J (1978) Weather compilations as a source of data for the reconstruction of European climate during the Medieval Period. *Climatic Change*, **1**, 331–348.

Bennett, K D (1989) A provisional map of forest types for the British Isles 5000 years ago. *Journal of Quaternary Science*, **4**, 141–144.

Bennett, K D, Fossitt, J A, Sharp, M J and Switsur, V R (1990) Holocene vegetation and environmental history at Loch Lang, South Uist, Western Isles, Scotland. *New Phytologist*, **114**, 281–298.

Berendsen, H J A and Zagwijn, W H (1984) Geological changes in the western Netherlands during the period 1000–1300 AD. *Geologie en Mijnbouw*, **3**, 225–336.

Beresford, G (1981) Climatic change and its effects upon the settlement and desertion of medieval villages in Britain. In *Consequences of Climatic Change* (edited by C Delano Smith and M Parry). Department of Geography, Nottingham, 30–39.

Berger, A (1977) Support for the astronomical theory of climatic change. *Nature*, **269**, 44–45.

Berger, A (1978a) Long-term variations of daily insolation and Quaternary climatic changes. *Journal of Atmospheric Science*, **35**, 2362–2367.

Berger, A (1978b) Long-term variations of calorific insolation resulting from the earth's orbital elements. *Quaternary Research*, **9**, 139–167.

Berger, A (1980) The Milankovitch astronomical theory of palaeoclimates: a modern review. In *Vistas in Astronomy* (edited by A Beer, K Pounds and P Beer). Pergamon Press, Oxford, 103–122.

Berglund, B E (1985) Early agriculture in Scandinavia: Research problems related to pollen analytical studies. *Norwegian Archaeological Review*, **18**, 77–105.

Berglund, B E (1986) *Handbook of Holocene Palaeoecology and Palaeohydrology*. John Wiley, Chichester.

Berglund, B E, Lemdahl, G, Liedberg-Jonsson, B and Persson, T (1984) Biotic response to climatic changes during the time span 13,000–10,000 BP – a case study from SW Sweden. In *Climatic Changes on a Yearly to Millennial Basis* (edited by N-A Mörner and W Karlén). Reidel, Dordrecht, 25–36.

Bergthórsson, P (1969) An estimate of drift ice and temperature in Iceland in 10,000 years. *Jökull*, **19**, 94–101.

Berti, A A (1975) Palaeobotany of Wisconsinan Interstadials, eastern Great Lakes region, North America. *Quaternary Research*, **5**, 591–619.

Billard, A and Orombelli, G (1986) Quaternary glaciations in the French and Italian Piedmonts of the Alps. *Quaternary Science Reviews*, **5**, 407–411.

Binford, L R (1968) Post-Pleistocene adaptations. In *New Perspectives in Archaeology* (edited by S R Binford and L R Binford). Aldine, Chicago, 313–341.

Binford, L R (1981) *Bones: ancient men and modern myths*. Academic Press, New York.

Binford, L R (1983) *In Pursuit of the Past*. Thames and Hudson, London.

Bintliff, J (1977) *Natural Environment and Human Settlement in Prehistoric Greece*. British Archaeological Reports, Oxford, IS 28.

Birchfield and Weertman, J (1983) Topography, albedo-temperature feedback, and climatic sensitivity. *Science*, **219**, 284–285.

Birkenmajer, K (1981) Lichenometric dating of raised beaches at Admiralty Bay, King George Island, South Shetland Islands, West Antarctica. *Bulletin of the Polish Academy of Sciences*, Ser. Sci. Terre, **29**, 119–128.

Birks, H H (1980) Plant macrofossils in Quaternary lake sediments. *Archives für Hydrobiologie und Limnologie*, **15**, 1–60.

Birks, H H, Birks, H J B, Kaland, P E and Moe, D (eds) (1988) *The Cultural Landscape: Past, Present and Future*. Cambridge University Press, Cambridge.

Birks, H J B (1981) The use of pollen analysis in the reconstruction of past climates: a review. In *Climate and History* (edited by T M L Wigley, M J Ingram and G Farmer). Cambridge University Press, Cambridge, 111–138.

Birks, H J B (1986) Late Quaternary biotic changes in terrestrial and lacustrine environments, with particular reference to north-west Europe. In *Handbook of Holocene Palaeoecology and Palaeohydrology* (edited by B E Berglund). John Wiley, Chichester and New York, 3–65.

Birks, H J B (1989) Holocene isochrone maps and patterns of tree spreading in the British Isles. *Journal of Biogeography*, **16**, 503–540.

Birks, H J B and Birks, H H (1980) *Quaternary Palaeoecology*. Edward Arnold, London.

Bishop, P (1980) Popper's principle of falsifiability and the Davisian cycle. *Professional Geographer*, **32**, 310–315.

Blong, R J (1982) *The Time of Darkness*. Australian National University Press, Canberra.

Blong, R J (1984) *Volcanic Hazards*. Academic Press, London.

Bloom, A L (1983) Sea level and coastal morphology of the United States through the Late Wisconsin glacial maximum. In *Late Quaternary Environments of the United States. Volume 1: The Pleistocene* (edited by S C Porter). Longman, London, 215–229.

Bloom, A L (1984) Sea level and coastal changes. In *Late Quaternary Environments of the United States. Volume 2*.

The Holocene (edited by H E Wright Jr). Longman, London, 42–51.

Boal, F W and Livingstone, D N (eds) (1989) *The Behavioural Environment: Essays in reflection, application and re-evaluation*. Routledge, London.

Boardman, J (1990a) Soil erosion on the South Downs: a review. In *Soil Erosion on Agricultural Land* (edited by J Boardman, I D L Foster and J A Dearing). John Wiley, Chichester, 87–105.

Boardman, J (1990b) *Soil Erosion in Britain: Costs, attitudes and policies*. Education network for environment and development, Sussex University, Falmer.

Boardman, J and Robinson, D A (1985) Soil erosion, climatic vagary and agricultural change on the Downs around Lewes and Brighton, autumn 1982. *Applied Geography*, **5**, 243–258.

Boer, M M, Koster, E A and Lundberg, H (1990) Greenhouse impact in Fennoscandia – preliminary findings of a European workshop on the effects of climatic change. *Ambio*, **19**, 2–16.

Bohnke, S J P (1988) Vegetation and habitation history of the Callanish area, Isle of Lewis, Scotland. In *The Cultural Landscape: Past, present and future* (edited by H H Birks, H J B Birks, P E Kaland and D Moe). Cambridge University Press, Cambridge, 445–461.

Bohnke, S, Vandenberghe, J, Coope, G R and Reiling, R (1987) Geomorphology and palaeoecology of the Mark Valley (southern Netherlands): palaeoecology, palaeohydrology and climate during the Weichselian Late Glacial. *Boreas*, **16**, 69–85.

Bolin, B, Jäger, J and Döös, B R (1986) The greenhouse effect, climatic change and ecosystems: a synthesis of present knowledge. In *SCOPE 29, The Greenhouse Effect, Climatic Change and Ecosystems* (edited by B Bolin, B R Döös, J Jäger and R A Warrick). John Wiley, Chichester and New York. 1–32.

Bolle, H-J, Seiler, W and Bolin, B (1986) Other greenhouse gases and aerosols. In *SCOPE 29: The Greenhouse Effect, Climatic Change and Ecosystems* (edited by B Bolin, B R Döös, J Jäger and R A Warrick). John Wiley, Chichester and New York, 157–203.

Bonavia, D and Grobman, A (1989) Andean maize: its origins and domestication. In *Foraging and Farming* (edited by D R Harris and G C Hillman). Unwin Hyman, London, 456–470.

Bonnischen, R, Stanford, D and Fastook, J L (1987) Environmental change and the developmental history of human adaptation patterns: the Palaeoindian case. In *North America and Adjacent Oceans During the Last Deglaciation* (edited by W F Ruddiman and H E Wright Jr). Geological Society of America, Boulder, Colorado, 403–424.

Bonsall, C, Sutherland, D, Tipping, R and Cherry, J (1986). The Eskmeals Project 1981–5: an interim report. *Northern*

Archaeology, **7**(1), 3–30.

Bottema, S (1988) Back to nature? Objectives of nature management in view of archaeological research. In *Archeologie en Landschap* (edited by H van Bierma, O H Harsema, W van Zeist). Biologisch-Archeologisch Instituut Rijkuniversiteit, Groningen, 185–206.

Boulton, G S, Smith, G D, Jones, A S and Newsome, J (1985) Glacial geology and glaciology of the last mid–latitude ice sheets. *Journal of the Geological Society, London*, **142**, 447–474.

Bowdler, S (1988) Tasmanian Aborigines in the Hunter Islands in the Holocene: island resource use and seasonality. In *The Archaeology of Prehistoric Coastlines* (edited by G Bailey and J Parkington). Cambridge University Press, 42–52.

Bowen, D Q (1978) *Quaternary Geology*. Pergamon, Oxford.

Bowen, D Q, Rose, J, McCabe, A M and Sutherland, D G (1986) Correlation of Quaternary glaciations in England, Ireland, Scotland and Wales. *Quaternary Science Reviews*, 5, 299–340.

Bowman, S (1990) *Radiocarbon Dating*. British Museum Publications, London.

Bradbury, J P (1975) Diatom stratigraphy and human settlement in Minnesota. *Geological Society of America, Special Paper*, 171.

Bradley, R (1984) *The Social Foundations of Prehistoric Europe*. Longman, London.

Bradley, R S (1985) *Quaternary Palaeoclimatology*. Allen and Unwin, London.

Bradley, R S, Diaz, H F, Eischeid, J K, Jones, P D, Kelly, P M and Goodess, C M (1987) Precipitation fluctuation over Northern Hemisphere land areas since the mid-19th century. *Science*, **237**, 171–175.

Brain, C K (1981) Hominid evolution and climatic change. *South African Journal of Science*, **77**, 104–105.

Brandt, R, Groenman–van Waateringe, W and van der Leeuw, S E (1987) *Assendelver Polder Papers I*. Van Giffen Instituut, Amsterdam.

Brandt, R and van der Leeuw, S E (1987) The Assendelver Polders of the Netherlands and a wet perspective on the European Iron Age. In *European Wetlands in Prehistory* (edited by J M Coles and A J Lawson). Clarendon Press, Oxford, 203–226.

Brasier, M D (1980) *Microfossils*. George Allen and Unwin, London.

Bray, J R (1970) Temporal patterning of post–Pleistocene glaciation. *Nature*, **228**, 353–354.

Bray, J R (1974) Volcanism and glaciations during the past 40 millennia. *Nature*, **252**, 679–680.

Bray, J R (1979) Surface albedo increase following massive Pleistocene explosive eruptions in western North America. *Quaternary Research*, **12**, 204–211.

Bray, J R (1982) Alpine glacier advance in relation to a proxy summer temperature index based mainly on wine harvest dates. *Boreas*, **11**, 1–10.

Brenninkmeijer, C A M, van Geel, B and Mook, W G (1982) Variations in the D/H and $^{18}O/^{16}O$ ratios in cellulose extracted from a peat bog core. *Earth and Planetary Science Letters*, **61**, 283–290.

Briffa, K R, Bartholin, T S, Eckstein, D, Jones, D D, Karlén, W, Schweingruber, F H and Zetterberg, P (1990) A 1,400-year tree-ring record of summer temperatures in Fennoscandia. *Nature*, **346**, 343–439.

Brimblecombe, P and Stedman, D (1982) Historical evidence for a dramatic increase in the nitrate component of acid rain. *Nature*, **298**, 460–462.

Broadbent, N D (1979) *Coastal Resources and Settlement Stability. A critical analysis of a Mesolithic site complex in northern Sweden*. Aun 3. Archaeological Studies Institute of North European Archaeology, University of Uppsala, Uppsala. Borgstroms tryckeri, 268pp.

Broadbent, N D and Bergqvist, K I (1986) Lichenometric chronology and archaeological features on raised beaches: preliminary results from the Swedish North Bothnian coastal region. *Arctic and Alpine Research*, **18**, 297–306.

Broecker, W (1984) Terminations. In *Milankovitch and Climate* (edited by A Berger, J Imbrie, J Hays, G Kukla and B Saltzman). Reidel, Dordrecht, 687–698.

Broecker. W S, Peteet, D M and Rind, D (1985) Does the ocean–atmosphere system have more than one stable mode of operation? *Nature*, **315**, 21–25.

Broecker, W S, Andree, J, Wolfli, W, Oeschger, H, Bonani, G, Peteet, D and Kennett, J P (1988) The chronology of the last deglaciation: implications to the cause of the Younger Dryas event. *Paleoceanography*, **3**, 1–29.

Broecker, W S, Kennett, J P, Flower, B P, Teller, J T, Trumboe, S, Bonani, G and Wolfli, W (1989) Routing of meltwater from the Laurentide Ice Sheet during the Younger Dryas cold episode. *Nature*, **341**, 318–321.

Broecker, W S and Denton, G H (1990) The role of ocean-atmosphere reorganisation in glacial cycles. *Quaternary Science Review*, **9**, 305–341.

Brown, A G (1987) Longterm sediment storage in the Severn and Wye Catchments. In *Palaeohydrology in Practice* (edited by K J Gregory, J Lewin and J B Thornes). John Wiley, Chichester, 307–332.

Brown, A G and Barber, K E (1985) Late Holocene palaeoecology and sedimentary history of a small lowland catchment in central England. *Quaternary Research*, **24**, 87–102.

Brown, T A, Nelson, D E, Mathewes, R W, Vogel, J S and Southon, J R (1989) Radiocarbon dating of pollen by Accelerator Mass Spectrometry. *Quaternary Research*, **32**, 205–212.

Brubaker, L B and Cook, E R (1984) Tree-ring studies of Holocene environments. In *Late-Quaternary Environments of the United States. Volume 2. The Holocene* (edited by H E Wright Jr). Longman, London, 222–235.

Bruckner, H (1986) Man's impact on the evolution of the physical environment in the Mediterranean region in historical times. *GeoJournal*, **13**(1), 7–17.

Brugam, R B (1975) *The human disturbance history of Linsley Pond, North Bramford*. Connecticut. Yale University PhD thesis.

Brugam, R B (1980) Postglacial diatom stratigraphy of Kirschner Marsh, Minnesota. *Quaternary Research*, **13**, 133–146.

Brunsden, D and Thornes, J B (1979) Landscape sensitivity and change. *Institute of British Geographers Transactions*, **4**, 463–484.

Bryant, I D (1983a) Facies sequences associated with some braided river deposits of late Pleistocene age from southern Britain. In *Modern and Ancient Fluvial Systems* (edited by J D Collinson and J Lewin). Blackwell, Oxford, 267–275.

Bryant, I D (1983b) The utilisation of Arctic river analogue studies in the interpretation of periglacial river sediments from southern Britain. In *Background to Palaeohydrology* (edited by K J Gregory). John Wiley, Chichester and New York, 413–431.

Bryant. I D, Holyoak. D T and Moseley. K A (1983) Late Pleistocene deposits at Brimpton, Berkshire, England. *Proceedings of the Geologists' Association*, **94**, 321–343.

Bryson, R A (1988) Civilisation and rapid climatic change. *Environmental Conservation*, **15**, 7–15.

Bryson, R A and Murray, T (1977) *Climates of Hunger*. University of Wisconsin Press, Madison.

Buckland, P C (1975) Synanthropy and the deathwatch: a discussion. *The Naturalist*, **100**(993), 37–42.

Buckland, P C (1976) The use of insect remains in the interpretation of archaeological environments. In *Geoarchaeology* (edited by D A Davidson and M L Shackley). Duckworth, London, 369–391.

Buckland, P C (1979) *Thorne Moors: a palaeoecological study of a Bronze Age site*. University of Birmingham, Department of Geography.

Buckland, P C and Coope, G R (1991) *A Bibliography and Literature Review of Quaternary Entomology*. J R Collis, Sheffield.

Buckland, P C and Edwards, K J (1984) The longevity of pastoral episodes of clearance activity in pollen diagrams: the role of post–occupation grazing. *Journal of Biogeography*, **11**, 243–249.

Buckland, P C, Sveinbjarnardottir, G, Savory, D, McGovern, T, Skidmore, P and Andreasen, C (1983) Norsemen at Nipáitsoq, Greenland: a palaeoecological investigation. *Norwegian Archaeological Review*, **16**(2), 86–98.

Buckland, P C, Dugmore, A J and Sadler, J (1991) Faunal change or taphonomic problem? A comparison of modern and fossil insect faunas from south–east Iceland. In *Environmental Change in Iceland: Past and present* (edited by J K Maizels and C J Caseldine). Kluwer, Dordrecht, 127–146.

Burckle, L H (1978) Marine diatoms. In *Introduction to Marine Micropalaeontology* (edited by B U Haq and A Boersma). Elsevier, Rotterdam, 246–266.

Burgess, C (1974) The Bronze Age. In *British Prehistory* (edited by C Renfrew). Duckworth, London, 223–232.

Burgess, C (1980) *The Age of Stonehenge*. Dent, London.

Burgess, C (1985) Prehistoric climate and upland settlement. In *Upland Settlement in Britain* (edited by D Spratt and C Burgess). British Archaeological Reports, Oxford, BS 143, 195–230.

Burgess, C (1989) Volcanoes, catastrophe and the global crisis of the late second millenium BC. *Current Archaeology*, **117**, 325–329.

Burk, R L and Stuiver, M (1981) Oxygen isotope ratios in trees reflect mean annual temperature and humidity. *Science*, **211**, 1417–1419.

Burke, R M and Birkeland, P W (1984) Holocene glaciation in the mountain ranges of the Western United States. In *Late Quaternary Environments of the United States. Volume 2: The Holocene* (edited by H E Wright Jr). Longman, London, 3–11.

Burleigh, R and Kerney, M P (1982) Some chronological implications of a fossil Molluscan assemblage from a Neolithic site at Brook, Kent, England. *Journal of Archaeological Science*, **9**, 29–38.

Burrin, P J and Scaife, R G (1988) Environmental thresholds, catastrophe theory and landscape sensitivity: their relevance to the impact of man on valley alluviation. In *Conceptual issues in Environmental Archaeology* (edited by J L Bintliff, D A Davidson and E G Grant). Edinburgh University Press, 211–232.

Bush, M B (1988) Early Mesolithic disturbance: a force in the Landscape. *Journal of Archaeological Science*, **15**, 453–462.

Butzer, K W (1964) Pleistocene cold-climate phenomena of the island of Mallorca. *Zeitschrift für Geomorphologie*, **8**, 7–31.

Butzer, K W (1976) *Geomorphology from the Earth*. Harper and Row, New York.

Butzer, K M (1977a) Environmental culture and human evolution. *American Scientist*, **65**, 572–584.

Butzer, K W (1977b) *Geomorphology of the Lower Illinois Valley as a Spatial–Temporal Context for the Koster Archaic site. Reports of Investigations of the Illinois State Museum*, no. 34.

Butzer, K W (1980) Holocene alluvial sequences: problems of dating and correlation. In *Timescales in Geomorphology* (edited by R A Cullingford, D A Davidson and J Lewin). John Wiley, Chichester and New York, 131–142.

Butzer, K W (1981) Cave sediments, Upper Pleistocene stratigraphy and Mousterian facies in Cantabrian Spain. *Journal of Archaeological Science*, **8**, 133–183.

Butzer, K W (1982) *Archaeology as Human Ecology*. Cambridge University Press, Cambridge.

Butzer, K W (1990) The indian legacy in the American landscape. In *The Making of the American Landscape* (edited by M P Conzen). Unwin Hyman, Boston, 27–50.

Byers, D S (ed.) (1967) *The Prehistory of the Tehuacan Valley*. University of Texas Press, Texas.

Carbonel, P, Colin, J–P, Danielopol, D L, Löffler, H and Neustreva, I (1988) Palaeoecology of limnic ostracodes: a review of some major topics. *Palaeogeography, Palaeoecology, Palaeoclimatology*, **62**, 413–461.

Caseldine, C J (1984) Pollen analysis of a buried arctic–alpine brown soil from Vestre Memurubreen, Jotunheimen, Norway: evidence for postglacial high-altitude vegetation change. *Arctic and Alpine Research*, **16**, 423–430.

Casparie, W A and Groenman–van Waateringe, W (1982) Palynological analysis of Dutch barrows. *Palaeohistoria*, **22**, 7–65.

Cassels, R (1984) Faunal extinction and prehistoric man in New Zealand and the Pacific Islands. In *Quaternary Extinctions* (edited by P Martin and R Klein). University of Arizona Press, 741–767.

Catchpole, A J W and Faurer, M–A (1983) Summer sea ice severity in Hudson Strait, 1751–1870. *Climatic Change*, **5**, 115–139.

Catchpole, A J W and Moodie, D W (1978) Archives and the environmental scientist. *Archivaria*, **6**, 113–136.

Catt, J A (1979) Soils and Quaternary geology in Britain. *Journal of Soil Science*, **30**, 607–642.

Catt, J A (1986) *Soils and Quaternary Geology*. Clarendon Press, Oxford.

Catt, J A (1988) Soils of the Plio-Pleistocene: do they distinguish types of interglacial? *Philosophical Transactions of the Royal Society, London*, **B318**, 539–557.

Catt, J A and Staines, S J (1982) Loess in Cornwall. *Proceedings of the Ussher Society*, **5**, 368–375.

Caulfield, S (1983) The Neolithic settlement of N. Connaught. In *Landscape Archaeology in Ireland* (edited by T Reeves–Smyth and F Hamond). British Archaeological Reports, Oxford, BS 116, 195–215.

Chalavoux, J (1988) *Ferme archéologique de Melrand*. Association de la ferme archéologique de Melrand.

Chalmers, A F (1982) *What is this thing called Science?* Open University, Milton Keynes.

Chamberlin, T C (1895) Glacial phenomena of North America. In *The Great Ice Age* (edited by J Geikie) 3rd edition. New York, 724–755.

Chamberlin, T C (1965) The method of multiple working

hypotheses. *Science*, **148**, 745–759. (Reprinted from *Science*, 1897.)

Champion, T, Gamble, C, Shennan, S and Whittle, A (1984) *Prehistoric Europe*. Academic Press, London.

Chappell, J and Shackleton, N J (1986) Oxygen isotopes and sea level. *Nature*, **324**, 137–139.

Chernosky, E J (1966) Double sunspot-cycle variation in terrestrial magnetic activity, 1884–1963. *Journal of Geophysical Research*, **71**, 965–974.

Childe, V G (1928) *The Most Ancient East: the oriental prelude to European prehistory*. Routledge and Kegan Paul, London.

Chorley, R J and Kennedy, B A (1971) *Physical Geography: a systems approach*. Prentice-Hall, Englewood Cliffs.

Chorley, R J, Dunn, A J and Beckinsale, R P (1964) *The History of the Study of Landforms. Volume 1: Geomorphology before Davis*. Methuen, London.

Church, M and Ryder, J M (1972) Paraglacial sedimentation: a consideration of fluvial processes conditioned by glaciation. *Geological Society of America Bulletin*, **83**, 3059–3072.

Clark, A J, Tarling, D H and Noël, M (1988) Developments in archaeomagnetic dating in Britain. *Journal of Archaeological Science*, **15**, 645–647.

Clark, J D and Harris, J W K (1985) Fire and its roles in early hominid lifeways. *African Archaeological Review*, **3**, 3–27.

Clark, J G D (1972) *Star Carr: a case study in bioarchaeology*. Addison-Wesley Modular Publications 10, Reading, Massachusetts.

Clark, J S (1988) Effect of climatic change on fire regimes in northwestern Minnesota. *Nature*, **334**, 233–235.

Clark, R L (1983) Pollen and charcoal evidence for the effects of aboriginal burning on the vegetation of Australia. *Archaeology in Oceania*, **18**, 32–37.

Clark, R M (1975) A calibration curve for radiocarbon dates. *Antiquity*, **49**, 251–259.

Clarke, G K C, Mathews, W H and Pack, R T (1984) Outburst floods from glacial Lake Missoula. *Quaternary Research*, **22**, 289–299.

Clayton, K M (1977) River terraces. In *British Quaternary Studies: Recent Advances* (edited by F W Shotton). Clarendon Press, Oxford, 153–168.

Cleere, H F (1984) *Approaches to the Archaeological Heritage*. Cambridge University Press, Cambridge.

Cleere, H F (1989) Introduction: the rationale of archaeological Heritage. In *Archaeological Heritage Management in the Modern World* (edited by H F Cleere). Unwin Hyman, London, 1–19.

Cloutman, E W (1988). Palaeoenvironments in the Vale of Pickering: Part 2: Environmental history at Seamer Carr. *Proceedings of the Prehistoric Society*, **54**, 21–36.

Cloutman, E W and Smith, A G (1988) Palaeoenvironments in the Vale of Pickering. Part 3:

Environmental history at Star Carr. *Proceedings of the Prehistoric Society*, **54**, 37–58.

Clutton-Brock, J (ed.) (1989) *The Walking Larder*. Unwin Hyman, London.

Cohen, M D (1977) *The Food Crisis in Prehistory*. Yale University Press, New Haven.

Coles, B and Coles, J (1986) *Sweet Track to Glastonbury*. Thames and Hudson, London.

Coles, B and Coles, J (1989) *People of the Wetlands*. Thames and Hudson, London.

Coles, J M (1987) Meare Village East: the excavations of A. Bulleid and H. St George Gray, 1932–1956. *Somerset Levels Papers*, **13**, 1–254.

Coles, J M and Lawson, A J (1987) *European Wetlands in Prehistory*. Clarendon Press, Oxford.

Coles, J and Orme, B (1983) *Homo sapiens* or *Castor fiber*? *Antiquity*, **57**, 95–102.

Collinson, J D and Lewin, J (eds) (1983) *Modern and Ancient Fluvial Systems*. Blackwell, Oxford.

Comani, S (1987) The historical temperature series of Bologna (Italy): 1716–1774. *Climatic Change*, **11**, 375–390.

Conchon, O (1978) Quaternary studies in Corsica (France). *Quaternary Research*, **9**, 41–53.

Conzen, M P (1990) *The Making of the American Landscape*. Unwin Hyman, Boston.

Cooke, H B S (1981) Age control of Quaternary sedimentary/climatic record from deep boreholes in the Great Hungarian Plain. In *Quaternary Palaeoclimatology* (edited by W C Mahaney). Geobooks, Norwich, 1–12.

Coope, G R (1975) Climatic fluctuations in north-west Europe since the last Interglacial indicated by fossil assemblages of Coleoptera. In *Ice Ages: Ancient and Modern* (edited by A E Wright and F Moseley). Seel House Press, Liverpool, 153–168.

Coope, G R (1977a) Quaternary Coleoptera as aids in the interpretation of environmental history. In *British Quaternary Studies – Recent Advances* (edited by F W Shotton). Oxford University Press, Oxford, 55–68.

Coope, G R (1977b) Fossil coleopteran assemblages as sensitive indicators of climatic changes during the last (Devensian) cold stage. *Philosophical Transactions of the Royal Society, London*, **B280**, 313–340.

Coope, G R (1986) Coleoptera analysis. In *Handbook of Holocene Palaeoecology and Palaeohydrology* (edited by B E Berglund). John Wiley, Chichester and New York, 703–713.

Cornwall, I W (1953) Soil science and archaeology with illustrations from some British Bronze Age monuments. *Proceedings of the Prehistoric Society*, **2**, 129.

Costa, J E (1978) Colorado Big Thompson Flood: geologic evidence of a rare hydrologic event. *Geology*, **6**, 617–620.

Courty, M A, Goldberg, P and Macphail, R (1989) *Soils and Micromorphology in Archaeology*. Cambridge University Press.

Coveney, P and Highfield, R (1990) *The Arrow of Time.* W.H. Allen, London.

Cowling, E B (1989) Recent changes in chemical climate and related effects on forests in North America and Europe. *Ambio,* **18,** 167–171.

Craddock, J M (1976) Annual rainfall in England since 1726. *Journal of the Royal Meteorological Society,* **102,** 823–840.

Crawford, G W (1987) Evidence for anthropogenic environmental change in the Green River Archaic. In *Man and the Holocene Optimum* (edited by N A McKinnon and G S L Stuart). *Proc. of the 17th Annual Chacmool Conference, Calgary University,* 303–307.

Crisman, T L (1978) Reconstruction of past lacustrine environments based on the remains of aquatic invertebrates. In *Biology and Quaternary Environments* (edited by D Walker and J C Guppy). Australian Academy of Sciences, Canberra, 69–101.

Crosby, A W (1986) *Ecological Imperialism: The Biological Expansion of Europe 900–1100.* Cambridge University Press, Cambridge.

Cubasch, U and Cess, R D (1990) Processe and modelling. In *Climate Change: The IPCC Scientific Assessment* (edited by J T Houghton, G J Jenkins and J J Ephraums). Cambridge University Press, Cambridge, 69–92.

Cwynar, L C (1978) Recent history of fire and vegetation from laminated sediment of Greenleaf Lake, Algonquin Park, Ontario. *Canadian Journal of Botany,* **56,** 10–21.

Daniel, G (1975) *One Hundred and Fifty Years of Archaeology.* Duckworth, London.

Dansgaard, W, Johnsen, S J, Reeh, N, Gundestrup, N, Clausen, H B and Hammer, C U (1975) Climatic changes, Norsemen and modern man. *Nature,* **255,** 24–28.

Dansgaard, W, Clausen, H B, Gundestrup, N, Hammer, CU, Johnsen, S F, Kristinsdottir, P M and Reeh, N (1982) A new Greenland deep ice core. *Science,* **218,** 1273–1277.

Dansgaard, W, White, J W C and Johnsen, S J (1989) The abrupt termination of the Younger Dryas climatic event. *Nature,* **339,** 532–534.

Darvill, T (1987) *Ancient Monuments in the Countryside.* English Heritage, London.

Darwin, C (1859) *The Origin of Species by Means of Natural Selection.* Murray, London.

Davidson, D A (1980) Erosion in Greece during the first and second millennium BC. In *Timescales in Geomorphology* (edited by R A Cullingford, D A Davidson and J Lewin). John Wiley, Chichester, 143–158.

Davidson, D A and Shackley, M L (eds) (1976) *Geoarchaeology; Earth Science and the Past.* Duckworth, London.

Davies, K H and Keen, D H (1985) The age of Pleistocene marine deposits at Portland, Dorset. *Proceedings of the Geologists' Association,* **96,** 217–225.

Davis, M B (1976a) Pleistocene biogeography of the temperate deciduous forests. *Geoscience and Man,* **13,** 13–26.

Davis, M B (1976b) Erosion rates and land use history in southern Michigan. *Environmental Conservation,* **3,** 139–148.

Davis, M B (1981) Outbreaks of forest pathogens in Quaternary History. *Proceedings of IV International Conference on Palynology, Luchnow, India,* **3,** 216–227.

Davis, M B (1984) Holocene vegetational history of the eastern United States. In *Late-Quaternary Environments of the United States. Volume 2. The Holocene* (edited by H E Wright Jr). Longman, London, 166–181.

Davis, P T (1988) Holocene glacier fluctuations in the American cordillera. *Quaternary Science Reviews,* **7,** 129–158.

Davis, P T and Osborn, G (eds) (1988) Holocene glacier fluctuations. *Quaternary Science Reviews,* **7,** 113–242.

Davis, R B and Norton, S A (1978) Paleolimnologic studies of human impact on lakes in the United States, with emphasis on recent research in New England. *Polskie Archiwum Hydrobiologii,* **25**(1/2), 99–115.

Davis, S J M (1987) *The Archaeology of Animals.* Batsford, London.

Dawson, M (1987) Sedimentological aspects of periglacial terrace aggradations: a case study from the English Midlands. In *Periglacial Processes and Landforms in Britain and Ireland* (edited by J Boardman). Cambridge University Press, 265–275.

Dawson, M R and Gardiner, V (1987) River terraces: the general model and a palaeohydrological and sedimentological interpretation of the terraces of the lower Severn. In *Palaeohydrology in Practice* (edited by K J Gregory, J Lewin and J B Thornes). John Wiley, Chichester and New York, 269–305.

Day, G M (1953) The Indian as an ecological factor in the North-Eastern Forest. *Ecology,* **34**(2), 329–346.

Dean, J S (1986) Dendrochronology. In *Dating and Age Determination of Biological Materials* (edited by M R Zimmerman and J L Angell). Croom Helm, Beckenham, 126–165.

Dean, J S (1988) Dendrochronology and palaeoenvironmental reconstruction on the Colorado Plateaus. In *The Anasazi in a Changing Environment* (edited by G J Gumerman). Cambridge University Press, 119–167.

Dean, J S, Euler, R C, Gumerman, G J, Plog, F, Hevly, R H and Karlstrom, T (1985) Human behaviour, demography and palaeoenvironment on the Colorado Plateaus. *American Antiquity,* **50,** 537–554.

Dearing, J A, Alstrom, K, Bergman, A, Reynell, J and Sandgren, P (1990) Recent and long-term records of soil erosion from southern Sweden. In *Soil Erosion on Agricultural Land* (edited by J Boardman, I D L Foster and

J A Dearing). John Wiley, Chichester, 173–191.

de Beaulieu, J–L and Reille, M (1984) A long Upper Pleistocene pollen record from Les Echets, near Lyon, France. *Boreas*, **13**, 111–132.

de Jong, J (1988) Climatic variability during the past three million years, as indicated by vegetational evolution in northwest Europe and with emphasis on data from The Netherlands. *Philosophical Transactions of the Royal Society, London*, **B318**, 603–617.

Delcourt, H R (1987) The impact of prehistoric agriculture and land occupation on natural vegetation. *Tree*, **2**, 39–44.

Delcourt, H R and Delcourt, P A (1987) *Long-term Forest Dynamics of the Temperate Zone with Particular Reference to Late-Quaternary Forest History*. Springer, New York.

Delcourt, P A and Delcourt, H R (1984) Late Quaternary palaeoclimates and biotic responses in eastern North America and the western North Atlantic ocean. *Palaeogeography, Palaeoclimatology, Palaeoecology*, **48**, 263–284.

Delorme, L D and Zoltai, S C (1984) Distribution of an arctic ostracod fauna in space and time. *Quaternary Research*, **21**, 65–73.

Denton, G H and Hughes, T J (1981) *The Last Great Ice Sheets*. John Wiley, Chichester and New York.

Department of the Environment (1990) *Archaeology and Planning: a consultative document*. Department of the Environment, London.

Derbyshire, E, Love, M A and Edge, M J (1985) Fabrics of probable segregated ground-ice origin in some sediment cores from the North Sea basin. In *Soils and Quaternary Landscape Evolution* (edited by J Boardman). John Wiley, Chichester and New York, 261–280.

Devoy, R J (1977) Flandrian sea-level changes and vegetational history of the lower Thames estuary. *Philosophical Transactions of the Royal Society, London*, **B285**, 355–410.

Devoy, R J (1983) Late Quaternary shorelines in Ireland: an assessment of their implications for isostatic land movement and relative sea-level change. In *Shorelines and Isostasy* (edited by D E Smith and A G Dawson). Academic Press, London and New York, 227–254.

Devoy, R J (1985) The problems of a Late Quaternary landbridge between Britain and Ireland. *Quaternary Science Reviews*, **4**, 43–58.

Devoy, R J (1987) Sea-level changes during the Holocene: the North Alantic and Arctic Oceans. In *Sea Surface Studies* (edited by R J N Devoy). Croom Helm, Beckenham, 294–347.

de Vries, J (1981) Measuring the impact of climate on history: the search for appropriate methodologies. In *Climate and History: Studies in Interdisciplinary History* (edited by R I Rotberg and T K Rabb). Princeton University

Press, Princeton, 19–50.

Diamond, J M (1989) Quaternary megafaunal extinctions: variations on a theme by Paganini. *Journal of Archaeological Science*, **16**, 167–175.

Diaz, H F, Andrews, J T and Short, S K (1989) Climate variations in northern North America (6000 BP to present) reconstructed from pollen and tree-ring data. *Arctic and Alpine Research*, **21**, 45–59.

di Castri, F, Hansen, A J and Debussche, M (eds) (1990) *Biological Invasions in Europe and the Mediterranean Basin*. Kluwer, Dordrecht.

Dickinson, R E (1986) How will climate change? In *SCOPE 29, The Greenhouse Effect, Climatic Change, and Ecosystems* (edited by B Bolin, B R Döös, J Jöger and R A Warrick). John Wiley, Chichester and New York, 206–270

Dickinson, R E and Cicerone, R J (1986) Future global warming from atmospheric trace gases. *Nature*, **319**, 109–115.

Digerfeldt, G (1988) Reconstruction and regional correlation of Holocene lake-level fluctuations in Lake Bysjon, South Sweden. *Boreas*, **17**, 165–182.

Dimbleby, G W (1962) *The Development of British Heathlands and their Soils*. Oxford Forestry Memoirs, 23.

Dimbleby, G W (1976) Climate, soil and man. *Philosophical Transactions of the Royal Society, London*, **B275**, 197–208.

Dimbleby, G W (1977) *Ecology and Archaeology*. Edward Arnold, London.

Dimbleby, G W (1985) *The Palynology of Archaeological Sites*. Academic Press, London.

Dimbleby, G W, Greig, J and Scaife, R (1981) Vegetational history of the Isles of Scilly. In *Environmental Aspects of Coasts and Islands* (edited by D Brothwell and G W Dimbleby). British Archaeological Reports, Oxford, IS 94, 127–144.

Dincauze, D F (1984) An archaeological evaluation of the case for pre-Clovis occupations. *Advances in World Archaeology*, **3**, 275–321.

Dincauze, D F (1987) Strategies for palaeoenvironmental reconstruction in archaeology. *Advances in Archaeological Method and Theory*, **11**, 255–336.

Dodge, R E, Fairbanks, R G, Benninger, L K and Maurasse, F (1983) Pleistocene sea levels from raised coral reefs in Haiti. *Science*, **219**, 1423–1425.

Dodson, J R (1989) Late Pleistocene vegetation and environmental shifts in Australia and their bearing on faunal extinctions. *Journal of Archaeological Science*, **16**, 207–217.

Dolukhanov, P M and Khotinskiy, N A (1984) Human cultures and natural environment in the USSR during the Mesolithic and Neolithic. In *Late Quaternary Environments of the Soviet Union* (edited by A A Velichko). Longman, London.

Dool, H M van den, Krijnen, H J and Schuurmans, C J E

(1978) Average winter temperatures at De Bilt (The Netherlands), 1634–1977. *Climatic Change*, **1**, 319–330.

Drew, D P (1982) Environmental archaeology and karstic terrains: the example of the Burren, Co Clare, Ireland. In *Archaeological Aspects of Woodland Ecology* (edited by M Bell and S Limbrey). British Archaeological Reports, Oxford, IS 146, 115–128.

Drewett, P L (1982) *The Archaeology of Bullock Down, Eastbourne, East Sussex: The Development of a Landscape*. Sussex Archaeological Society Monograph 1, Lewes.

Drewett, P, Rudling, D and Gardiner, M (1988) *The South–East to AD 1000*. Longman, London.

Dubois, A D and Ferguson, D K (1985) The climatic history of pine in the Cairngorms based on radiocarbon dates and stable isotope analysis, with an account of events leading up to its colonisation. *Review of Palaeobotany and Palynology*, **46**, 55–80.

Duchaufour, P (1982) *Pedology*. Allen and Unwin, London.

Dugmore, A J (1989a) Tephrochronological studies of glacier fluctuations in Iceland. In *Glacier Fluctuations and Climatic Change* (edited by J Oerlemans). Kluwer, Dordrecht, 37–57.

Dugmore, A J (1989b) Icelandic volcanic ash in Scotland. *Scottish Geographical Magazine*, **105**, 168–172.

Dumond, D (1979) People and pumice on the Alaska Peninsula. In *Volcanic Activity and Human Ecology* (edited by P D Sheets and D K Grayson). Academic Press, London, 373–392.

Dunnell, R C (1980) Evolutionary theory and archaeology. In *Advances in Archaeological Method and Theory*, 3 (edited M B Schiffer). Academic Press, London, 35–99.

Duplessy, J–C and Shackleton, N J (1985) Response of global deep-water circulation to Earth's climatic change 135,000–107,000 years ago. *Nature*, **316**, 500–507.

Dupont, L M (1986) Temperature and rainfall variations in the Holocene based on comparative palaeoecology and isotope geology of hummock and hollow (Bourtangerveen, The Netherlands). *Review of Palaeobotany and Palynology*, **48**, 71–159.

Durham, W H (1978) The coevaluation of human biology and culture. In *Human Behaviour and Adaptation* (edited by V Reynolds and N Blurton Jones). Taylor & Francis, London, 11–32.

Eddy, J A (1976) The maunder minimum. *Science*, **191**, 1189–1202.

Eddy, J A (1977) Climate and the changing sun. *Climatic Change*, **1**, 172–190.

Eddy, J A (1980) Climate and the role of the sun. *Journal of Interdisciplinary History*, **10**, 725–747.

Eddy, J A (1981) Climate and the role of the sun. In *Climate and History* (edited by R I Rotberg and T K Rabb).

Princeton University Press, Princeton, New Jersey, 145–167.

Edwards, K J (1982) Man, space and the woodland edge – speculations on the detection and interpretation of human impact in pollen profiles. In *Archaeological Aspects of Woodland Ecology* (edited by M Bell and S Limbrey). British Archaeological Reports, Oxford, IS 146, 5–22.

Edwards, K J (1983) Quaternary palynology: consideration of a discipline. *Progress in Physical Geography*, **7**, 113–125.

Edwards, K J (1988) The hunter–gatherer agricultural transition and the pollen record in the British Isles. In *The Cultural Landscape: Past, Present and Future* (edited by H H Birks, H J B Birks, P E Kaland and D Moe). Cambridge University Press, 255–266.

Edwards, K J (1989) Meso-Neolithic vegetational impact in Scotland and beyond: palynological considerations. In *The Mesolithic in Europe* (edited by C Bonsall). Donald, Edinburgh, 143–155.

Edwards, K J (1990) Fire and the Scottish Mesolithic: evidence from microscopic charcoal. In *Contributions to the Mesolithic in Europe* (edited by P M Vermeersch and P van Peer). Leuven University Press, 71–79.

Edwards, K J and Ralston, I (1984) Post-glacial hunter–gatherers and vegetational history in Scotland. *Proceedings of the Society of Antiquaries of Scotland*, **114**, 15–34.

Edwards, K J and Rowntree, K M (1980) Radiocarbon and palaeoenvironmental evidence for changing rates of erosion at a Flandrian stage site in Scotland. In *Timescales in Geomorphology* (edited by R A Cullingford, D A Davidson and J Lewin). John Wiley, Chichester and New York, 207–223.

Ehlers, J (1983) The glacial history of north-west Germany. In *Glacial Deposits in North–West Europe* (edited by J Ehlers). Balkema, Rotterdam, 229–247.

Eicher, U and Siegenthaler, U (1976) Palynological and oxygen isotope investigations on Late-Glacial sediment cores from Switzerland. *Boreas*, **5**, 109–117.

Einarsson, Th (1963) Pollen-analytical studies on the vegetation and climatic history of Iceland in late and postglacial time. In *North Atlantic Biota and their History* (edited by A Lowe and D Love). Pergamon Press, Oxford, 355–365.

Einarsson, Th (1986) Tephrochronology. In *Handbook of Holocene Palaeoecology and Palaeohydrology* (edited by B E Berglund). John Wiley, Chichester and New York, 329–342.

Eliade, M (1973) *Australian Religions: an Introduction*. Cornell University Press, Ithaca.

Elsworth, S (1984) *Acid Rain*. Pluto, London.

Emanuel, W R, Shugart, H H and Stevenson, M P (1985) Response to comment: climatic change and the broad-scale distribution of terrestrial ecosystem complexes. *Climatic Change*, **7**, 457–460.

Embleton, C and King, C A M (1975) *Glacial Geomorphology*. Edward Arnold, London.

Engstrom, D R and Wright, H E Jr (1984) Chemical stratigraphy of lake sediments as a record of environmental change. In *Lake Sediments and Environmental History* (edited by E Y Haworth and J W G Lund). Leicester University Press, 11–67.

Erikstad, L and Sollid, L (1986) Neoglaciation in South Norway using lichenometric methods. *Norsk Geografisk Tidsskrift*, **40**, 85–105.

Eronen, M (1983) Late Weichselian and Holocene shorelines in Finland. In *Shorelines and Isostasy* (edited by D E Smith and A G Dawson). Academic Press, London and New York, 183–208.

Evans, J G (1972) *Land Snails in Archaeology*. Seminar Press, London.

Evans, J G (1975) *The Environment of Early Man in the British Isles*. Paul Elek, London.

Evans, J G (1979) The palaeoenvironment of coastal blown-sand deposits in western and northern Britain. *Scottish Archaeological Forum*, **9**, 16–26.

Evans, J G and Jones, H (1979) Mount Pleasant and Woodhenge: the land Mollusca. In *Mount Pleasant, Dorset: Excavations 1970–1971* (edited by G J Wainwright). Society of Antiquaries, London, 190–213.

Evans, J G, French, C and Leighton, D (1978) Habitat change in two Late-glacial and Post-glacial sites in southern Britain: the molluscan evidence. In *The Effect of Man on the Landscape: The Lowland Zone* (edited by S Limbrey and J G Evans). CBA Research Report 21, London, 63–75.

Evans, J G, Limbrey, S, Mate, I and Mount, R J (1988) Environmental change and land-use history in a Wiltshire River Valley in the last 14,000 years. In *The Archaeology of Context in the Neolithic and Bronze Age* (edited by J C Barrett and I A Kinnes). Department of Archaeology and Prehistory, University of Sheffield, 97–103.

Evans, R (1990) Soil erosion: its impact on the English and Welsh Landscape since woodland clearance. In *Soil Erosion on Agricultural Land* (edited by J Boardman, I D L Foster and J A Dearing). John Wiley, Chichester, 231–254.

Fagan, B M (1991) *Ancient North America*. Thames and Hudson, London.

Fairbanks, R G (1989) A 17,000-year glaciomarine sea-level record: influence of glacial melting rates on the Younger Dryas event in deep-ocean circulation. *Nature*, **342**, 637–642.

Fairbridge, R W (1983) Isostasy and eustasy. In *Shorelines and Isostasy* (edited by D E Smith and A G Dawson). Academic Press, London and New York, 3–28.

Fairbridge, R W (1984) Planetary periodicities and terrestrial climatic stress. In *Climatic Changes on a Yearly to Millennial Basis* (edited by N–A Mörner and W Karlén), Reidel. Dordrecht, 509–520.

Falkengren–Grerup, U (1989) Soil acidification and its impact on ground vegetation. *Ambio*, **18**, 179–183.

Farman, J C, Gardiner, B G and Shanklin, J D (1985) Large seasonal losses of ozone in Antarctica reveal seasonal ClO_x/NO_x interactions. *Nature*, **315**, 207–210.

Farmer, C B, Toon, G C, Schaper, P W, Blavier, J–F and Lowes, L L (1987) Stratospheric trace gases in the spring 1986 Antarctic atmosphere. *Nature*, **329**, 126–130.

Farrand, W R (1979) Chronology and palaeoenvironment of Levantine prehistoric sites as seen from sediment studies. *Journal of Archaeological Science*, **6**, 369–392.

Faustova, M A (1984) Late Pleistocene glaciation of European USSR. In *Late Quaternary Environments of the Soviet Union* (edited by A A Velichko). Longman, London, 3–12.

Fink, J and Kukla, G J (1977) Pleistocene climates in central Europe: at least 17 interglacials after the Olduvai event. *Quaternary Research*, **7**, 363–371.

Fisher, D A (1982) Carbon-14 production compared to oxygen isotope records from Camp Century, Greenland and Devon Island, Canada. *Climatic Change*, **4**, 419–426.

Fisher, D A and Koerner, R M (1980) Some aspects of climatic change in the High Arctic during the Holocene as deduced from ice cores. In *Quaternary Palaeoclimate* (edited by W C Mahaney). Geobooks, Norwich, 349–371.

Flannery, K V (1969) Origins and ecological effects of early domestication in Iran and the Near East. In *The Domestication and Exploitation of Plants and Animals* (edited by P J Ucko and G W Dimbleby). Duckworth, London, 73–100.

Flannery, K V (1973) The origins of agriculture. *Annual Review of Anthropology*, **2**, 271–310.

Flannery, K V (ed.) (1986) *Guilá Naquitz: Archaic Foraging and Early Agriculture in Oaxaca, Mexico*. Academic Press, London.

Fleming, A (1988) *The Dartmoor Reaves*. Batsford, London.

Flint, R F (1971) *Glacial and Quaternary Geology*. John Wiley, New York and London.

Flohn, H (1979) On timescales and causes of abrupt palaeoclimatic events. *Quaternary Research*, **12**, 135–149.

Flohn, H (1984) A possible mechanism of abrupt climatic changes. In *Climatic Changes on a Yearly to Millennial Basis* (edited by N–A Mörner and W Karlén). Reidel, Dordrecht, 521–532.

Flood, J (1989) Tread softly for you tread on my bones: the development of cultural resource management in Australasia. In *Archaeological Resource Management in the Modern World* (edited by H F Cleere). Unwin Hyman, London, 79–101.

Foley, R (1987) *Another Unique Species*. Longman, London.

Foley, R (1989) The ecological conditions of speciation: a comparitive approach to the origins of anatomically-modern humans. In *The Human Revolution* (edited by P Mellars and C Stringer). Edinburgh University Press, 298–318.

Fowler, D D (1982) Cultural resources management. *Advances in Archaeological Method and Theory*, **5**, 1–50.

Fowler, P J (1987) The contemporary past. In *Landscape and Culture* (edited by J M Wagstaff). Blackwell, Oxford, 173–191.

Fowler, P J (1989) The experimental earthworks 1958–88. *Council for British Archaeology Annual Report*, **39**, 83–98.

Frankel, H (1988) From continental drift to plate tectonics. *Nature*, **335**, 127–130.

Fredskild, B (1988) Agriculture in a marginal area – South Greenland from the Norse *Landnam* (985 AD) to the present (1985 AD). In *The Cultural Landscape: Past, Present and Future* (edited by H H Birks, H J B Birks, P E Kaland and D Moe). Cambridge University Press, Cambridge, 381–393.

French, H M (1976) *The Periglacial Environment*. Longman, London.

Frenzel, B (1966) Climatic change in the Atlantic/sub–Boreal transition in the Northern Hemisphere: botanical evidence. In *World Climate from 8000–0 BC* (edited by J S Sawyer). Royal Meteorological Society, London, 99–123.

Frey, D G (1986) *Cladocera* analysis. In *Handbook of Holocene Palaeoecology and Palaeohydrology* (edited by B E Berglund). John Wiley, Chichester and New York, 667–692.

Friedman, I, Carrara, P and Gleason, J (1988) Climatic change in the San Juan Mountains, Colorado. *Quaternary Research*, **30**, 350–353.

Fritts, H C (1976) *Tree rings and climate*. Academic Press, London and New York.

Fuji, N (1988) Palaeovegetation and palaeoclimate changes around Lake Biwa, Japan, during the last *ca.* 3 million years. *Quaternary Science Reviews*, **7**, 21–28.

Fullen, M A (1985) Wind erosion of arable soils in east Shropshire (England) during spring 1983. *Catena*, **12**, 111–120.

Fullerton, D S (1986) Chronology and correlation of glacial deposits in the Sierra Nevada, California. *Quaternary Science Reviews*, **5**, 161–169.

Galloway, J N (1989) Atmospheric acidification: projections for the future. *Ambio*, **18**, 161–166.

Geel, B van and Middeldorp, A A (1988) Vegetational history of Carbury Bog (Co. Kildare, Ireland) during the last 850 years and a test of the temperature indicator value of $^2H/^1H$ measurements of peat samples in relation to historical sources and meteorological data. *New Phytologist*, **109**, 377–392.

Genthon, C, Barnola, J M, Raynaud, D, Lorius, C, Jouzel, J, Barkov, N I, Korotkevich, Y S and Kotlyakov, V M (1987) Vostok ice core: climatic response to CO_2 and orbital forcing changes over the last climatic cycle. *Nature*, **329**, 414–418.

Gérard, J–C (1990) Modelling the climatic response to solar variability. *Philosophical Transactions of the Royal Society, London*, **A330**, 561–574.

Gerrard, A J (1984) Multiple working hypotheses and equifinality in geomorphology: comments on the recent article by Haines-Young and Petch. *Transactions of the Institute of British Geographers, New Series*, **9**, 364–366.

Gibbard, P L (1985) *The Pleistocene History of the Middle Thames Valley*. Cambridge University Press.

Gijn, A L van and Waterbolk, H T (1984) The colonisation of the salt marshes of Friesland and Groningen: the possibility of a transhument prelude. *Palaeohistoria*, **26**, 101–122.

Gilliland, R L (1982) Solar, volcanic and CO_2 forcing of recent climatic changes. *Climatic Change*, **4**, 111–131.

Gilman, A and Thornes, J B (1985) *Land Use and Prehistory in South-east Spain*. Allen and Unwin, London.

Gimingham, C H (1972) *Ecology of Heathlands*. Chapman and Hall, London.

Gimingham, C H and de Smidt, J T (1983) Heaths as natural and semi-natural vegetation. In *Man's Impact on Vegetation* (edited by W Holzner, M J A Werger and I Ikusima). W. Junk, The Hague, 185–200.

Girling, M (1982) Fossil insect faunas from forest sites. In *Archaeological Aspects of Woodland Ecology* (edited by M Bell and S Limbrey). British Archaeological Reports, Oxford, IS 146, 129–146.

Girling, M A (1984) Investigations of a second insect assemblage from the Sweet Track. *Somerset Levels Papers*, **10**, 79–91.

Girling, M A (1988) The bark beetle *Scolytus scolytus* (Fabricius) and the possible role of elm disease in the early Neolithic. In *Archaeology and the Flora of the British Isles* (edited by M Jones). Oxford University Committee for Archaeology Monograph, **14**, 34–38.

Gleeson, P and Grosso, G (1976) Ozette site. In *The Excavation of Water–saturated Archaeological Sites (Wet Sites) on the North-West Coast of North America* (edited by D R Croes). National Museum of Man, Mercury Series, Archaeological Survey of Canada, Paper 50, Ottawa.

Gleick, J (1987) *Chaos*. Sphere, London.

Godwin, H (1967) Strip lynchets and soil erosion. *Antiquity*, **41**, 66–67.

Godwin, H (1975) *The History of the British Flora*, 2nd edition. Cambridge University Press, Cambridge.

Godwin, H (1981) *The Archives of the Peat Bogs*. Cambridge University Press, London.

Golson, J (1989) The origins and development of New Guinea agriculture. In *Foraging and Farming* (edited by

D R Harris and G C Hillman). Unwin Hyman, London, 678–687.

Göransson, H (1986) Man and the forests of nemoral broad-leaved trees during the Stone Age. *Striae*, **24**, 143–152.

Gordon, D, Smart, P L, Ford, D C, Andrews, J N, Atkinson, T C and Reeve, P J (1989) Dating of Late Pleistocene interglacial and interstadial periods in the United Kingdom from speleothem growth frequencies. *Quaternary Research*, **31**, 14–26.

Goreau, T J (1990) Balancing atmospheric carbon dioxide. *Ambio*, **19**, 230–236.

Gornitz, V, Lebedeff, S and Hansen, J (1982) Global sea-level trend in the past century. *Science*, **215**, 1611–1614.

Goudie, A (1981) *The Human Impact: Man's Role in Environmental Change*. Basil Blackwell, Oxford.

Goudie, A S (1983) *Environmental Change*. 2nd edition. Clarendon Press, Oxford.

Goudie, A S (1987) Geography and archaeology: the growth of a relationship. In *Landscape and Culture* (edited by J M Wagstaff). Blackwell, Oxford, 11–25.

Goudie, A S (1989) *The Nature of the Environment*, 2nd edition. Blackwell, Oxford.

Gould, S J (1965) Is uniformitarianism necessary? *American Journal of Science*, **263**, 223–228.

Gould, S J and Eldredge, N (1972) Punctuated equilibria: the tempo and mode of evolution reconsidered. *Paleobiology*, **3**, 115–51.

Graf, W L (1983) The arroyo problem – palaeohydrology and palaeohydraulics in the short term. In *Background to Palaeohydrology* (edited by K J Gregory). John Wiley, Chichester, 279–302.

Graham, R W and Mead, J I (1987) Environmental fluctuations and evolution of mammalian faunas during the last deglaciation of North America. In *The Geology of North America, Vol K–3: North America and Adjacent Oceans During the Last Deglaciation* (edited by W F Ruddiman and H E Wright Jr). Geological Society of America, Boulder, Colorado, 371–402.

Gray, J (1981) The use of stable isotope data in climatic reconstruction. In *Climate and History* (edited by T M L Wigley, M J Ingram and B Farmer). Cambridge University Press, London, 53–81.

Gray, J M and Coxon, P (1991) The Loch Lomond Stadial glaciation in Britain and Ireland. In *Glacial Deposits in Great Britain and Ireland* (edited by J Ehlers, P L Gibbard and J Rose). Balkema, Rotterdam, 89–105.

Gray, J M and Thompson, P (1976) Climatic information from $^{18}O/^{16}O$ ratios of cellulose in tree rings. *Nature*, **262**, 481–482.

Grayson, D K (1979) Mount Mazama, climatic change, and Fort Rock Basin archaeofaunas. In *Volcanic Activity and Human Ecology* (edited by P D Sheets and D K Grayson). Academic Press, London, 427–457.

Grayson, D K (1987) An analysis of the chronology of late Pleistocene mammalian extinctions in North America. *Quaternary Research*, **28**(2), 281–289.

Grayson, D K (1989) The chronology of North America Late Pleistocene extinctions. *Journal of Archaeological Science*, **16**, 153–166.

Greeley, W B (1925) The relation of geography to timber supply. *Economic Geography*, **1**, 1–11.

Greensmith, J T and Tooley, M J (eds) (1982) IGCP Project 61, Sea-level movements during the last deglacial hemicycle (about 15,000 years) final report of the UK working group. *Proceedings of the Geologists' Association*, **93**, 3–125.

Greeves, T (1989) Archaeology and the Green Movement: a case for *perestroika*. *Antiquity*, **63**, 659–665.

Gregory, K J (ed.) (1983) *Background to Palaeohydrology*. John Wiley, Chichester and New York.

Gregory, K J, Lewin, J and Thornes, J B (eds) (1987) *Palaeohydrology in Practice*. John Wiley, Chichester and New York.

Greig, J (1982) Past and present limewoods of Europe. In *Archaeological Aspects of Woodland Ecology* (edited by M Bell and S Limbrey). British Archaeological Reports, Oxford, IS 146, 23–55.

Griffin, K O (1975) Vegetation studies and modern pollen spectra from Red lake peatland, northern Minnesota. *Ecology*, **56**, 531–546.

Grigson, C and Mellars, P A (1987) The mammalian remains from the middens. In *Excavations on Oronsay* (edited by P Mellars). Edinburgh University Press, 243–289.

Groenman-van Waateringe, W (1979) Nogle aspekter af jernalderens agerbrug i Holland og NV Tyskland. In *Fra Jernalder till Middelalder* (edited by H Thrane). Odense.

Groenman-van Waateringe, W (1983) The early agricultural utilization of the Irish landscape: the last word on the elm decline. In *Landscape Archaeology in Ireland* (T Reeves-Smyth and F Hamond). British Archaeological Reports, Oxford, BS 116, 217–232.

Groenman-van Waateringe, W (1988) New trends in palynoarchaeology in Northwest Europe or the frantic search for local pollen data. In *Recent Developments in Environmental Analysis in Old and New World Archaeology* (edited by R E Webb). British Archaeological Reports, Oxford, IS 416, 1–19.

Groenman-van Waateringe, W and Wijngaarden-Bakker, L H van (eds) (1987) *Farm Life in a Carolingian Village*. Van Gorcum, Assen.

Groenman-van Waateringe, W and Robinson, M (1988) *Man–made Soils*. British Archaeological Reports, Oxford, IS 410.

Grootes, P M (1984) Radioactive isotopes in the Holocene. In *Late-Quaternary Environments of the United States. Volume 2. The Holocene* (edited by H E Wright Jr). Longman, London, 86–105.

Grosse-Brauckmann, G (1986) Analysis of vegetative plant macrofossils. In *Handbook of Holocene Palaeoecology and Palaeohydrology* (edited by B E Berglund). John Wiley, Chichester and New York, 591–618.

Grove, J M (1979) The glacial history of the Holocene. *Progress in Physical Geography*, **3**, 1–54.

Grove, J M (1988) *The Little Ice Age*. Methuen, London.

Grove, J M and Battagel, A (1983) Tax records from western Norway, as an index of little Ice Age environmental and economic deterioration. *Climatic Change*, **5**, 265–282.

Guillet, B (1982) Study of the turnover of soil organic matter using radio-isotopes (^{14}C). In *Constituents and Properties of Soils* (edited by M Bonneau and B Souchier). Academic Press, London and New York, 238–260.

Guiot, J (1987a) Late Quaternary climatic change in France estimated from multivariate pollen time series. *Quaternary Research*, **28**, 100–118.

Guiot, J (1987b) Reconstruction of seasonal temperatures in central Canada since AD 1700 and detection of the 18.6 and 22-year signals. *Climatic Change*, **10**, 249–268.

Guiot, J, Pons, A, de Beaulieu, J L and Reille, M (1989) A 140,000-year continental climate reconstruction from two European pollen records. *Nature*, **338**, 309–313.

Gumerman, G J (ed.) (1988) *The Anasazi in a Changing Environment*. Cambridge University Press, Cambridge.

Gupt, S K, Sharma, P, Juyal, N and Agrawal, D P (1991) Loess–palaeosol sequence in Kashmir: correlation of mineral magnetic stratigraphy with the marine palaeoclimatic record. *Journal of Quaternary Science*, **6**, 3–12.

Hafsten, U (1983) Biostratigraphical evidence for late Weichselian and Holocene sea-level changes in southern Norway. In *Shorelines and Isostasy* (edited by D E Smith and A G Dawson). Academic Press, London and New York, 161–182.

Haines-Young, R and Petch, J (1980) The challenge of critical rationalism for physical geography. *Progress in Physical Geography*. **4**, 63–77.

Haines-Young, R and Petch, J (1983) Multiple working hypotheses, equifinality and the study of landforms. *Transactions of the Institute of British Geographers, New Series*, **8**, 458–466.

Haines-Young, R and Petch, J (1986) *Physical Geography: its Nature and Methods*. Harper and Row, London.

Halstead, P and O'Shea, J (eds) (1989) *Bad Year Economics: Cultural Responses to Risk and Uncertainty*. Cambridge University Press, Cambridge.

Hammen, T, van der Maarleveld, G C, Vogel, J C and Zagwijn, W H (1967) Stratigraphy, climatic succession and radiocarbon dating of the last glacial in the Netherlands. *Geologie en Mijnbouw*, **46**, 79–95.

Hammer, C U (1980) Acidity of polar ice cores in relation to absolute dating, past volcanism and radio-echoes. *Journal of Glaciology*, **25**, 359–372.

Hammer, C U, Clausen, H B and Dansgaard, W (1980) Greenland ice sheet evidence of postglacial volcanism and its climatic impact. *Nature*, **288**, 230–235.

Hammer, C U, Clausen, H B and Dansgaard, W (1981) Past volcanism and climate revealed by Greenland ice cores. *Journal of Volcanology and Geothermal Research*, **11**, 3–11.

Hammer, C U, Clausen, H B, Friedrich, W L and Tauber, H (1987) The Minoan eruption of Santorini in Greece dated to 1645 BC? *Nature*, **328**, 517–519.

Handler, P (1989) The effects of volcanic aerosols on global climate. *Journal of Volcanology and Geothermal Processes*, **37**, 233–249.

Handmer, J W (ed.) (1987) *Flood Hazard Management: British and International Perspectives*. Geobooks, Norwich.

Hansen, H-O (1982) *Lejre Research Center*. Historical–Archaeological Research Centre, Lejre.

Hansen, J P H, Meldgaard, J and Nordqvist, J (1985) The mummies of Qilakitsoq. *National Geographic*, **167**, 191–207.

Hansen, J P H, Meldgaard, J and Nordqvist, J (1991) *The Greenland Mummies*. British Museum Publications, London.

Hardesty, D L (1977) *Ecological Anthropology*. John Wiley, New York.

Harmon, R S, Thompson, P, Schwarcz, H P and Ford, D C (1978) Late Pleistocene palaeoclimates of North America as inferred from stable isotope studies of speleotherms. *Quaternary Research*, **9**, 54–70.

Harris, C (1987) Solifluction and related periglacial deposits in England and Wales. In *Periglacial Processes and Landforms in Britain and Ireland* (edited by J Boardman). Cambridge University Press, 209–223.

Harris, D R (1969) Agricultural systems, ecosystems and the origins of agriculture. In *The Domestication and Exploitation of Plants and Animals* (edited by P J Ucko and G W Dimbleby). Duckworth, London, 3–15.

Harris, D R (ed.) (1980) *Human Ecology in Savanna Environments*. Academic Press, London.

Harris, D R (1987) The impact on archaeology of radiocarbon dating by accelerator mass spectrometry. *Philosophical Transactions of the Royal Society, London*, **A323**, 23–43.

Harris, D R (1989) An evolutionary continuum of people–plant interaction. In *Foraging and Farming* (edited by D R Harris and G C Hillman). Unwin Hyman, London, 11–26.

Harris, D R and Hillman G C (1989) Introduction. In *Foraging and Farming* (edited by D R Harris and G C Hillman). Unwin Hyman, London, 1–8.

Harvey, L D D (1980) Solar variability as a contributing factor to Holocene climatic change. *Progress in Physical Geography*, **4**, 487–530.

Harvey, L D D (1989) Modelling the Younger Dryas. *Quaternary Science Reviews*, **8**, 137–149.

Haworth, E Y (1972) Diatom succession in a core from Pickerel Lake, Northeastern, South Dakota. *Bulletin of the Geological Society of America*, **83**, 157–172.

Hays, J D, Imbrie, J and Shackleton, N J (1976) Variations in the earth's orbit: pacemaker of the Ice Ages. *Science*, **194**, 1121–1132.

Heden, L O, Likens, G E and Borman, F H (1987) Decrease in precipitation acidity resulting from decreasing SO_{2-4} concentration. *Nature*, **325**, 244–246.

Heidinga, H A (1987) *Medieval Settlement and Economy North of the Lower Rhine. Archaeology and History of Kootwijk and the Veluwe*. Cingula 9, Assen.

Heiser, C B (1989) Domestication of Cucurbitaceae: *Cucurbita* and *Lagenaria*. In *Foraging and Farming* (edited by D R Harris and G C Hillman). Unwin Hyman, London, 471–480.

Hendy, C H and Wilson, A T (1968) Palaeoclimatic data from speleothems. *Nature*, **219**, 48–51.

Hennig, G J, Grun, R and Brunnacker, K (1983) Speleothems, travertines and palaeoclimates. *Quaternary Science Reviews*, **2**, 1–29.

Henry, D O, Leroi-Gourhan, A and Davis, S (1981) The excavation of Hayonim Terrace: an examination of terminal Pleistocene climatic and adaptive changes. *Journal of Archaeological Science*, **8**, 33–58.

Herity, M (1971) Prehistoric fields in Ireland. *Irish University Review*, spring, 258–265.

Hevly, R H, Kelly, R E, Anderson, G A and Olsen, S J (1979) Comparative effects of climatic change, cultural impact and volcanism in the palaeoecology of Flagstaff, Arizona, AD 900–1300. In *Volcanic Activity and Human Ecology* (edited by P D Sheets and D K Grayson). Academic Press, London, 487–523.

Heyworth, A and Kidson, C (1982) Sea-level changes in southwest England and in Wales. *Proceedings of the Geologists' Association*, **93**, 91–112.

Hillaire-Marcel, C (1980) Multiple component postglacial emergence, eastern Hudson Bay, Canada. In *Earth Rheology, Isostasy and Eustasy* (edited by N-A Mörner). John Wiley, Chichester and New York, 215–230.

Hillaire-Marcel, C and Occhietti, S (1980) Chronology, palaeogeography and palaeoclimatic significance of the late and post-glacial events in eastern Canada. *Zeitschrift für Geomorphologie*, **24**, 373–392.

Hillam, J, Groves, C M, Brown, D M, Baillie, M G L, Coles, J M and B J (1990) Dendrochronology of the English Neolithic. *Antiquity*, **64**, 210–220.

Hillman, G C (1989) Late Palaeolithic plant foods from Wadi Kubbaniya in Upper Egypt: dietary diversity, infant weaning, and seasonality in a riverine environment. In *Foraging and Farming* (edited by D R Harris and G C Hillman). Unwin Hyman, London, 207–239.

Hillman, G C, Colledge, S M and Harris, D R (1989) Plant-food economy during the Epipalaeolithic period at Tell Abu Hureyra, Syria: dietary diversity, seasonality and modes of exploitation. In *Foraging and Farming* (edited by D R Harris and G C Hillman). Unwin Hyman, London, 240–268.

Hirons, K R (1990) The Post-glacial environment. In *Rhum: Mesolithic and Later Sites at Kinloch, Excavations 1984–86* (edited by C R Wickham Jones). Society of Antiquaries of Scotland Monograph 7, 137–143.

Hirons, K R and Edwards, K J (1986) Events at and around the first and second *Ulmus* declines: palaeoecological investigations in Co. Tyrone, Northern Ireland. *New Phytologist*, **104**, 131–153.

Hirons, K R and Edwards, K J (1990) Pollen and related studies at Kinloch, Isle of Rhum, Scotland, with particular reference to possible early human impacts on vegetation. *New Phytologist*, **116**, 715–727.

Hodder, I (1986) *Reading the Past*. Cambridge University Press, Cambridge.

Hodder, I (1990) *The Domestication of Europe*. Blackwell, Oxford.

Hofmann, D J, Deshler, T L, Aimedieu, P, Matthews, W A, Johnston, P V, Kondo, Y, Sheldon, W R, Byrne, G J and Benbrook, J R (1989) Stratospheric clouds and ozone depletion in the Arctic during January 1989. *Nature*, **340**, 117–121.

Hoffman, J S, Keyes, D and Titus, J G (1983) *Projecting Future Sea Level Rise*. 2nd edition. Government Printing Office, Washington DC.

Hoffman, J S, Wells, J B and Titus, J G (1985) Future global warming and sea-level rise. In *Icelandic Symposium September 1985* (edited by P Brunn). National Energy Authority, Reykjavik, 53–71.

Hole, F and Flannery, K V (1967) The prehistory of South-West Iran: a preliminary report. *Proceedings of the Prehistoric Society*, **33**, 147–206.

Hollin, J T and Schilling, D H (1981) Late Wisconsin–Weichselian mountain glaciers and small ice caps. In *The Last Great Ice Sheets* (edited by G H Denton and T J Hughes). John Wiley, Chichester and New York, 179–206.

Holyoak, D (1983) The colonisation of Berkshire, England, by land and freshwater Mollusca since the Late Devensian. *Journal of Biogeography*, **10**, 483–498.

Holyoak, D and Preece, R C (1983) Evidence of a high Middle Pleistocene sea-level from estuarine deposits at Bembridge, Isle of Wight. *Proceedings of the Geologists' Association*, **94**, 231–244.

Hooghiemstra, H (1984) *Vegetation and Climatic History of the High Plain of Bogota, Colombia: a Continuous Record of the Last 3.5 Million Years*. Cramer, Vaduz.

Hooykas, R (1963) *The Principle of Uniformity in Geology, Biology and Theology*. Brill, Leiden.

Horie, S (ed.) (1979) *International Project on Palaeolimnology and Late Cenozoic Climate*. No. 2,

Contribution to the Palaeolimnology of Lake Biwa and the Japanese Pleistocene, No. 263, Kyoto University, Otsu.

Horton, D R (1982) The burning question: aborigines, fire and Australian ecosystems. *Mankind*, **13**, 237–251.

Horton, D R (1984) Red kangaroos: last of the Australian megafauna. In *Quaternary Extinctions* (edited by P S Martin and R G Klein). University of Arizona Press, Tucson, 639–680.

Hoskins, W G (1955) *The Making of the English Landscape*. Hodder and Stoughton, London.

Houghton, J T, Jenkins, G J and Ephraums, J J (1990) *Climate Change. The IPCC Scientific Assessment*. Cambridge University Press, Cambridge.

Housley, R A (1988) The environmental context of Glastonbury Lake Village. *Somerset Levels Papers*, **14**, 63–82.

Hovgaard, W (1925) The Norsemen in Greenland. *Geographical Review*, **15**, 605–616.

Hubbert, M K (1967) Critique of the principle of uniformity. In *Uniformity and Simplicity* (edited by C C Albritton). *Geological Society of America, Special Paper*, **89**, 3–23.

Hudson, J C (1990) Settlement of the American grassland. In *The Making of the American Landscape* (edited by M P Conzen). Unwin Hyman, Boston, 169–185.

Hughes, M K, Kelly, P M, Pilcher, J R and LaMarche, V C (1982) *Climate from Tree Rings*. Cambridge University Press, London.

Hughes, T (1983) The stability of the west Antarctic ice sheet: what has happened and what will happen. In *Proceedings: Carbon Dioxide Research Conference: Carbon Dioxide, Science and Consensus*. Department of Energy Conference 820970, Washington DC, IV 51–73.

Hughes, T (1987) Ice dynamics and deglaciation models when ice sheets collapsed. In *The Geology of North America. Volume K–3. North America and Adjacent Oceans During the Last Deglaciation* (edited by W F Ruddiman and H E Wright Jr). Geological Society of America, Boulder, Colorado, 183–220.

Hughes, T J, Denton, G H, Andersen, B G, Schilling, D H, Fastook, J L and Lingle, C S (1981) The last great ice sheets: a global view. In *The Last Great Ice Sheets* (edited by G H Denton and T J Hughes). John Wiley, Chichester and New York, 275–317.

Huntley, B (1990) European vegetation history: palaeovegetation maps from pollen data – 13,000 yr BP to present. *Journal of Quaternary Science*, **5**, 103–122.

Huntley, B and Birks, H J B (1983) *An Atlas of Past and Present Pollen Maps of Europe 0–13,000 years ago*. Cambridge University Press, London.

Huntley, B and Webb, T (eds) (1988) *Vegetation History*. Kluwer Academic Publishers, Dordrecht.

Hutton, J (1788) Theory of the earth, or the investigation of the laws observable in the composition, dissolution and restoration of land upon the globe. *Transactions of the Royal Society of Edinburgh*, **1**, 209–304.

Hutton, J (1795) *Theory of the Earth*, 2 volumes. William Creech, Edinburgh.

Idso, S B (1980) The climatological significance of a doubling of the earth's atmospheric carbon dioxide concentration. *Science*, **207**, 1462–1463.

Idso, S B (1984) What if increases in atmospheric CO_2 have an *inverse* greenhouse effect? 1. Energy balance considerations related to surface albedo. *Journal of Climatology*, **4**, 399–409.

Ijzereef, G F (forthcoming) The presentation of environmental archaeology in Archeon, a plan for an archaeological theme park in the Netherlands. In *Presenting Environmental Archaeology* (edited by J Rackham and N Balaam). Institute of Archaeology, London.

Imbrie, J (1985) A theoretical framework for the Pleistocene ice ages. *Journal of the Geological Society*, **142**, 417–432.

Imbrie, J and Imbrie, K P (1979) *Ice Ages: Solving the Mystery*. Macmillan, London.

Imbrie, J, Hays, J D, Martinson, A, McIntyre, A, Mix, A C, Morley, J J, Pisias, W L, Prell, W L and Shackleton, N J (1984) The orbital theory of Pleistocene climate: support from a revised chronology of the $\delta^{18}O$ record. In *Milankovitch and Climate* (edited by A Berger, J Imbrie, J Hays, G Kukla and B Saltzman). Reidel, Dordrecht, 269–306.

Imeson, A C, Kwaad, F J P M and Mücher, H J (1980) Hillslope processes and deposits in forested areas of Luxembourg. In *Timescales in Geomorphology* (edited by R A Cullingford, D A Davidson and J Lewin). John Wiley, Chichester, 31–42.

Ingold, T (1986) *The Appropriation of Nature*. Manchester University Press, Manchester.

Ingold, T (1990) *Environment and Culture in Ecological Anthropology*. British Association, Science 90, Lecture Aa9, 20–24 August 1990.

Ingram, M J, Underhill, D J and Farmer, G (1981) The use of documentary sources for the study of past climate. In *Climate and History* (edited by T M L Wigley, M J Ingram and G Farmer). Cambridge University Press, London, 180–213.

Ingstad, A S (1977) *The Discovery of a Norse Settlement in America: Excavations at L'Anse aux Meadows, Newfoundland 1961–1968. Volume 1*. Universitetsforlaget, Oslo.

Innes, J B and Simmons, I G (1988) Disturbance and diversity: floristic changes associated with pre-elm decline woodland recession in north-east Yorkshire. In *Archaeology and the Flora of the British Isles* (edited by M Jones). Oxford University Committee for Archaeology

Monograph 14, 7–20.

International Union for Conservation of Nature and Nature Reserves (1980) *A World Conservation Strategy*. International Union for Conservation of Nature and Natural Resources, Gland.

Ivanovich, M and Harmon, R S (1982) *Uranium Series Disequilibrium: Applications to Environmental Problems*. Clarendon Press, Oxford.

Iversen, J (1941) Land occupation in Denmark's stone age. *Danmarks Geologiske Undersøgelse*, Raekke 2, no 66, 7–68.

Iversen, J (1944) *Viscum, Hedera* and *Ilex* as climatic indicators. *Geologiska Föreningens Stockholm Förhandlingar*, **66**, 463–483.

Iversen, J (1954) The late-glacial flora of Denmark and its relation to climate and soil. *Danmarks Geologiske Undersøgelse*. Series 2, **80**, 87–119.

Iversen, J (1958) The bearing of glacial and interglacial epochs on the formation and extinction of plant taxa. *Uppsala Universiteit Årssk*, **6**, 210–215.

Iversen, J (1964) Retrogressive vegetational succession in the Post-glacial. *Journal of Ecology*, **52**, 59–70.

Iversen, J (1969) Retrogressive development of a forest ecosystem demonstrated by pollen diagrams from a fossil mor. *Oikos Suppl*, **12**, 35–49.

Iversen, J (1973) The development of Denmark's nature since the last glacial. *Danmarks Geologiske Undersøgelse*, V Raekke, 7–C, 1–126.

Jacobi, R M (1980) The early Holocene settlements of Wales. In *Culture and Environment in Prehistoric Wales* (edited by J A Taylor). British Archaeological Reports, Oxford, BS 76, 131–206.

Jacobi, R M, Tallis, J H and Mellars, P A (1976) The southern Pennine Mesolithic and the ecological record. *Journal of Archaeological Science*, **3**, 307–320.

Jacobson, G L, Webb, T III and Grimm, E C (1987) Patterns and rates of vegetational change during the deglaciation of eastern North America. In *The Geology of North America. Volume K–3. North America and Adjacent Oceans During the Last Deglaciation* (edited by W F Ruddiman and H E Wright Jr). Geological Society of America, Boulder, Colorado, 277–288.

Jardine, W G (1982) Sea-level changes in Scotland during the last 18,000 years. *Proceedings of the Geologists' Association*, **93**, 25–41.

Jarman, M R, Bailey, G N and Jarman, H N (eds) (1982) *Early European Agriculture: its Foundation and Development*. Cambridge University Press, Cambridge.

Jashemski, W F (1979) Pompeii and Mount Vesuvius AD 79. In *Volcanic Activity and Human Ecology* (edited by P D Sheets and D K Grayson). Academic Press, London, 587–622.

Jelgersma, S (1979) Sea-level changes in the North Sea basin. In *The Quaternary History of the North Sea* (edited by E Oele, R T E Schuttenenhelm and A J Wiggers). Acta Univ. Ups. Symp. Univ. Ups Annum Quingentesimum Celebrantis 2, Uppsala, 233–248.

Jelgersma, S, de Jong, J, Zagwijn, W H and van Regteren Altena, J F (1970) The coastal dunes of the western Netherlands; geology, vegetational history and archaeology. *Mededelingen Rijks Geologische Dienst*, NS 21, 93–167.

Jensen, J (1982) *The Prehistory of Denmark*. Methuen, London.

Johnsen, S J, Dansgaard, W, Clausen, H B and Langway, C C (1970) Climatic oscillations 1200–2000 A.D. *Nature*, **227**, 482.

Johnsen, S J, Dansgaard, W, Clausen, H B and Langway, C C Jr (1972) Oxygen isotope profiles through the Antarctic and Greenland ice sheets. *Nature*, **235**, 429–434.

Jones, M (1981) The development of crop husbandry. In *The Environment of Man: the Iron Age to the Anglo–Saxon Period* (edited by M Jones and G Dimbleby). British Archaeological Report, Oxford, BS 87, 95–127.

Jones, M (1986) *England Before Domesday*. Batsford, London.

Jones, M H D and Henderson-Sellars, A (1990) History of the Greenhouse Effect. *Progress in Physical Geography*, **14**, 1–18.

Jones, P D (1988) Large-scale precipitation fluctuations: a comparison of grid-based and areal precipitation estimates. In *Recent Climatic Change* (edited by S Gregory). Belhaven Press, London and New York, 30–40.

Jones, P D, Raper, S C B, Bradley, R S, Diaz, H F, Kelly, P M and Wigley, T M L (1986a) Northern Hemisphere surface air temperature variations 1851–1984. *Journal of Climatology and Applied Meteorology*, **25**, 161–179.

Jones, P D, Wigley, T M L and Wright, P B (1986b) Global temperature variations between 1861 and 1984. *Nature*, **322**, 430–434.

Jones, P D, Wigley, T M L, Folland, C K, Parker, D E, Angell, J K, Lebedeff, S and Hansen, J E (1988) Evidence for global warming in the past decade. *Nature*, **332**, 790.

Jones, R (1968) The geographical background to the arrival of man in Australia and Tasmania. *Archaeology and Physical Anthropology in Oceania*, **3**, 186–215.

Jones, R (1969) Fire-stick farming. *Australian Natural History*, September, 224–228.

Jones, R (1989) East of Wallace's Line: Issues and problems in the colonisation of the Australian continent. In *The Human Revolution* (edited by P Mellars and C Stringer). Edinburgh University Press, 743–782.

Jones, R and Meehan, B (1989) Plant foods of the Gidjingali: ethnographic and archaeological perspectives from northern Australia on tuber and seed exploitation. In *Foraging and Farming* (edited by D R Harris and G C Hillman). Unwin Hyman, London, 120–135.

Jones, R, Benson-Evans, K and Chambers, F M (1985) Human influence upon sedimentation in Llangorse Lake, Wales. *Earth Surface Processes and Landforms*, **10**, 227–235.

Jones, V J, Stevenson, A C and Battarbee, R W (1989) Acidification of lakes in Galloway, south-west Scotland: a diatom and pollen study of the postglacial history of the Roundloch of Glenhead. *Journal of Ecology*, **77**, 1–23.

Jong, de J (1988) Climatic variability during the past three million years, as indicated by vegetational evolution in northwest Europe and with emphasis on data from The Netherlands. *Philosophical Transactions of the Royal Society, London*, **B318**, 603–617.

Joos, M (1982) Swiss midland lakes and climatic changes. In *Climatic Change in Later Prehistory* (edited by A F Harding). Edinburgh University Press, 44–51.

Jouzel, J, Lorius, C, Petit, J R, Genthon, C, Barkov, N I, Kotlyakov, V M and Petrov, V M (1987) Vostok ice core: a continuous isotope temperature record over the last climatic cycle (160,000 years). *Nature*, **329**, 403–408.

Kaland, P E (1986) The origin and management of Norwegian coastal heaths as reflected by pollen analysis. In *Anthropogenic Indicators in Pollen Diagrams* (edited by K E Behre). Balkema, Rotterdam, 19–36.

Kaland, P E (1988) The development of blanket mires in western Norway. In *The Cultural Landscape. Past, Present and Future* (edited by H H Birks, H J B Birks, P E Kaland and D Moe). Cambridge University Press, 475.

Kana, T W, Michel, J M, Hayes, M O and Jensen, J R (1984) The physical impact of sea-level rise in the area of Charleston, North Carolina. In *Greenhouse Effect and Sea Level Rise* (edited by M C Barth and J G Titus). Van Nostrand Reinhold, New York, 105–150.

Kanari, S, Fuji, N and Horie, S (1984) The palaeoclimatological constituents of palaeotemperature in Lake Biwa. In *Milankovitch and Climate* (edited by A Berger, J Hays, J Imbrie, G Kukla and B Saltzman). Reidel, Dordrecht, 405–414.

Karlén, W (1983) Holocene fluctuations of the Scandinavian alpine tree limit. In *Tree-line Ecology. Proceedings of the Northen Quebec Tree-line Conference* (edited by P Morisset and S Payette), *Nordicana*, **47**, 55–59.

Karlén, W (1984) Dendrochronology, mass balance and glacier front fluctuations in northern Sweden. In *Climatic Changes on a Yearly to Millennial Basis* (edited by N-A Mörner and W Karlén). Reidel, Dordrecht, 263–271.

Karlén, W (1988) Scandinavian glacial and climatic fluctuations during the Holocene. *Quaternary Science Reviews*, **7**, 199–209.

Karlén, W and Denton, G H (1976) Holocene glacier variations in Salek National Park, northen Sweden. *Boreas*, **5**, 25–56.

Karlstrom, T N V (1988) Alluvial chronology and hydrologic change of Black Mesa and nearby regions. In *The Anasazi in a Changing Environment* (edited by G J Gumerman). Cambridge University Press, 45–91.

Karte, J (1987) Pleistocene periglacial conditions and geomorphology on north central Europe. In *Periglacial Processes and Landforms in Britain and Ireland* (edited by J Boardman). Cambridge University Press, 66–75.

Karte, J and Liedtke, H (1981) The theoretical and practical definition of the term 'periglacial' in its geographical and geological meaning. *Biuletyn Peryglacjalny*, **28**, 123–135.

Keating, G M (1978) Relation between monthly variations in global ozone and solar activity. *Nature*, **274**, 873–874.

Keepin, W, Mintzer, I and Kristoferson, L (1986) Emission of CO_2 into the atmosphere. In *SCOPE 29, The Greenhouse Effect, Climatic Change and Ecosystems* (edited by B Bolin, B R Döös, J Jäger and R A Warrick). John Wiley, Chichester and New York, 35–91.

Keigwin, L D (1978) Pliocene closing of the Isthmus of Panama, based on biostratigraphic evidence from nearby Pacific Ocean and Caribbean Sea cores. *Geology*, **6**, 630–634.

Keller, J (1981) Quaternary tephrochronology in Mediterranean regions. In *Tephra Studies* (edited by S Self and R J S Sparks). Reidel, Dordrecht, 227–244.

Kellogg, T B (1984) Late-glacial-Holocene high-frequency climatic changes in deep-sea cores from the Denmark Strait. In *Climatic Changes on a Yearly to Millennial Basis* (edited by N-A Mörner and W Karlén). Reidel, Dordrecht, 123–133.

Kellogg, W W (1987) Mankind's impact on climate: the evolution of an awareness. *Climatic Change*, **10**, 113–136.

Kelly, P M and Sear, C B (1984) Climatic impact of explosive volcanic eruptions. *Nature*, **311**, 740–743.

Kemp, R A (1986) Pre-Flandrian Quaternary soils and pedogenic processes in Britain. In *Palaeosols: Their Recognition and Interpretation* (edited by V P Wright). Blackwell, Oxford, 242–262.

Kenward, H K (1982) Insect communities and death assemblages past and present. In *Environmental Archaeology in the Urban Context* (edited by A R Hall and H K Kenward). Council for British Archaeology Research Report 43, 71–78.

Kerney, M P (1966) Snails and Man in Britain. *Journal of Conchonology*, **26**, 3–14.

Kerney, M P (1968) Britain's fauna of land Mollusca and its relation to the postglacial climatic optimum. *Symposium of the Zoological Society of London*, **22**, 273–291.

Kerney, M P (1976) *Atlas of the Non-Marine Mollusca of the British Isles*. Institute of Terrestrial Ecology, Cambridge.

Kerney, M P (1977) British Quaternary non-marine

Mollusca: a review. In *British Quaternary Studies – Recent Advances* (edited by F W Shotton). Oxford University Press, Oxford, 31–42.

Kerney, M P and Stubbs, A (1980) *The Conservation of Snails, Slugs and Freshwater Molluscs*. Nature Conservancy Council, Shrewsbury.

Khotinskiy, N A (1984a) Holocene vegetational history. In *Late Quaternary Environments of the Soviet Union* (edited by A A Velitchko). Longman, London, 179–200.

Khotinskiy, N A (1984b) Holocene climatic changes. In *Late Quaternary Environments of the Soviet Union* (edited by A A Velitchko). Longman, London, 305–309.

Kidson, C (1953) The Exmoor storm and the Lynmouth floods. *Geography*, **38**, 1–9.

Kidson, C (1982) Sea level changes in the Holocene. *Quaternary Science Reviews*, **1**, 121–151.

Kidson, C, Gilbertson, D D, Haynes, J R, Heyworth, A, Hughes, C E and Whatley, R C (1978) Interglacial marine deposits of the Somerset Levels, South West England. *Boreas*, **7**, 215–228.

Kincer, J B (1941) Climate and weather data for the United States. In *Climate and Man 1941 Yearbook of Agriculture*. United States Department of Agriculture, Superintendent of Documents, Washington DC, 685–699.

Klein, R G (1984) Mammalian extinctions and stone age people in Africa. In *Quaternary Extinctions* (edited by P S Martin and R G Klein). University of Arizona Press, Tucson, 553–573.

Klimanov, V A (1984) Palaeoclimatic reconstructions based on the information statistical method. In *Late Quaternary Environments of the Soviet Union* (edited by A A Velitchko). Longman, London, 287–296.

Knox, J C (1984) Responses of river systems to Holocene climates. In *Late Quaternary Environments of the United States. 2. The Holocene* (edited by H E Wright Jr). Longman, London, 26–41.

Knox, J C (1985) Responses of floods to Holocene climatic change in the Upper Mississippi valley. *Quaternary Research*, **23**, 287–300.

Koerner, R M and Fisher, D A (1981) Studying climatic change from Canadian High Arctic ice cores. *Syllogeus*, **33**, *Climatic Change in Canada* (edited by C R Harington). National Museum of Man, Ottawa, Canada, 195–218.

Koerner, R M and Fisher, D A (1986) The Devon Island ice core and the glacial record. In *Quaternary Environments. Eastern Canadian Arctic, Baffin Bay and Western Greenland* (edited by J T Andrews). George Allen and Unwin, London, 309–327.

Koehler, J and Hajost, S A (1990) The Montreal Protocol: a dynamic agreement for protecting the ozone layer. *Ambio*, **19**, 82–86.

Kolstrup, E (1988) Late Atlantic and early Subboreal vegetational development at Trundholm, Denmark. *Journal of Archaeological Science*, **15**, 503–513.

Koster, E A (1988) Ancient and modern cold-climate aeolian sand deposition: a review. *Journal of Quaternary Science*, **3**, 69–83.

Kristiansen, K (1989) Perspectives on the archaeological heritage: history and future. In *Archaeological Heritage Management in the Modern World* (edited by H F Cleere). Unwin Hyman, London.

Kroeber, A L (1939) *Cultural and Natural Areas of Native North America*. University of California Press, Berkeley.

Kukla, G (1987) Loess stratigraphy in Central China and correlation with an extended oxygen isotope stage scale. *Quaternary Science Reviews*, **6**, 191–219.

Kullman, L (1987) Sequences of Holocene forest history in the Scandes, inferred from megafossil *Pinus sylvestris*. *Boreas*, **16**, 21–26.

Kullman, L (1988) Holocene history of the forest–alpine tundra ecotone in the Scandes Mountains (central Sweden). *New Phytologist*, **108**, 101–110.

Kuhn, J, Libbrecht, K G and Dicke, R (1988) The surface temperature of the sun and changes in the solar constant. *Science*, **242**, 908–911.

Kutzbach, J E (1987) Model simulations of the climatic patterns during the deglaciation of North America. In *The Geology of North America. Volume K–3. North America and Adjacent Oceans During the Last Deglaciation* (edited by W F Ruddiman and H E Wright Jr). Geological Society of America, Boulder, Colorado, 425–446.

Kutzbach, J E and Street-Perrott, F A (1985) Milankovitch forcing of fluctuations in the level of tropical lakes from 18–0 k yr BP. *Nature*, **317**, 130–134.

Kwaad, F (1977) Measurement of rainsplash erosion and the formation of colluvium beneath deciduous woodland in the Luxembourg Ardennes. *Earth Surface Processes*, **2**, 161–173.

Laidlaw, R (1989) Cultural resource planning and management in a mutiple-use agency. In *Archaeological Heritage Management in the Modern World* (edited by H F Cleere). Unwin Hyman, London, 232–235.

Lal, D and Revelle, R (1984) Atmospheric PCO_2 changes recorded in lake sediments. *Nature*, **308**, 344–346.

LaMarche, V C and Hirschboeck, K K (1984) Frost rings in trees as records of major volcanic eruptions. *Nature*, **307**, 121–126.

Lamb, H F, Eicher, U and Switsur, V R (1989) An 18,000-year record of vegetation, lake-level and climatic change from Tigalmamine, Middle Atlas, Morocco. *Journal of Biogeography*, **16**, 65–74.

Lamb, H H (1968) The climatic background to the birth of civilisation. *Advancement of Science*, **25**, 103–120.

Lamb, H H (1970) Volcanic dust in the atmosphere, with a chronology and assessment of its meteorological significance. *Philosophical Transactions of the Royal Society of London*, **A266**, 425–533.

Lamb, H H (1972) *Climate: Present, Past and Future*, Volume 1. Methuen, London.

Lamb, H H (1977) *Climate: Present, Past and Future*, Volume 2. Methuen, London.

Lamb, H H (1979) Climatic variation and changes in the wind and oceanic circulation: The Little Ice Age in the Northeast Atlantic. *Quaternary Research*, **11**, 1–20.

Lamb, H H (1981) An approach to the study of the development of climate and its impact on human affairs. In *Climate and History* (edited by T M L Wigley, M J Ingram and G Farmer). Cambridge University Press, London, 291–309.

Lamb, H H (1982) *Climate, History and the Modern World.* Methuen, London.

Lambrick, G (1985) *Archaeology and Nature Conservation.* Oxford University Department for External Studies, Oxford.

Lambrick, G and McDonald, A (1985) The archaeology and ecology of Port Meadow and Wolvercote Common, Oxford. In *Archaeology and Nature Conservation* (edited by G Lambrick). Oxford University Department for External Studies, Oxford. 95–109.

Landscheidt, T (1984) Cycles of solar flares and weather. In *Climatic Changes on a Yearly to Millennial Basis* (edited by N-A Mörner and W Karlén). Reidel, Dordrecht, 473–481.

Lang, G and Schlüchter, C (eds) (1988) *Lake Mire and River Environments*. Balkema, Rotterdam.

Larson, E, Eide, F, Longva, O and Mangerud, J (1984) Allerød-Younger Dryas climatic inferences from the cirque glaciers and vegetational development in the Nordfjord area, western Norway. *Arctic and Alpine Research*, **16**, 137–160.

Laville, H (1976) Deposits in calcareous rock shelters: analytical methods and climatic interpretation. In *Geoarchaeology* (edited by D A Davidson and M L Shackley). Duckworths, London, 137–155.

Lean, J L (1984) Solar ultraviolet irradiance variations and the earth's atmosphere. In *Climatic Changes on a Yearly to Millennial Basis* (edited by N-A Mörner and W Karlén). Reidel, Dordrecht, 449–471.

Lee, R B and DeVore, I (1968) *Man the Hunter.* Aldine, Chicago.

Lemdahl, G (1991) A rapid climatic change at the end of the Younger Dryas in south Sweden – palaeoclimatic and palaeoenvironmental reconstructions based on fossil insect assemblages. *Palaeogeography, Palaeoecology, Palaeoclimatology*, **83**, 313–331.

Leopold, L B (1963) Some climatic indications in the period AD 1200–1400 in New Mexico. In *Changes of Climate: Proceedings of the Rome Symposium organised by UNESCO and the World Meteorological Organisation.* UNESCO, 265–270.

Le Roy Ladurie, E (1972) *Times of Feast, Times of Famine.*

Allen and Unwin, London.

Le Roy Ladurie, E and Baulant, M (1981) Grape harvests from the fifteenth through the nineteenth centuries. In *Climate and History: Studies in Interdisciplinary History* (edited by R I Rotberg and T K Rabb). Princeton University Press, Princeton, 259–269.

Likens, G E (1989) Some aspects of air pollutant effects on terrestrial ecosystems and prospects for the future. *Ambio*, **18**, 172–178.

Lill, G O and Smalley, I J (1978) Distribution of loess in Britain. *Proceedings of the Geologists' Association*, **89**, 57–65.

Limbrey, S (1975) *Soil Science and Archaeology.* Academic Press, London.

Limbrey, S (1983) Archaeology and Palaeohydrology. In *Background to Palaeohydrology* (edited by K J Gregory). John Wiley, Chichester and New York, 189–212.

Limbrey, S and Robinson, S (1988). Dry land to wetland: soil resources in the Upper Thames Valley. In *The Exploitation of Wetlands* (edited by P Murphy and C French). British Archaeological Reports, Oxford, BS 186, 129–144.

Linick, T W, Damon, P E, Donahue, T J and Jull, A J T (1989) Accelerator mass spectrometry: the new revolution in radiocarbon dating. *Quaternary International*, **1**, 1–6.

Lipe, W D (1984) Value and meaning in cultural resources. In *Approaches to the Archaeological Heritage* (edited by H Cleere). Cambridge University Press, 1–10.

Liu Tungsheng, Zhang Shouxin and Han Jiamao (1986) Stratigraphy and palaeoenvironmental changes in the loess of central China. *Quaternary Science Reviews*, **5**, 489–496.

Lockley, R M (1970) *The Naturalist in Wales.* David and Charles, Newton Abbot.

Lockwood, J G (1983) Modelling climatic change. In *Background to Palaeohydrology* (edited by K J Gregory). John Wiley, Chichester, 25–50.

Löffler, H (1986) Ostracod analysis. In *Handbook of Holocene Palaeoecology and Palaeohydrology* (edited by B E Berglund). John Wiley, Chichester and New York, 693–702.

Logan, J A (1985) Tropospheric ozone: seasonal behaviour, trends and anthropogenic effect. *Journal of Geophysical Research*, **90**, 10 463–10 482.

Long, A (1982) The study of isotopic parameters. In *Climate from Tree Rings* (edited by M K Hughes, P M Kelly, J Pilcher and V LaMarche Jr). Cambridge University Press, London, 50–56.

Long, D and Stoker, M S (1986) Valley asymmetry: evidence for periglacial activity in the central North Sea. *Earth Surface Processes and Landforms*, **11**, 525–532.

Longmore, M E, O'Leary, B M and Rose, C W (1983) Caesium-137 profiles in the sediment of a partial-meromictic lake on Great Sandy Island (Fraser

Island), Queensland, Australia. *Hydrobiologia*, **103**, 21–27.

Loosli, H H (1983) A dating method with ^{39}A. *Earth and Planetary Science Letters*, **63**, 51–62.

Lorius, C, Jouzel, J, Ritz, C, Merlivat, L, Barkov, N I, Korotkevich, Y S and Kotlyakov, V M (1985) A 150,000-year climatic record from Antarctic ice. *Nature*, **316**, 591–596.

Lorius, L, Jouzel, J, Raynaud, D, Hansen, J and Le Treut, H (1990) The ice-core record: climatic sensitivity and future greenhouse warming. *Nature*, **347**, 139–145.

Lough, J M and Fritts, H C (1987) An assessment of the possible effects of volcanic eruptions on North American climate using tree-ring data, 1602–1900 AD. *Climatic Change*, **10**, 219–239.

Louwe Kooijmans, L P (1974) *The Rhine Meuse Delta, Analecta Praehistorica Leidensia* VII. Leiden University Press, Leiden.

Louwe Kooijmans, L P (1980) Archaeology and coastal change in the Netherlands. In *Archaeology and Coastal Change* (edited by F H Thompson). Society of Antiquaries, London, 106–133.

Louwe Kooijmans, L P (1985) *Sporen in het land: De Nederlandse delta in de prehistorie*. Meulenhoff Informaticf, Amsterdam.

Louwe Kooijmans, L P (1987) Neolithic settlement and subsistence in the wetlands of the Rhine–Meuse Delta of the Netherlands. In *European Wetlands in Prehistory* (edited by J M Coles and A J Lawson). Clarendon Press, Oxford, 227–252.

Louwe Kooijmans, L P (1991) Wetland exploitation and upland relations of prehistoric communities in the Netherlands. *East Anglian Archaeology*, **50**.

Lowe, J J and Walker, M J C (1984) *Reconstructing Quaternary Environments*. Longman, London.

Lowenthal, D (1976) The place of the past in the American landscape. In *Geographies of the Mind* (edited by D Lowenthal and M J Bowden). Oxford University Press, New York.

Lowenthal, D (1985) *The Past is a Foreign Country*. Cambridge University Press.

Ložek, V (1986) Mollusca analysis. In *Handbook of Holocene Palaeoecology and Palaeohydrology* (edited by B E Berglund). John Wiley, 729–740.

Lu, Y C (1981) Pleistocene climatic cycles and variation of $CaCO_3$ contents in a loess profile. *Scientia Geologica Sinica*, **2**, 122–131.

Lundelius, E L (1989) The implications of disharmonious assemblages for Pleistocene extinctions. *Journal of Archaeological Science*, **16**, 407–417.

Lundqvist, J (1985) The 1984 symposium on clay varve chronology in Stockholm. *Boreas*, **14**, 97–100.

Lundqvist, J (1986) Late Weichselian glaciation and deglaciation in Scandinavia. *Quaternary Science Reviews*, **5**, 269–292.

Lyell, C (1830–33) *Principles of Geology*, 3 volumes. Murray, London.

Lynch, A (1981) *Man and Environment in Southwest Ireland*. British Archaeological Reports, Oxford, BS 85.

Lynch, F (1980) Bronze Age Monuments in Wales. In *Culture and Environment in Prehistoric Wales* (edited by J A Taylor). British Archaeological Reports, Oxford, BS 76, 233–241.

Maarleveld, G C (1976) Periglacial phenomena and the mean annual temperature during the last glacial time in the Netherlands. *Biuletyn Peryglacjalny*, **26**, 57–78.

MacDonald, G M, Larsen, C P S, Szeicz, J M and Moser, K A (1991) The reconstruction of boreal forest fire history from lake sediments: a comparison of charcoal, pollen, sedimentological and geochemical indices. *Quaternary Science Reviews*, **10**, 53–71.

Macklin, M and Lewin, J (1991) Holocene river alluviation in Britain. In *Proceedings of the Second International Geomorphology Conference*. Frankfurt.

MacNeish, R S (1967) A summary of the subsistence. In *The Prehistory of the Tehuacan Valley* (edited by D S Byers), Vol 1. University of Texas Press, Austin, 290–309.

Macphail, R I (1986) Palaeosols in archaeology: their role in understanding Flandrian pedogenesis. In *Palaeosols: Their Recognition and Interpretation* (edited by V P Wright). Blackwell, Oxford, 263–290.

Macphail, R I (1987) A review of soil science in archaeology in England. In *Environmental Archaeology: A regional review*, Vol II (edited by H C M Keeley). Historic Buildings and Monuments Commission for England, London, 332–379.

Macphail, R I, Courty, M A and Gebhardt, A (1990) Soil micromorphological evidence of early agriculture in north-west Europe. *World Archaeology*, **22**(1), 53–69.

Macpherson, J (1985) The postglacial development of vegetation in Newfoundland and eastern Labrador–Ungava: synthesis and climatic implications. *Syllogeus*, **55**, 267–280.

Magnuson, J J, Baker, J P and Rahel, E J (1984) A critical assessment of effects of acidification on fisheries in North America. *Philosophical Transactions of the Royal Society, London*, **B305**, 501–516.

Maizels, J K (1983a) Palaeovelocity and palaeodischarge determination for coarse gravel deposits. In *Background to Palaeohydrology* (edited by K J Gregory). John Wiley, Chichester and New York, 101–139.

Maizels, J K (1983b) Proglacial channel systems: change and thresholds for change over long, intermediate and short time-scales. In *Modern and Ancient Fluvial Systems* (edited by J D Collinson and J Lewin). Blackwell, Oxford, 251–266.

Manabe, S, Wetherald, R T and Stauffer, R J (1981) Summer dryness due to an increase of atmospheric CO_2

concentration. *Climatic Change*, **4**, 347–385.

Mangerud, J, Andersen, S Th, Berglund, B E and Donner, J J (1974) Quaternary stratigraphy of Norden; a proposal for terminology and classification. *Boreas*, **3**, 109–126.

Manley, G (1974) Central England temperatures: monthly means 1659–1973. *Quarterly Journal of the Royal Meteorological Society*, **100**, 389–405.

Marinatos, S (1939) The volcanic destruction of Minoan Crete. *Antiquity*, **13**, 425–439.

Martin, P S (1967) Prehistoric overkill. In *Pleistocene Extinctions: the Search for a Cause* (edited by P S Martin and H E Wright). Yale University Press, New Haven, 75–120.

Martin, P S (1984) Prehistoric overkill: The global model. In *Quaternary Extinctions* (edited by P S Martin and R G Klein). University of Arizona Press, Tucson, 354–403.

Martin, P S. and Klein, R G (1984) *Quaternary Extinctions*, University of Arizona Press, Tucson.

Martin, P S and Plog, F (1973) *The Archaeology of Arizona*. Doubleday, New York.

Martin, P S and Wright, H E (1967) *Pleistocene Extinctions: the Search for a Cause*. Yale University Press, New Haven.

Martin, P S, Thompson, R S and Long, A (1985) Shasta ground sloth extinction: a test of the blitzkreig model. In *Environments and Extinctions: Man in Late-glacial North America* (edited by J I Mead and D J Meltzer). Center for the Study of Early Man, Orono.

Martinson, D G, Pisias, N G, Hays, J D, Imbrie, J, Moore, T C Jr and Shackleton, N J (1987) Age dating and the orbital theory of Ice Ages: development of a high resolution 0–300,000 year chronology. *Quaternary Research*, **27**, 1–29.

Masson, V M (1989) Archaeological heritage management in the USSR. In *Archaeological Heritage Management in the Modern World* (edited by H F Cleere). Unwin Hyman, London, 195–206.

Masters, P M and Fleming, N C (1983) *Quaternary Coastlines and Marine Archaeology*. Academic Press, London.

Matthews, J A (1974) Families of licheometric dating curves from the Storbreen gletschervorfeld, Jotunheimen, Norway. *Norsk Geografisk Tidsskrift*, **28**, 215–235.

Matthews, J A (1978) Plant colonisation patterns on a gletschervorfeld, southern Norway: a meso-scale geographical approach to phytometric dating. *Boreas*, **7**, 155–178.

Maxwell, J B and Barrie, L A (1989) Atmospheric and climatic change in the Arctic and Antarctic. *Ambio*, **18**, 42–49.

Mayewski, P A, Denton, G H and Hughes, T J (1981) Late Wisconsin ice sheets of North America. In *The Last Great Ice Sheets* (edited by G H Denton and T J Hughes). John Wiley, Chichester and New York, 67–178.

McAndrews, J H (1988) Human disturbance of North American forests and grasslands: the fossil pollen record. In *Vegetation History* (edited by B Huntley and T Webb). Kluwer, Dordrecht, 674–697.

McCracken Peck, R (1990) *Land of the Eagle: a Natural History of North America*. BBC Books, London.

McElroy, M (1983) Marine biological controls on atmospheric CO_2 and climate. *Nature*, **302**, 328–329.

McGimsey, C R and Davis, H A (1984) United States of America. In *Approaches to the Archaeological Heritage* (edited by H Cleere). Cambridge University Press, Cambridge.

McGovern, T H (1981) The economics of extinction in Norse Greenland. In *Climate and History* (edited by T M L Wigley, M J Ingram and G Farmer). Cambridge University Press, London, 404–433.

McGovern, T H, Buckland, P C, Savory, D, Sveinbjarnardottir, G, Andreasen, C and Skidmore, P (1983) A study of the faunal and floral remains from two Norse farms in the Western settlement, Greenland. *Arctic Anthropology*, **20**, 93–120.

McGovern, T H, Bigelow, G, Amorosi, T and Russell, D (1988) Northern Islands, human error and environmental degradation: a view of social and ecological change in the Medieval North Atlantic. *Human Ecology*, **16**, 225–270.

Mead, J I and Meltzer, D J (1984). North American late Quaternary extinctions and the ^{14}C record. In *Quaternary Extinctions* (edited by P S Martin and R G Klein). University of Arizona Press, Tucson, 440–450.

Mead, J I and Meltzer, D J (eds) (1985) *Environments and Extinctions: Man in Late-glacial North America*. Center for the Study of Early Man, Orono.

Mellaart, J (1967) *Çatal Hüyük: a Neolithic town in Anatolia*. Thames and Hudson, London.

Mellars, P (1976) Fire, ecology, animal populations and man: a study of some ecological relationships in prehistory. *Proceedings of the Prehistoric Society*, **42**, 15–45.

Mellars, P (1987) *Excavations on Oronsay: Prehistoric Human Ecology on a Small Island*. Edinburgh University Press, Edinburgh.

Meltzer, D J and Mead, J I (1985) Dating late Pleistocene extinctions: theoretical issues, analytical bias and substantive results. In *Environments and Extinctions: Man in Late-glacial North America* (edited by J I Mead and D J Meltzer). Center for the Study of Early Man, Orono.

Mercer, J H (1969) The Allerød oscillation: a European climatic anomaly? *Arctic and Alpine Research* **1**, 227–234.

Mercer, J H (1978) West Antarctic ice sheet and CO_2 greenhouse effect: a threat of disaster. *Nature*, **271**, 321–325.

Merryfield, D L and Moore, P D (1974) Prehistoric human activity and blanket peat initiation on Exmoor. *Nature*,

250, 439–441.

Mesolella, K J, Matthews, R K, Broecker, W S and Thurber, D L (1969) The astronomical theory of climatic change: Barbados data. *Journal of Geology*, **77**, 250–274.

Messerli, B, Messerli, P, Pfister, C and Zumbuhl, H J (1978) Fluctuations of climate and glaciers in the Bernese Oberland, Switzerland, and their geological significance, 1600 to 1975. *Arctic and Alpine Research*, **10**, 247–260.

Mickleson, D M, Clayton, L, Fullerton, D S and Borns, H W Jr (1983) The late Wisconsin glacial record of the Laurentide ice sheet in the United States. In *Late Quaternary Environments of the United States. Volume 1: The Late Pleistocene* (edited by S C Porter). Longman, London, 3–37.

Mikkelsen, E and Høeg, H I (1979) A reconsideration of Neolithic agriculture in Eastern Norway. *Norwegian Archaeological Review*, **12**(1), 33–47.

Minc, L and Smith, K (1989) The spirit of survival: cultural responses to resource variability in North Alaska. In *Bad Year Economics: Cultural Responses to Risk and Uncertainty* (edited by P Halstead and J O'Shea). Cambridge University Press, Cambridge, 8–39.

Mitchell, G F (1956) Post-Boreal pollen diagrams from Irish raised bogs. *Proceedings of the Royal Irish Academy*, **B57**, 185–251.

Mitchell, F (1986) *Reading the Irish Landscape*. Country House, Dublin.

Mitchell, F (1989) *Man and Environment in Valencia Island*, Royal Irish Academy, Dublin.

Mitchell, J F B (1989) The 'Greenhouse' effect and climatic change. *Reviews of Geophysics*, **27**, 115–139.

Mitchell, J F B and Warrilow, D A (1987) Summer dryness in northern mid-latitudes due to increased CO_2. *Nature*, **330**, 238–240.

Mitchell, J F B, Manabe, S, Tokioka, T and Meleshko, V (1990) Equilibrium climatic change. In *Climate Change. The IPCC Scientific Assessment* (edited by J T Houghton, G J Jenkins and J J Ephraums). Cambridge University Press, Cambridge, 131–172.

Mithen, S (1990) *Thoughtful Foragers: a Study of Prehistoric Decision-making*. Cambridge University Press, Cambridge.

Moffet, L, Robinson, M A and Straker, V (1989) Cereals, fruit and nuts. In *The Beginnings of Agriculture* (edited by A Milles, D Williams and N Gardner). British Archaeological Reports, Oxford, IS 496, 243–261.

Moghissi, A A (1986) Potential public health impacts of acidic deposition. *Water Quality Bulletin*, **11**, 3–5.

Molfino, B, Heusser, L H and Woillard, G M (1984) Frequency components of a Grande Pile pollen record: evidence of precessional orbital forcing. In *Milankovitch and Climate* (edited by A Berger, J Hays, J Imbrie, G Kukla and B Saltzman). Reidel, Dordrecht, 391–404.

Molina, M J and Rowland, F S (1974) Stratospheric sink for chlorofluoromethanes: chlorine catalyzed destruction of ozone. *Nature*, **249**, 812–814.

Molnar, G (1981) A survey of possible effects of long-lasting absence of solar activity on climate and some speculations on possible mechanisms. *Climatic Change*, **33**, 189–207.

Moore, A M T (1985) The development of Neolithic societies in the Near East. *Advances in World Archaeology*, **4**, 1–69.

Moore, A M T (1989) The transition from foraging to farming in Southwest Asia: present problems and future directions. In *Foraging and Farming* (edited by D R Harris and G C Hillman). Unwin Hyman, London, 620–631.

Moore, P D (1975) Origin of blanket mire. *Nature*, **256**, 267–269.

Moore, P D (1985) Forests, man and water. *International Journal of Environmental Studies*, **25**, 159–166.

Moore, P D (1986) Hydrological changes in mires. In *Handbook of Holocene Palaeoecology and Palaeohydrology* (edited by B E Berglund). John Wiley, Chichester and New York, 92–107.

Moore, P D and Webb, J A (1978) *An Illustrated Guide to Pollen Analysis*. Hodder and Stoughton, London.

Morgan, A V (1987) Late Wisconsin and early Holocene palaeoenvironments of east-central North America based on assemblages of fossil Coleoptera. In *The Geology of North America. Volume K–3. North America and Adjacent Oceans During the Last Deglaciation* (edited by W F Ruddiman and H E Wright Jr). Geological Society of America, Boulder, Colorado, 353–370.

Morgan, R A, Litton, C D and Salisbury, C R (1987) Trackways and tree trunks – dating Neolithic oaks in the British Isles. *Tree-ring Bulletin*, **47**, 61–69.

Morgan, R P C (1986) *Soil Erosion and Conservation*. Longman, London.

Mörner, N-A (1980a) The Fennoscandian uplift: geological data and their geodynamical implications. In *Earth Rheology, Isostasy and Eustasy* (edited by N-A Mörner). John Wiley, Chichester and New York, 251–284.

Mörner, N-A (1980b) The Northwest European sea-level laboratory and regional Holocene eustasy. *Palaeogeography, Palaeoclimatology, Palaeoecology*, **29**, 281–300.

Mörner, N-A (1984) Planetary, solar, atmospheric, hydrospheric and endogene processes as origin of climatic changes on the earth. In *Climatic Changes on a Yearly to Millennial Basis* (edited by N-A Mörner and W Karlén). Reidel, Dordrecht, 483–507.

Mörner, N-A and Wallin, B (1977) A 10,000 year temperature record from Gotland, Sweden. *Palaeogeography, Palaeoclimatology, Palaeoecology*, **21**, 113–138.

Morrison, I (1985) *Landscape with Lake Dwellings: the Crannogs of Scotland*. Edinburgh University Press, Edinburgh.

Moss, R P (1977) Deductive strategies in geographical generalisation. *Progress in Physical Geography*, **10**, 23–39.

Mott, R J, Grant, D R, Stea, R and Occhietti, S (1986) Late-glacial climatic oscillation in Atlantic Canada equivalent to the Allerød-Younger Dryas event. *Nature*, **323**, 247–250.

Mottershead, D N and White, I D (1972) The lichenometric dating of glacier recession, Tunbergalbre, southern Norway. *Geografiska Annaler*, **54A**, 47–52.

Mulholland, M T (1988) Territoriality and horticulture: a perspective for prehistoric southern New England. In *Holocene Human Ecology in Northeastern North America* (edited by G P Nicholas). Plenum Press, New York.

Munaut, A V (1986) Dendrochronology applied to mire environments. In *Handbook of Holocene Palaeoecology and Palaeohydrology* (edited by B E Berglund). John Wiley, Chichester and New York, 371–385.

Muniz, I P (1984) The effects of acidification on Scandinavian freshwater fish fauna. *Philosophical Transactions of the Royal Society, London*, **B305**, 517–528.

Nagy, J and Ofstad, K (1980) Quaternary Foraminifera and sediments in the Norwegian Channel. *Boreas*, **9**, 39–52.

Neftel, A, Oeschger, H, Schwander, J, Stauffer, B and Zumbrunn, R (1982) Ice core sample measurements give atmospheric CO_2 content during the past 40 000 years. *Nature*, **295**, 220–223.

Neftel, A, Moor, E, Oeschger, H and Stauffer, B (1985) Evidence from polar ice cores for the increase in atmospheric CO_2 in the past two centuries. *Nature*, **315**, 45–47.

Newell, R E (1981) Further studies of the atmospheric temperature change produced by Mt Agung volcanic eruption in 1963. *Journal of Volcanology and Geothermal Research*, **11**, 61–66.

Newell, R E and Doplick, T G (1979) Questions concerning the possible influence of anthropogenic CO_2 on atmospheric temperature. *Journal of Applied Meteorology*, **18**, 822–825.

Newell, R E and Hsiung, J (1984) Sea-surface temperature, atmospheric CO_2 and the global energy budget: some comparisons between the past and present. In *Climatic Changes on a Yearly to Millennial Basis* (edited by N-A Mörner and W Karlén). Reidel, Dordrecht, 533–561.

Newman, W S, Cinquemani, L J, Pardi, R R and Marcus, L F (1980) Holocene delevelling of the United States east coast. In *Earth Rheology, Isostasy and Eustasy* (edited by N-A Mörner). John Wiley, Chichester and New York, 449–463.

Nicholas, G P (1988) Ecological leveling: the archaeology and environmental dynamics of early post-glacial land use. In *Holocene Human Ecology of Northeastern North America* (edited by G P Nicholas). Plenum, New York, 257–296.

Norton, P E P (1977) Marine Mollusca in the East Anglian Pleistocene. In *British Quaternary Studies – Recent Advances* (edited by F W Shotton). Oxford University Press, Oxford, 43–53.

O'Connell, M (1990) Origins of Irish lowland blanket bog. In *Ecology and Conservation of Irish Peatlands* (edited by G J Doyle). Royal Irish Academy, Dublin, 49–71.

Odgaard, B V (1988) Heathland history in western Jutland, Denmark. In *The Cultural Landscape: Past, Present and Future* (edited by H H Birks, H J B Birks, P E Kaland and D Moe). Cambridge University Press, 311–319.

Odgaard, B V and Rostholm, H (1987) A single grave barrow at Harreskor, Jutland. *Journal of Danish Archaeology*, **6**, 87–100.

Oerlemans, J (1982) Response of the Antarctic ice sheet to a climatic warming: a model study. *Journal of Climatology*, **2**, 1–11.

Oeschger, H and Langway, C C Jr (eds) (1989) *The Environmental Record in Glaciers and Ice Sheets*. John Wiley, Chichester and New York.

Oeschger, H, Welten, M, Eicher, U, Moll, M, Riesen, T, Siegenthaler, U and Wegmuller, S (1980) ^{14}C and other parameters during the Younger Dryas cold phase. *Radiocarbon*, **22**, 299–310.

Ogilvie, A E J (1981) Climate and economy in eighteenth century Iceland. In *Consequences of Climatic Change* (edited by C Delano Smith and M Parry), Department of Geography, University of Nottingham, 54–69.

Ogley, B (1988) *In the Wake of the Hurricane*. Froglets Publications, Westerham.

O'Hear, A (1989) *An Introduction to the Philosophy of Science*. Clarendon, Oxford.

Oldane, R N and O'Hara, C J (1980) New radiocarbon dates from the inner Continental Shelf off southeastern Massachusetts and a local sea-level curve for the past 12,000 years. *Geology*, **8**, 102–106.

Oldfield, F (1978) Lakes and their drainage basins as units of sediment-based ecological study. *Progress in Physical Geography*, **1**, 460–504.

Oldfield, F, Appleby, P G, Cambray, R S, Eakins, J D, Barber, K E, Battarbee, R W, Pearson, G W and Williams, J M (1979) ^{210}Pb, ^{137}Cs and ^{239}Pu profiles in ombrotrophic peat. *Oikos*, **33**, 40–45.

Olsson, I U (1986) Radiometric dating. In *Handbook of Holocene Palaeoecology and Palaeohydrology* (edited by B E Berglund). John Wiley, Chichester and New York, 273–312.

Orme, B J (1982). The use of radiocarbon dates from the Somerset Levels. *Somerset Levels Papers*, **8**, 9–25.

Osborn, G and Luckman, B H (1988) Holocene glacier

fluctuations in the Canadian cordillera (Alberta and British Columbia). *Quaternary Science Reviews*, **7**, 115–127.

Osborne, P J (1974) An insect assemblage of early Flandrian age from Lea Marston, Warwickshire, and its bearing on the contemporary climate and ecology. *Quaternary Research*, **4**, 471–486.

O'Shea, J M (1989) The role of wild resources in small-scale agricultural systems: tales from the lakes and the plains. In *Bad Year Economics* (edited by P Halstead and J O'Shea). Cambridge University Press, 57–67.

O'Sullivan, P E (1983) Annually-laminated lake sediments and the study of Quaternary environmental changes. *Quaternary Science Reviews*, **1**, 245–312.

Paludan-Müller, C (1978) High Atlantic food gathering in northwestern Zealand: ecological conditions and spatial representation. *Studies in Scandinavian Prehistory and Early History*, **1**, 120–157.

Park, C C (1987) *Acid Rain: Rhetoric and Reality*. Methuen, London.

Parry, M L (1975). Secular climatic change and marginal agriculture. *Transactions of the Institute of British Geographers*, **64**, 1–13.

Parry, M L (1978) *Climatic Change, Agriculture and Settlement*. Dawson, Folkstone.

Parry, M L (1981) Evaluating the impact of climatic change. In *Consequences of Climatic Change* (edited by C Delano Smith and M Parry). Department of Geography, University of Nottingham, 3–16.

Parry, M L (1990) *Climatic Change and World Agriculture*. Earthscan, London.

Parry, M L and Carter, T R (1985) The effect of climatic variations on agricultural risk. *Climatic Change*, **7**, 95–110.

Paterson, W S B and Hammer, C U (1987) Ice core and other glaciological data. In *The Geology of North America. Volume K–3. North America and Adjacent Oceans During the Last Deglaciation* (edited by W F Ruddiman and H E Wright Jr). Geological Society of America, Boulder, Colorado, 91–110.

Paterson, W S B, Koerner, R M, Fisher, D, Johnsen, S J, Clausen, H B, Dansgaard, W, Bucher, P and Oeschger, H (1977) An oxygen isotope climatic record from the Devon Island Ice cap, Arctic Canada. *Nature*, **266**, 508–511.

Patterson, W A and Backman, A E (1988) Fire and disease history of forests. In *Vegetation History* (edited by B Huntley and T Webb). Kluwer, Dordrecht, 603–632.

Patterson, W A and Sassaman, K E (1988) Indian fires in the prehistory of New England. In *Holocene Human Ecology in Northeastern North America* (edited by G P Nicholas). Plenum, New York, 107–135.

Patterson, W A III, Edwards, K J and McGuire, D J (1987) Microscopic charcoal as a fossil indicator of fire. *Quaternary Science Reviews*, **6**, 3–23.

Peacock, J D (1989) Marine molluscs and Late Quaternary environmental studies with particular reference to the Late-Glacial period in Northwest Europe. *Quaternary Science Reviews*, **8**, 179–192.

Pearce, F (1989) *Climate and Man*. Vision Books, London.

Pears, N V (1968) Postglacial tree-lines of the Cairngorm Mountains, Scotland. *Transactions of the Botanical Society of Edinburgh*, **40**, 361–394.

Pearsall, W H (1950) *Mountains and Moorlands*, New Naturalist, Collins, London.

Pearson, G W, Pilcher, J R and Baillie, M G L (1983) High-precision ^{14}C measurement of Irish oaks to show the natural ^{14}C variations from 200 BC to 4000 BC. *Radiocarbon*, **25**, 179–186.

Pearson, M G (1973) Snowstorms in Scotland 1782–1786. *Weather*, **28**, 195–201.

Peglar, S, Fritz, S C, Alapieti, T, Saarnisto, M and Birks, H J B (1984) The composition and formation of laminated lake sediments in Diss Mere, Norfolk, England. *Boreas*, **13**, 13–28.

Peltier, W R (1987) Mechanisms of relative sea-level change and the geophysical responses to ice-water loading. In *Sea Surface Studies* (edited by R J N Devoy). Croom Helm, Beckenham, 57–96.

Peltier, W R and Andrews, J T (1983) Glacial geology and glacial isostasy of the Hudson Bay region. In *Shorelines and Isostasy* (edited by D E Smith and A G Dawson). Academic Press, London and New York, 285–319.

Penck, A and Brückner, E (1909) *Die Alpen im Eiszeitalter*. Tachnitz, Leibnitz.

Penney, D N (1987) Application of Ostracoda to sea-level studies. *Boreas*, **16**, 237–247.

Pennington, W (1965) The interpretation of some Postglacial vegetation diversity at different Lake District sites. *Proceedings of the Royal Society, London*, **B161**, 293–375.

Pennington, W (1978) The impact of man on some English lakes: rates of change. *Polskie Archiwum Hydrobiologii*, **25**(1/2), 429–437.

Pennington, W, Haworth, E Y, Bonny, A P and Lishman, J P (1972) Lake sediments in Northern Scotland. *Philosophical Transactions of the Royal Society, London*, **B264**, 191–294.

Pennington, W, Cambray, R S, Eakins, J D and Harkness, D D (1976) Radionuclide dating of the recent sediments of Blelham Tarn. *Freshwater Biology*, **6**, 317–331.

Perry, I and Moore, P D (1987) Dutch elm disease as an analogue of Neolithic elm decline. *Nature*, **326**, 72–73.

Peteet, D M (1987) Younger Dryas in North America – modelling, data analysis and re-evaluation. In *Abrupt Climatic Change* (edited by W H Berger and L D Labeyrie). Reidel, Dordrecht, 185–193.

Péwé, T L (1973) Ice wedge casts and past permafrost distribution in North America. *Geoforum*, **15**, 15–26.

Péwé, T L (1983a) The periglacial environment in North America during Wisconsin time. In *Late Quaternary Environments of the United States. Volume 1: The Late Pleistocene* (edited by S C Porter). Longman, London, 157–189.

Péwé, T L (1983b) Alpine permafrost in the contiguous United States: a review. *Arctic and Alpine Research*, **15**, 145–156.

Pfister, C (1981) An analysis of the Little Ice Age climate in Switzerland and its consequences for agricultural production. In *Climate and History* (edited by T M L Wigley, M J Ingram and G Farmer). Cambridge University Press, London, 214–248.

Pfister, C (1984) The potential of documentary data for the reconstruction of past climates. Early 16th to 19th century. Switzerland as a case study. In *Climatic Changes on a Yearly to Millennial Basis* (edited by N-A Mörner and W Karlén). Reidel, Dordrecht, 331–337

Pickersgill, B (1989) Cytological and genetical evidence on the domestication and diffusion of crops within the Americas. In *Foraging and Farming* (edited by D R Harris and G C Hillman). Unwin Hyman, London, 426–439.

Pilcher, J R, Baillie, M G L, Schmidt, B and Becker, B (1984) A 7272-year tree-ring record for western Europe. *Nature*, **312**, 150–152.

Pilles, P J (1979) Sunset Crater and the Sinague: a new interpretation. In *Volcanic Activity and Human Ecology* (edited by P D Sheets and D K Grayson). Academic Press, London, 459–485.

Pisias, N G and Shackleton, N J (1984) Modelling the global climatic response to orbital forcing and atmospheric carbon dioxide changes: a frequency domain approach. *Nature*, **310**, 757–759.

Pissart, A (1987) Weichselian periglacial structures and their environmental significance: Belgium, the Netherlands, and northern France. In *Periglacial Processes and Landforms in Britain and Ireland* (edited by J Boardman). Cambridge University Press, 77–85.

Plassche van de, O (1982) Sea-level change and water-level movements in the Netherlands during the Holocene. *Mededelingen Rijks Geologische Dienst*, 36–1.

Playfair, J (1802) *Illustrations of the Huttonian Theory of the Earth*. Cadell, Davies and Creech, Edinburgh.

Plog, F, Gumerman, G J, Evler, R C, Dean, J S, Hevly, R H and Karlstrom, T N V (1988) Anasazi adaptive strategies: the model, predictions and results. In *The Anasazi in a Changing Environment* (edited by G J Gumerman). Cambridge University Press, 230–276.

Pollack, J B, Toon, O B, Sagan, C, Summers, A, Baldwin, B and van Camp, W (1976) Volcanic explosions and climatic change: a theoretical assessment. *Journal of Geophysical Research*, **81**, 1071–1083.

Ponel, P and Coope, G R (1990) Lateglacial and early Flandrian Coleoptera from La Taphanel, Massif Central, France: climatic and ecological implications. *Journal of Quaternary Science*, **5**, 235–249.

Popper, K R (1972) *Objective Knowledge*. Oxford University Press, Oxford.

Popper, K R (1974) *Conjectures and Refutations*. 5th edition, Routledge and Kegan Paul, London.

Porter, S C (1986) Pattern and forcing of Northern Hemisphere glacier variations during the last millennium. *Quaternary Research*, **26**, 27–48.

Porter, S C, Pierce, K L and Hamilton, T D (1983) Late Wisconsin mountain glaciation in the western United States. In *Late-Quaternary Environments of the United States. Volume 1. The Late-Pleistocene* (edited by S C Porter). University of Minnesota Press, Minneapolis, 71–111.

Pott, R (1986) Der pollenanalytische Nachweis extensiver Waldbewirtschaftungen in den Haubergen des Siegerlandes. In *Anthropological Indicators in Pollen Diagrams* (edited by K-E Behre). Balkema, Rotterdam, 125–134.

Potter, T W (1976) Valleys and settlements: some new evidence, *World Archaeology*, **8**, 207–219.

Preece, R C (1986) Faunal remains from radiocarbon-dated soils within landslip debris from the Undercliff, Isle of Wight, England. *Journal of Archaeological Science*, **13**, 189–200.

Probert-Jones, J R (1984) On the homogeneity of the annual temperature of central England since 1659. *Journal of Climatology*, **4**, 241–253.

Proffitt, M H, Margitan, J J, Kelly, K K, Lowenstein, M, Padolske, J R and Chan, K R (1990) Ozone loss in the Arctic polar vortex inferred from high-altitude aircraft measurements. *Nature*, **347**, 31–36.

Pyle, D M (1989) Ice core acidity peaks, retarded tree growth and putative eruptions. *Archaeometry*, **31**, 88–91.

Rackham, O (1980) *Ancient Woodland*, Arnold, London.

Rackham, O (1986) *The History of the Countryside*. Dent and Sons, London.

Rackham, O (1988) Trees and woodland in a crowded landscape – the cultural landscape of the British Isles. In *The Cultural Landscape: Past, Present and Future* (edited by H H Birks, H J B Birks, P E Kaland and D Moe). Cambridge University Press, 53–77.

Radley, J and Simms, C (1967) Wind erosion in east Yorkshire. *Nature*, **216**, 20–22.

Raftery, B (1990) *Trackways through Time: Archaeological Investigations on Irish Bog Roads 1985–1989*. Headline Publishing, Dublin.

Ramanathan, V (1975) Greenhouse effect due to chlorofluorocarbons: climatic implications. *Science*, **190**, 50–52.

Ramanathan, V, Cicerone, R J, Singh, H B and Kiehl, J T (1985) Trace gas trends and their potential role in climate change. *Journal of Geophysical Research*, **90**, 5547–5566.

Rampino, M R and Self, S (1982) Historic eruptions of Tambora (1815), Krakatau (1883) and Agung (1963): their stratospheric aerosols and climatic impact. *Quaternary Research*, **18**, 127–143.

Rawlence, D J (1988) The post-glacial history of Splan Lake, New Brunswick. *Journal of Palaeolimnology*, **1**, 51–60.

Raynaud, D and Barnola, J M (1985) An Antarctic ice core reveals atmospheric CO_2 variations over the past few centuries. *Nature*, **315**, 309–311.

Raynaud, D, Chappellaz, J, Barnola, J M, Korotkevich, Y S and Lorius, C (1988) Climatic and CH_4 cycle implications of glacial–interglacial CH_4 change in the Vostok ice core. *Nature*, **333**, 655–657.

Reichstein, J (1984) Federal Republic of Germany. In *Approaches to the Archaeological Heritage* (edited by H. Cleere). Cambridge University Press, 37–47.

Reiter, C (1984) The turbulent nature of a chaotic world. *New Scientist*, **1412**, 11.

Renberg, I (1981) Formation, structure and visual appearance of iron-rich, varved lake sediments. *Verhandlungen Internationalen Vereinigung für Limnologie*, **21**, 94–101.

Rendall, H, Worsley, P, Green, F and Parks, D (1991) Thermoluminescence dating of the Chelford Interstadial. *Earth and Planetary Science Letters*, **103**, 182–189.

Renfrew, C (1979) The eruption of Thera and Minoan Crete. In *Volcanic Activity and Human Ecology* (edited by P D Sheets and D K Grayson). Academic Press, London, 565–585.

Renfrew, C (1983) Geography, archaeology and environment. *Geographical Journal*, **149**, 316–322.

Renfrew, C and Bahn, P (1991) *Archaeology: Theories, Methods and Practice*. Thames and Hudson, London.

Renouf, M (1988) Sedentary coastal hunter–fishers: an example from the younger stone age of northern Norway. In *The Archaeology of Prehistoric Coastlines* (edited by G Bailey and J Parkington). Cambridge University Press, Cambridge, 102–115.

Revelle, R (1983) Probable future changes in sea level resulting from increasing atmospheric carbon dioxide. In *Carbon Dioxide Assessment Committee, Changing Climate*. National Academy Press, Washington DC.

Revelle, R (1985) Introduction: The scientific history of carbon dioxide. In *The Carbon Cycle and Atmospheric CO_2: Natural Variations Archaeoan to Present* (edited by E T Sundquist and W S Broecker). Geophysical Monograph 32, American Geophysical Union, Washington DC, 1–4.

Reynolds, P (1981) Deadstock and livestock. In *Farming Practice in British Prehistory* (edited by R Mercer). Edinburgh University Press, 97–122.

Reynolds, P J (1987) *Ancient Farming*. Shire, Aylesbury.

Ribes, E (1990) Astronomical determination of the solar variability. *Philosophical Transactions of the Royal Society, London*, **A33**, 487–497.

Rice. R J (1988) *Fundamentals of Geomorphology*, 2nd edition, Longman, London.

Richter, G (1980) On the soil erosion problem in the temperate humid area of Central Europe. *Geojournal*, **4**, 279–287.

Richter, G (1986) Investigations of soil erosion in Central Europe. In *Soil Erosion* (edited by C P Burnham and J I Pitman). SEESOIL, **3**, 14–27.

Rind, D, Peteet, D, Broecker, W, McIntyre, A and Ruddiman, W F (1986) The impact of cold North Atlantic sea surface temperatures on climate: implications for the Younger Dryas cooling (11–10 ka). *Climate Dynamics*, **1**, 3–33.

Rinsland, C P, Levine, J S and Miles, T (1985) Concentration of methane in the troposphere deduced from 1951 infrared solar spectra. *Nature*, **318**, 245–249.

Roberts, N (1984) Pleistocene environments in time and space. In *Hominid Evolution and Community Ecology* (edited by R Foley). Academic Press, London, 25–53.

Roberts, N (1989) *The Holocene: an Environmental History*. Blackwell, Oxford.

Robin, G de Q (1986) Changing the sea level. In *SCOPE 29. The Greenhouse Effect, Climatic Change and Ecosystems* (edited by B Bolin, B R Döös, J Jäger and R A Warrick). John Wiley, Chichester and New York, 323–359.

Robinson, D and Rasmussen, P (1989) Botanical investigations at the Neolithic Lake Village at Weier, north east Switzerland: Leaf, hay and cereals as animal fodder. In *The Beginnings of Agriculture* (edited by A Milles, D Williams and N Gardner). British Archaeological Reports, Oxford, IS 496, 149–163.

Robinson, J E (1980) The marine ostracod record from the Lateglacial period in Britain and NW Europe: a review. In *Studies in the Lateglacial of North West Europe* (edited by J J Lowe, J M Gray and J E Robinson). Pergamon Press, Oxford, 115–122.

Robinson, M A (1985) Nature conservation and environmental archaeology. In *Archaeology and Nature Conservation* (edited by G Lambrick). Oxford University Department for External Studies, 11–17.

Robinson, M A and Lambrick, G H (1984) Holocene alluviation and hydrology in the Upper Thames Basin. *Nature*, **308**, 809–814.

Rodhe, H (1989) Acidification in a global perspective. *Ambio*, **18**, 155–160.

Rose, F (1974) The epiphytes of oak. In *The British Oak: its History and Natural History* (edited by M G Morris and F H Perring). Classey, Faringdon, 250–273.

Rose, J (1981) Raised shorelines. In *Techniques in Geomorphology* (edited by A S Goudie). Allen and Unwin, London, 327–341.

Rose, J (1985) The Dimlington Stadial/Dimlington

Chronozone: a proposal for naming the main glacial episode of the Late Devensian in Britain. *Boreas*, **14**, 225–230.

Rose, J, Turner, C, Coope, G R and Bryan, M D (1980) Channel changes in a lowland river catchment over the last 13,000 years. In *Timescales in Geomorphology* (edited by R A Cullingford, D A Davidson and J Lewin). John Wiley, Chichester and New York, 159–175.

Rose, J, Boardman, J, Kemp, R A and Whiteman, C (1985) Palaeosols and the interpretation of the British Quaternary stratigraphy. In *Geomorphology and Soils* (edited by K Richards, R R Arnett and S Ellis). George Allen and Unwin, London, 348–375.

Rose, M R, Dean, J S and Robinson, W J (1981) *The Past Climate of Arroyo Hondo, New Mexico, Reconstructed from tree rings.* Arroyo Hondo Archaeological Series no 4. Santa Fe NM School of American Research Press.

Rottländer, R C A (1976) Variations in the chemical composition of bone as an indicator of climatic change. *Journal of Archaeological Science*, **3**, 83–86.

Rowley-Conwy, P A (1981) Slash and burn in the temperate European Neolithic. In *Farming Practice in British Prehistory* (edited by R Mercer). Edinburgh University Press, 85–96.

Rowley-Conwy, R (1982) Forest grazing and clearance in temperate Europe with special reference to Denmark: an archaeological view. In *Archaeological Aspects of Woodland Ecology* (edited by M Bell and S Limbrey). British Archaeological Reports, Oxford, IS 146, 199–216.

Rozanski, K, Harmata, K, Noryskiewicz, B, Ralska-Jasiewiczowa, M and Wcislo, D (1988) Palynological and isotope studies on carbonate sediments from some Polish lakes – preliminary results. In *Lake Mire River Environments* (edited by G Lang and C Schlüchter). Balkema, Rotterdam, 41–49.

Ruddiman, W F (1987) Synthesis: the ocean/ice sheet record. In *The Geology of North America. Volume K–3. North America and Adjacent Oceans During the Last Deglaciation* (edited by W F Ruddiman and H E Wright Jr). Geological Society of America, Boulder, Colorado, 463–478.

Ruddiman, W F and Duplessy, J-C (1985) Conference on the last deglaciation: timing and mechanism. *Quaternary Research*, **23**, 1–17.

Ruddiman, W F and McIntyre, A (1977) Late Quaternary surface ocean kinematics and climatic change in the high-latitude North Atlantic. *Journal of Geophysical Research*, **82**, 3877–3887.

Ruddiman, W F and McIntyre, A (1981a) The North Atlantic ocean during the last deglaciation. *Palaeogeography, Palaeoclimatology, Palaeoecology*, **35**, 145–214.

Ruddiman, W F and McIntyre, A (1981b) The mode and mechanism of the last deglaciation;: oceanic evidence.

Quaternary Research, **16**, 125–134.

Ruddiman, W F and Raymo, M (1988) Northern Hemisphere climate regimes during the past 3 Ma: possible tectonic connections. *Philosophical Transactions of the Royal Society, London*, **B318**, 411–430.

Ruddiman, W, Sancetta, C D and McIntyre, A (1977) Glacial–interglacial response rate of subpolar North Atlantic waters to climatic change: the record in oceanic sediments. *Philosophical Transactions of the Royal Society, London*, **B280**, 119–142.

Ruddiman, W F, McIntyre, A and Shackleton, N J (1986a) North Atlantic sea-surface temperatures for the last 1.1 million years. In *North Atlantic Palaeoceanography* (edited by C P Summerhayes and N J Shackleton). Geological Society of America Special Publication, 21, 155–173.

Ruddiman, W F, Raymo, M and McIntyre, A (1986b) Matuyama 41,000 year cycles: North Atlantic Ocean and northern hemisphere ice sheets. *Earth and Planetary Science Letters*, **80**, 117–129.

Ruddiman, W F, McIntyre, A and Raymo, M (1986c) Palaeoenvironmental results from North Atlantic Sites 607 and 609. *Initial Reports of the Deep-Sea Drilling Project*, Volume 94. US Government Printing Office, Washington DC, 855–878.

Ruhe, R V (1983) Depositional environment of Late Wisconsin loess in the midcontinental United States. In *Late Quaternary Environments of the United States. Volume 1 The Late Pleistocene* (edited by S C Porter). Longman, London, 130–137.

Rummery, T A (1983) The use of magnetic measurements in interpreting the fire histories of lake drainage basins. *Hydrobiologia*, **103**, 53–58.

Russell, B (1961) *A History of Western Philosophy*, 2nd edition. George Allen and Unwin, London.

Rymer, L (1978) The use of uniformitarianism and analogy in palaeoecology, particularly pollen analysis. In *Biology and Quaternary Environments* (edited by D Walker and J C Guppy). Australian Academy of Sciences, Canberra, 245–258.

Saarnisto, M (1986) Annually laminated lake sediments. In *Handbook of Holocene Palaeoecology and Palaeohydrology* (edited by B E Berglund). John Wiley, Chichester and New York, 343–370.

Sadler, J P (1991) *Archaeological and Biogeographical Implications of Palaeoentomological Studies in Orkney and Iceland.* Unpublished Ph.D. Thesis, University of Sheffield.

Salisbury, E (1964) *Weeds and Aliens.* Collins, London.

Scaife, R G (1987) A review of later Quaternary plant microfossil and macrofossil research in southern England. In *Environmental Archaeology: a Regional Review*, 2 (edited by H C M Keeley). Historic Buildings and

Monuments Commission for England, London.

Scaife, R G (1988) The elm decline in the pollen record of South East England and its relationship to early agriculture. In *Archaeology and the Flora of the British Isles* (edited by M Jones). Oxford University Committee for Archaeology Monograph 14, 21–33.

Scaife, R G and Burrin, P J (1983) Floodplain development in and the vegetational history of the Sussex High Weald and some archaeological implications. *Sussex Archaeological Collections*, **121**, 1–10.

Scaife, R G and Macphail, R I (1983) The Post-Devensian development of heathland soils and vegetation. In *Soils of the Heathlands and Chalklands* (edited by P Burnham). *Seesoil*, **1**, 70–99.

Schiffer, M B (1987) *Formation Processes of the Archaeological Record*. University of New Mexico Press, Albuquerque.

Schirmer, W (1983) Criteria for the differentiation of the Late Quaternary river terraces. *Quaternary Studies in Poland*, **4**, 199–205.

Schirmer, W (1988) Holocene valley development on the Upper Rhine and Main. In *Lake, Mire and River Environments* (edited by G Lang and C Schlüchter). Balkema, Rotterdam, 153–160.

Schlüchter, C (1986) The Quaternary glaciations of Switzerland, with special reference to the Northern Alpine Foreland. *Quaternary Science Reviews*, **5**, 413–420.

Schmidt, M (1986) Possible influences of solar radiation variations on the atmospheric circulation in the Northern Hemisphere of the earth. *Climatic Change*, **8**, 279–296.

Schönwiese, C-D (1988) Volcanism and air temperature variations in recent centuries. In *Recent Climatic Change* (edited by S Gregory). Belhaven Press, London and New York, 20–29.

Schönwiese, C-D and Malcher, J (1987) The CO_2 temperature response. A comparison of the results from General Circulation Models with statistical assessments. *Journal of Climatology*, **7**, 215–229.

Schove, D J (1983) *Sunspot Cycles*. (Benchmark Volume 38). Hutchinson and Ross, London.

Schove, D J (1984) Sunspot cycles and global oscillations. In *Climatic Changes on a Yearly to Millennial Basis* (edited by N-A Mörner and W Karlén). Reidel, Dordrecht, 257–259.

Schove, D J and Fairbridge, R W (eds) (1984) *Ice Cores, Varves and Tree-rings*. Balkema, Rotterdam.

Schumer, B (1984) *The Evolution of Wychwood to 1400: Pioneers, Frontiers and Forests*. Leicester University, Department of English Local History Occasional Paper, 3rd Ser no 6.

Schumm, S A (1979) Geomorphic thresholds: the concept and its applications. *Transactions of the Institute of British Geographers*, **4**, 485–515.

Schumm, S A and Brakenridge, G R (1987) River responses.

In *The Geology of North America. Volume K–3. North America and Adjacent Oceans During the Last Deglaciation* (edited by W F Ruddiman and H E Wright Jr). Geological Society of America, Boulder, Colorado, 221–240.

Schwarcz, H P (1989) Uranium series dating of Quaternary deposits. *Quaternary International*, **1**, 7–18.

Schweingruber, F H (1988) *Tree Rings. Basics and Applications of Dendrochronology*. Reidel, Dordrecht.

Scuderi, L A (1990) Tree-ring evidence for climatically effective volcanic eruptions. *Quaternary Research*, **34**, 67–85.

Sear, C B, Kelly, P M, Jones, P D and Goodess, C M (1987) Global surface temperature responses to major volcanic eruptions. *Nature*, **330**, 365–367.

Sejrup, H P, Aarseth, I, Ellingsen, K L, Reither, E, Jansen, E, Lovlie, R, Bent, A, Brigham-Grette, J, Larsen, E and Stoker, M (1987) Quaternary stratigraphy of the Fladen area, central North Sea: a multidisciplinary study. *Journal of Quaternary Science*, **2**, 35–58.

Sejrup, H P, Sjøholm, J, Furnes, H, Beyer, I, Eide, L, Jansen, E and Mangerud, J (1989) Quaternary tephrachronology In the Iceland Plateau, north of Iceland. *Journal of Quaternary Science*, **4**, 109–114.

Self, S, Rampino, M A and Borbera, J J (1981) The possible effects of large 19th and 20th century volcanic eruptions on zonal and hemispherical surface temperatures. *Journal of Volcanology and Geothermal Research*, **11**, 41–60.

Semken, H A Jr (1984) Holocene mammalian biogeography and climatic change in the eastern and central United States. In *Late Quaternary Environments of the United States. Volume 2. The Holocene* (edited by H E Wright Jr). Longman, London, 182–207.

Seppala, M (1987) Periglacial phenomena of northern Fennoscandia. In *Periglacial Processes and Landforms in Britain and Ireland* (edited by J Boardman). Cambridge University Press, 45–55.

Shackleton, N J (1977) The oxygen isotope stratigraphic record of the Late Pleistocene. *Philosophical Transactions of the Royal Society, London*, **B280**, 169–179.

Shackleton, N J and Opdyke, N D (1973) Oxygen isotope and palaeomagnetic stratigraphy of equatorial Pacific core V28–238: oxygen isotope temperatures and ice volumes on a 10^5 and 10^6 year scale. *Quaternary Research*, **3**, 39–55.

Shackleton, N J, Hall, M A, Line, J and Shuxi, C (1983) Carbon isotope data in core V19–30 confirm reduced carbon dioxide concentration in the ice age atmosphere. *Nature*, **306**, 319–322.

Shackleton, N J, Backman, J, Zimmerman, H, Kent, D V, Hall, M A, Roberts, D G, Schnitker, D, Baldauf, J G, Desprairies, A, Homrighausner, R, Huddlestone, P, Keen, J B, Kaltenbach, A J, Krumsiek, K A O, Morton, A C, Murray, J W and Westberg-Smith, J (1984) Oxygen isotope calibration of the onset of ice-rafting and history of glaciation in the North Atlantic region. *Nature*, **307**, 620–623.

Shaw, E M (1985) Some aspects of rainfall records with selected computational examples from northern England. In *The Climatic Scene* (edited by M J Tooley and G M Sheail). Allen and Unwin, London, 60–92.

Shaw, E M (1988) *Hydrology in Practice*. Van Nostrand Reinhold International, London.

Shea, J H (1983) Twelve fallacies of uniformitarianism. *Geology*, **10**, 455–460.

Shennan, I (1983) Flandrian and Late Devensian sea-level changes and crustal movements in England and Wales. In *Shorelines and Isostasy* (edited by D E Smith and A G Dawson). Academic Press, London and New York, 255–284.

Shennan, I (1986a) Flandrian sea-level changes in the Fenland. 1. The geographical setting and evidence of sea-level change. *Journal of Quaternary Science*, **1**, 119–154.

Shennan, I (1986b) Flandrian sea-level changes in the Fenland. II: Tendencies of sea-level movement, altitudinal changes, and local and regional factors. *Journal of Quaternary Science*, **1**, 155–179.

Shennan, I (1987) Global analysis and correlation of sea-level data. In *Sea Surface Studies* (edited by R J N Devoy). Croom Helm, Beckenham, 198–230.

Shennan, I (1989) Holocene crustal movements and sea-level changes in Great Britain. *Journal of Quaternary Science*, **4**, 77–85.

Shipek, F C (1981) A native American adaptation to drought: the Kumeyaay as seen in the San Diego Mission Records 1770–1798. *Ethnohistory*, **28**(4), 295–312.

Shipek, F C (1989) An example of intensive plant husbandry: the Kumeyaay of southern California. In *Foraging and Farming* (edited by D R Harris and G C Hillman). Unwin Hyman, London, 159–170.

Shotton, F W (1978) Archaeological inferences from the study of alluvium in the lower Severn–Avon valleys. In *The Effect of Man on the Landscape: the Lowland Zone* (edited by S Limbrey and J G Evans). CBA Research Report 21, London, 27–31.

Shroder, J F (1980) Dendrogeomorphology: a review of new techniques, of tree-ring dating. *Progress in Physical Geography*, **4**, 161–188.

Shugart, H H, Antonovksy, M Ya, Jarvis, P G and Sandford, A P (1986) CO_2, climatic change and forest ecosystems. In *SCOPE 29, The Greenhouse Effect, Climatic Change and Ecosystems* (edited by B Bolin, B R Döös, J Jäger and R A Warrick). John Wiley, Chichester and New York, 475–521.

Šibrava, V, Bowen, D Q and Richmond, G M (ed.) (1986) Quaternary glaciations of the Northern Hemisphere. *Quaternary Science Reviews*, **5**.

Siegenthaler, U and Eicher, U (1986) Stable oxygen and carbon isotope analysis. In *Handbook of Holocene Palaeoecology and Palaeohydrology* (edited by B E Berglund). John Wiley, Chichester and New York, 407–422.

Siegenthaler, U, Eicher, U and Oeschger, H (1984) Lake sediments as continental ^{18}O records from the transition Glacial–Postglacial. *Annals of Glaciology*, **5**, 149–152.

Simmons, I G (1975) Towards an ecology of Mesolithic man in the uplands of Great Britain. *Journal of Archaeological Science*, **2**, 1–15.

Simmons, I G (1979) Late Mesolithic societies and the environment of the uplands of England and Wales. *Bulletin of the Institute of Archaeology London*, **16**, 11–129.

Simmons, I G (1989) *Changing the Face of the Earth*. Blackwell, Oxford.

Simmons, I G and Dimbleby, G W (1974) The possible role of ivy (*Hedera helix* L.) in the Mesolithic economy of Western Europe. *Journal of Archaeological Science*, **1**, 291–296.

Simmons, I G and Tooley, M J (eds) (1981) *The Environment in British Prehistory*. Duckworth, London.

Simmons, I G, Dimbleby, G W and Grigson, C (1981) The Mesolithic. In *The Environment in British Prehistory* (edited by I G Simmons and M J Tooley). Duckworth, London, 82–124.

Simmons, I G, Atherden, M A, Cundill, P R, Innes, J B and Jones, R L (1982) Prehistoric environments. In *Prehistoric and Roman Archaeology of North East Yorkshire*. British Archaeological Reports, Oxford, BS 104, 33–99.

Simola, H (1977) Diatom succession in the formation of annually laminated lake sediments. *Annales Botanica Fennica*, **14**, 143–148.

Sissons, J B (1967) *The Evolution of Scotland's Scenery*. Oliver and Boyd, London.

Sissons, J B (1976) *The Geomorphology of the British Isles: Scotland*. Methuen, London.

Sissons, J B (1979) Palaeoclimatic inferences from former glaciers in Scotland and the Lake District. *Nature*, **278**, 518–521.

Sissons, J B (1983) Shorelines and isostasy in Scotland. In *Shorelines and Isostasy* (edited by D E Smith and A G Dawson). Aacademic Press, London and New York, 209–226.

Sleeman, D P, Devoy, R J and Woodman, P C (eds) (1983) *Proceedings of the Postglacial colonization Conference, University College, Cork, 15–16 October 1983*. Irish Biogeographical Society Occasional Publication I.

Smirnova, T Y and Nikonov, A A (1990) A revised lichenometric method and its application to dating past earthquakes. *Arctic and Alpine Research*, **22**, 375–388.

Smith, A G (1970) The influence of Mesolithic and Neolithic man on british vegetation. In *Studies in the Vegetational History of the British Isles* (edited by D

Walker and R G West). Cambridge University Press, London, 81–96.

Smith, A G (1981) The Neolithic. In *The Environment in British Prehistory* (edited by I G Simmons and M J Tooley). Duckworth, London, 125–209.

Smith, C J (1980) *Ecology of the English Chalk*. Academic Press, London.

Smith, D E, Cullingford, R A and Haggart, B A (1985) A major coastal flood during the Holocene in eastern Scotland. *Eiszeitalter und Gegenwart*, **35**, 109–118.

Smith, E A, Vonder Haar, T H, Hickey, J H and Maschhoff, R (1983) The nature of the short-period fluctuations in solar irradiance received by the earth. *Climatic Change*, **5**, 211–235.

Smith, R T and Taylor, J A (1969) The Postglacial development of vegetation and soils in north Cardiganshire. *Institute of British Geographers Transactions*, **48**, 75–96.

Sofia, S, Demarque, P and Endal, A (1985) From solar dynamo to terrestrial climate. *American Scientist*, **73**, 326–333.

Solem, T, (1986) Age, origin and development of blanket mires in Sør-Trøndelag, central Norway. *Boreas*, **15**, 101–115.

Solem, T (1989) Blanket mire formation at Haramsøy, Møre og Romsdal, western Norway. *Boreas*, **18**, 221–235.

Søllerød Museum (1985) *Vedbaekfundene*. Vedbaek, Denmark.

Sonett, C P and Finney, S A (1990) The spectrum of radiocarbon. *Philosophical Transactions of the Royal Society, London*, **A330**, 413–426.

Sonett, C P and Suess, H E (1984) Correlation of bristlecone pine ring-widths with atmospheric ^{14}C variations: a climate–sun relation. *Nature*, **307**, 141–143.

Sorensen, C J (1977) Reconstructed Holocene bioclimates. *Annals of the Association of American Geographers*, **67**, 214–222.

Spaulding, W G (1983) The overkill hypothesis as a plausible explanation for the extinctions of Late Wisconsin megafauna. *Quaternary Research*, **20**, 110–112.

Spaulding, W G, Leopold, E B and Van Devender, R (1983) Late Wisconsin palaeoecology of the American southwest. In *Late Quaternary Environments of the United States. 1: The Pleistocene* (edited by S C Porter). Longman, London, 259–293.

Speight, M C D (1991) *Saproxylic Invertebrates and their Conservation*. Council of Europe.

Spencer, P J (1975) Habitat change in coastal sand-dune areas: the molluscan evidence. In *The Effect of Man on the Landscape of the Highland Zone* (edited by J G Evans, S Limbrey and H Cleere). Council for British Archaeology, London, Research Report No. 11, 96–103.

Starkel, L (1983a) The reflection of hydrological changes in the fluvial environment of the temperate zone during the last 15,000 years. In *Background to Palaeohydrology*

(edited by K J Gregory). John Wiley, Chichester and New York, 213–235.

Starkel, L (1983b) Climatic change and fluvial response. In *Mega-Geomorphology* (edited by R Gardner and H Scoging). Clarendon Press, Oxford, 195–211.

Starkel, L (1985) Lateglacial and Postglacial history of river valleys in Europe as a reflection of climatic changes. *Zeitschrift für gletscherkunde und Glazialgeologie*, **21**, 159–164.

Starkel, L (1987) The evolution of European rivers – a complex response. In *Palaeohydrology in Practice* (edited by K J Gregory, J Lewin and J B Thornes). John Wiley, Chichester and New York, 333–339.

Starkel, L (1988) Tectonic, anthropogenic and climatic factors in the history of the Vistula river valley downstream of Cracow. In *Lake, Mire and River Environments* (edited by G Lang and C Schlüchter). Balkema, Rotterdam, 161–170.

Starkel, L, Alexandrowicz, SW, Klimek, K, Kowalkowski, A, Mamakowa, K, Niedzialkowska, E and Pazdur, M (1981) The evolution of the Wisloka Valley near Dębica during the late glacial and Holocene. *Folia Quaternaria*, **53**, 1–91.

Stauffer, B, Lochbronner, E, Oeschger, H and Schwander, J (1988) Methane concentration in the glacial atmosphere was only half that of the preindustrial Holocene. *Nature*, **332**, 812–814.

Stead, I M, Bourke, J B and Brothwell, D (1986) *Lindow Man. The Body in the Bog*. British Museum, London.

Steensberg, A (1979) *Draved: An Experiment in Stone Age Agriculture : Burning, Sowing and Harvesting*. National Museum of Denmark, Copenhagen.

Steensberg, A (1986) *Man the Manipulator*. National Museum of Denmark, Copenhagen.

Stephenson, F R (1990) Historical evidence concerning the sun: interpretation of sunspot records during the telescopic and pretelescopic eras. *Philosophical Transactions of the Royal Society, London*, **A339**, 499–512.

Stewart, I (1990) *Does God play Dice? The Mathematics of Chaos*. Penguin, London.

Stewart, O C (1956) Fire as the first great force employed by man. In *Man's Role in Changing the Face of the Earth* (edited by W L Thomas). University of Chicago Press, Chicago, 115–133.

Stewart, T G and England, J (1983) Holocene sea-ice variations and palaeoenvironmental change, northernmost Ellesmere Island, NWT, Canada. *Arctic and Alpine Research*, **15**, 1–17.

Stoddart, D R (1986) *On Geography*. Blackwell, Oxford.

Stolarski, R S (1988) The Antarctic Ozone Hole. *Scientific American*, **258**, 20–26.

Street, F A and Grove, A T (1979) Global maps of lake-level fluctuations since 30,000 yr BP. *Quaternary Research*, **12**, 83–118.

Stringer, C B and Grün, R (1991) Time for the last Neanderthals. *Nature*, **351**, 701–702.

Strong, A E (1989) Greater global warming revealed by satellite-derived sea-surface temperature trends. *Nature*, **338**, 642–645.

Stuart, A J (1979) Pleistocene occurrences of the European pond tortoise (*Emys orbicularis* L.) in Britain. *Boreas*, **8**, 359–371.

Stuart, A J (1982) *Pleistocene Vertebrates in the British Isles*. Longman, London.

Stuart, A J and van Wijngaarden-Bakker, L H (1985) Quaternary vertebrates. In *The Quaternary History of Ireland* (edited by K J Edwards and W P Warren). Academic Press, London, 221–249.

Stuiver, M (1980) Solar variability and climatic change during the current millennium. *Nature*, **286**, 868–871.

Stuiver, M and Braziunas, T F (1989) Atmospheric ^{14}C and century-scale solar oscillations. *Nature*, **338**, 405–408.

Stuiver, M and Quay, P D (1980) Changes in atmospheric carbon-14 attributed to a variable sun. *Science*, **207**, 11–19.

Sturlodottir, S A and Turner, J (1985) The elm decline at Pawlaw Mire: an anthropogenic interpretation. *New Phytologist*, **99**, 323–329.

Sturm, M (1979) Origin and composition of clastic varves. In *Moraines and Varves* (edited by C Schlüchter). Balkema, Rotterdam, 281–285.

Succow, M and Lange, E (1984) The mire types of the German Democratic Republic. In *European Mires* (edited by P D Moore). Academic Press, London, 149–175.

Summerhayes, C P and Shackleton, N J (1986) *North Atlantic Palaeoceanography*. Geological Society of America Special Publication, 21.

Sundquist, E T (1987) Ice core links CO_2 to climate. *Nature*, **329**, 389–390.

Sutcliffe, A J (1970) A section of an imaginary bone cave. *Studies in Speleology*, **2**, 79–80.

Sutherland, D G (1984a) Modern glacier characteristics as a basis for inferring former climates with particular reference to the Loch Lomond Stadial. *Quaternary Science Reviews*, **3**, 291–309.

Sutherland, D G (1984b) The Quaternary deposits and landforms of Scotland and the neighbouring shelves – a review. *Quaternary Science Reviews*, **3**, 157–254.

Swain, A M (1978) Environmental changes during the past 2000 years in north-central Wisconsin: analysis of pollen, charcoal and seeds from varved lake sediments. *Quaternary Research*, **10**, 55–68.

Sykora, K V (1990) History of the impact of man on the distribution of plant species. In *Biological Invasions in Europe and the Mediterranean Basin* (edited by F di Castri, A J Hansen and M Debussche). Kluwer, Dordrecht. 37–50.

Tallis, J H and Johnson, R H (1980) The dating of landslides in Longdendale, north Derbyshire, using pollen-analytical techniques. In *Timescales in Geomorphology* (edited by R A Cullingford, D A Davidson and J Lewin). John Wiley, Chichester and New York, 189–205.

Taylor, J A (1980) Environmental changes in Wales during the Holocene. In *Culture and Environment in Prehistoric Wales* (edited by J A Taylor). British Archaeological Reports, Oxford, BS 76, 101–130.

Taylor, R E (1987) *Radiocarbon Dating: an Archaeological Perspective*. Academic Press, San Diego.

Teller, J T (1987) Proglacial lakes and the southern margins of the Laurentide ice sheet. In *The Geology of North America. Volume K–3. North America and Adjacent Oceans During the Last Deglaciation* (edited by W F Ruddiman and H E Wright Jr). Geological Society of America, Boulder, Colorado, 39–69.

Terasmae, J and Weeks, N C (1979) Natural fires as an index of palaeoclimate. *Canadian Field Naturalist*, **93**, 116–125.

Therkorn, L L and Abbink, A A (1987) Seven levee sites B,C,D,G,H,F and P. In *Assendelver Polder Papers* I (edited by R W Brandt, W Groenman-van Waateringe and S E van der Leeuw). Van Giffen Instituut, Amsterdam, 115–167.

Thomas, C (1985) *Exploration of a Drowned Landscape*. Batsford, London.

Thomas, J (1988) Neolithic explanations revisited: the Mesolithic–Neolithic transition in Britain and south Scandinavia. *Proceedings of the Prehistoric Society*, **54**, 59–66.

Thomas, K D (1982) Neolithic enclosures and woodland habitats on the South Downs in Sussex, England. In *Archaeological Aspects of Woodland Ecology* (edited by M Bell and S Limbrey). British Archaeological Reports, Oxford, IS 146, 147–170.

Thomas, R H, Sanderson, T J O and Rose, K E (1979) Effect of climatic warming on the West Antarctic ice sheet. *Nature*, **277**, 355–358.

Thompson, A and Jones, A (1986) Rates and causes of proglacial river terrace formation on southeast Iceland: an application of lichenometric dating techniques. *Boreas*, **15**, 231–246.

Thompson, P, Schwarcz, H P and Ford, D C (1974) Continental Pleistocene climatic variations from speleothem and isotopic data. *Science*, **184**, 893–895.

Thompson, R (1986) Palaeomagnetic dating. In *Handbook of Holocene Palaeoecology and Palaeohydrology* (edited by B E Berglund). John Wiley, Chichester and New York, 313–327.

Thompson. R and Oldfield. F (1986) *Environmental Magnetism*. George Allen and Unwin, London.

Thompson, R and Turner, G M (1979) British geomagnetic master curve 10,000–0 BP from dating European sediments. *Geophysical Research Letters*, **6**, 249–252.

Thorarinsson, S (1971) Damage caused by tephra fall in some big Icelandic eruptions and its relation to the thichness of the tephra layers. In *Acta of the First International Scientific Congress on the Volcano of Thera* (September 1969) (edited by A Kaloyeropoyloy), 213–236.

Thorarinsson, S (1979) On the damage caused by volcanic eruptions with special reference to tephra and gases. In *Volcanic Activity and Human Ecology* (edited by P D Sheets and D K Grayson). Academic Press, New York, 125–159.

Thorarinsson, S (1981) Greetings from Iceland. Ash-fall and volcanic aerosols in Scandinavia. *Geografiska Annaler*, **63A**, 109–118.

Thorbahn, P F and Cox, D C (1988) The effect of estuary formation on prehistoric settlement in southern Rhode Island. In *Holocene Human Ecology in Northeastern North America* (edited by G P Nicholas). Plenum, New York, 167–182.

Thorley, A (1981) Pollen analytical evidence relating to the vegetation history of the chalk. *Journal of Biogeography*, **8**, 93–106.

Thorne, A (1980) The arrival of man in Australia. In *Cambridge Encyclopedia of Archaeology* (edited by A Sherratt). Cambridge University Press, Cambridge, 96–100.

Thornes, J B (1988) Erosional equilibria under grazing. In *Conceptual Issues in Environmental Archaeology* (edited by J L Bintliff, D A Davidson and E G Grant). Edinburgh University Press, Edinburgh, 193–210.

Thorp, P (1986) A mountain icefield of Loch Lomond Stadial age, western Grampians, Scotland. *Boreas*, **15**, 83–97.

Titus, J G (1987) The greenhouse effect, rising sea levels and society's response. In *Sea Surface Studies* (edited by R J N Devoy). Croom Helm, Beckenham, 499–528.

Tolonen, K (1986) Charred particle analysis. In *Handbook of Holocene Palaeoecology and Palaeohydrology* (edited by B E Berglund). John Wiley, Chichester and New York, 485–496.

Tooley, M J (1978) *Sea-level changes in Northern England during the Flandrian Stage*. Clarendon Press, Oxford.

Tooley, M J (1985) Climate, sea-level and coastal changes. In *The Climatic Scene* (edited by M J Tooley and G M Sheail). Allen and Unwin, London, 206–234.

Torrence, R, Specht, J and Fullagar, R (1990) Pompeiis in the Pacific. In *Australian Natural History* **23.6**, 457–463.

Trenhaile, A S (1980) Shore platforms: a neglected geomorphological resource. *Progress in Physical Geography*, **4**, 1–23.

Troels-Smith, J (1960) Ivy, mistletoe and elm. Climatic indicators – fodder plants. A contribution to the interpretation of the pollen zone border VII–VIII. *Danmarks Geologiske Undersøgelse II. Raekke*, **4**(4), 1–32.

Trotter, M M and McCullock, B (1984) Moas, men and middens . In *Quaternary Extinctions* (edited by P S Martin and R G Klein). University of Arizona Press, London, 708–727.

Turner, C (1970) The Middle Pleistocene deposits at Marks Tey, Essex. *Philosophical Transactions of the Royal Society, London*, **B257**, 373–440.

Turner, C and Hannon, G E (1988) Vegetational evidence for late Quaternary climatic changes in southwest Europe in relation to the influence of the North Atlantic Ocean. *Philosophical Transactions of the Royal Society, London*, **B318**, 451–485.

Turner, C and West, R G (1968) The subdivision and zonation of interglacial periods. *Eiszeitalter und Gegenwart*, **19**, 93–101.

Turner, J (1962) The *Tilia* decline: an anthropogenic interpretation. *New Phytologist*, **61**, 328–341.

Turner, J (1979) The Environment of North East England during Roman times as shown by pollen analysis. *Journal of Archaeological Science*, **6**, 285–290.

Turner, J (1981) The Iron Age. In *The Environment in British Prehistory* (edited by I Simmons and M Tooley). Duckworth, London, 250–281.

Valentine, K W G and Dalrymple, J B (1975) The identification, lateral variation and chronology of two buried palaeocatenas at Woodland Spa and West Runton, England. *Quaternary Research*, **5**, 551–591.

Vandenberghe, J, Bohnke, S, Lammers, W and Zilverberg, L (1987) Geomorphology and palaeoecology of the Mark Valley (southern Netherlands): geomorphological valley development during the Weichselian and Holocene. *Boreas*, **16**, 55–68.

Vayda, A P and McCay, B J (1978). New directions in ecology and ecological anthropology. In *Human Behaviour and Adaptation* (edited by V Reynolds and N Blurton Jones). Taylor and Francis, London, 33–51.

Vereshschagin, N K and Baryshnikov, G F (1984) Quaternary mammalian extinctions in North Eurasia. In *Quaternary Extinctions* (edited by P S Martin and R G Klein). University of Arizona Press, Tucson, 483–516.

Viles, H A (1989) The greenhouse effect, sea-level rise and coastal geomorphology. *Progress in Physical Geography*, **13**, 452–461.

Vita-Finzi, C (1969) *The Mediterranean Valleys*. Cambridge University Press, London.

Vita-Finzi, C (1975) Related territories and alluvial sediments. In *Palaeoeconomy* (edited by E S Higgs). Cambridge University Press, London, 225–231.

Voytek, M A (1990) Addressing the biological effects of decreased ozone on the Antarctic environment. *Ambio*, **19**, 52–61.

Vuorela, I (1986) Palynological and historical evidence of slash and burn cultivation in south Finland. In *Anthropogenic Indicators in Pollen Diagrams* (edited by K E Behre). Balkema, Rotterdam, 53–64.

Wagner, P L (1977) The concept of environmental determinism in cultural evolution. In *Origins of Agriculture* (edited by C A Reed). Mouton, The Hague, 49–74.

Wagstaff, J M (1981) Buried assumptions: some problems in the interpretation of the 'Younger Fill' raised by recent data from Greece. *Journal of Archaeological Science*, **8**, 247–264.

Wagstaff, J M (1987) *Landscape and Culture: Geographical and Archaeological Perspectives*. Blackwell, Oxford.

Walker, D (1970) Direction and rate in some British post-glacial hydroseres. In *Studies in the Vegetational History of the British Isles* (edited by D Walker and R G West). Cambridge University Press, London, 117–139.

Walker, M J C (1984) Pollen analysis and Quaternary research in Scotland. *Quaternary Science Reviews*, **3**, 369–404.

Walters, J C (1978) Polygonal patterned ground in central New Jersey. *Quaternary Research*, **10**, 42–54.

Warrick, R A (1988) Carbon dioxide, climatic change and agriculture. *Geographical Journal*, **1154**, 221–233.

Warrick, R A and Oerlemans, H (1990) Sea-level rise. In *Climate Change. The IPCC Scientific Assessment* (edited by J T Houghton, G J Jenkins and J J Ephraums). Cambridge University Press, Cambridge, 257–282.

Warrick, R A, Shugart, H H, Antonovsky, M Ja, Tarrant, J A and Tucker, C J (1986a) The effects of increased CO_2 and climatic change on terrestrial ecosystems. In *SCOPE 29, The Greenhouse Effect, Climatic Change and Ecosystems* (edited by B Bolin, B R Döös, J Jäger and R A Warrick). John Wiley, Chichester and New York, 363–392.

Warrick, R A, Gifford, R M and Parry, M L (1986b) CO_2, climatic change and agriculture. In *SCOPE 29. The Greenhouse Effect, Climatic Change and Ecosystems* (edited by B Bolin, B R Döös, J Jäger and R A Warrick). John Wiley, Chichester and New York, 393–473.

Washburn, A L (1979) *Geocryology*. Edward Arnold, London.

Washburn, A L (1980) Permafrost features as evidence of climatic change. *Earth Science Reviews*, **15**, 327–402.

Wasylikowa, K (1986) Analysis of fossil seeds and fruits. In *Handbook of Holocene Palaeoecology and Palaeohydrology* (edited by B E Berglund). John Wiley, Chichester and New York, 571–590.

Waters, M R (1983) Late Holocene lacustrine chronology and archaeology of ancient Lake Cahuilla, California. *Quaternary Research*, **19**, 373–387.

Watkins, N D, Sparks, R S J, Sigurdsson, H, Huang, T C, Federman, A, Carey, S and Ninkovich, D (1978) Volume and extent of the Minoan tephra from Santorini Volcano new evidence from deep-sea cores. *Nature*, **271**, 122–126.

Waton, P V (1982) Man's impact on the chalklands: some new pollen evidence. In *Archaeological Aspects of Woodland Ecology* (edited by M Bell and S Limbrey). British Archaeological Reports, Oxford, IS 146, 75–91.

Watson, E (1977) The periglacial environment of Great Britain during the Devensian. *Philosophical Transactions of the Royal Society, London*, **B280**, 183–198.

Watson, P J (1989) Early plant cultivation in the eastern woodlands of North America. In *Foraging and Farming* (edited by D R Harris and G C Hillman). Unwin Hyman, London, 555–571.

Watson, R T, Rodhe, H, Oeschger, H and Siegenthaler, U (1990) Greenhouse gases and aerosols. In *Climate Change. The IPCC Scientific Assessment* (edited by J T Houghton, G J Jenkins and J J Ephraums). Cambridge University Press, Cambridge, 1–40.

Watts, R G (1985) Global climatic variation due to fluctuations in the rate of deep water formation. *Journal of Geophysical Research*, **90**, 8067–8070.

Watts, W A (1978) Plant macrofossils and Quaternary palaeoecology. In *Biology and Quaternary Environments* (edited by D Walker and J C Guppy). Australian Academy of Sciences, Canberra, 53–67.

Watts, W A (1979) Late Quaternary vegetation of central Appalachia and the New Jersey coast plain. *Ecological Monographs*, **49**, 427–469.

Watts, W A (1980) Regional variations in the response of vegetation to late-glacial climatic events in Europe. In *Studies in the Lateglacial of Northwest Europe* (edited by J J Lowe, J M Gray and J E Robinson). Pergamon Press, Oxford, 1–21.

Watts, W A (1983) Vegetational history of the eastern United States 25,000 to 10,000 years ago. In *Late Quaternary Environments of the United States. 1. The Late Pleistocene* (edited by S C Porter). Longman, London, 294–310.

Webb, N (1986) *Heathlands*. Collins, London.

Webb, T III, Richard, P and Mott, R J (1983) A mapped history of Holocene vegetation in southern Quebec. *Syllogeus*, **49**, 273–336.

Webb, T III, Cushing, E J and Wright, H E Jr (1984) Holocene changes in the vegetation of the Midwest. In *Late Quaternary Environments of the United States. Volume 2. The Holocene* (edited by H E Wright Jr). Longman, London, 142–165.

Webb, T III, Bartlein, P J and Kutzbach, J E (1987) Climatic change in eastern North America during the past 18,000 years: comparisons of pollen data with model results. In *The Geology of North America. Volume K–3. North America and Adjacent Oceans During the Last Deglaciation* (edited by W F Ruddiman and H E Wright Jr). Geological Society of America, Boulder, Colorado, 447–462.

Welinder, S (1983) Ecosystems change at the Neolithic transition. *Norwegian Archaeological Review*, **16**(2), 99–105.

Welinder, S (1989) Mesolithic forest clearance in Scandinavia. In *The Mesolithic in Europe* (edited by C Bonsall). John Donald, Edinburgh, 362–366.

West, R G (1970) Pleistocene history of the British flora. In *Studies in the Vegetational History of the British Isles* (edited by D Walker and R G West). Cambridge University Press, London, 1–11.

West, R G (1977a) *Pleistocene Geology and Biology*, 2nd edition. Longman, London.

West, R G (1977b) Early and Middle Devensian flora and vegetation. *Philosophical Transactions of the Royal Society*, **B280**, 229–246.

West, R G (1980) *The Pre-glacial Pleistocene of the Norfolk and Suffolk Coasts*. Cambridge University Press, London.

West, R G (1988) The record of the cold stages. *Philosophical Transactions of the Royal Society, London*, **B318**, 505–522.

Westeringh, W van de (1988) Man-made soils in the Netherlands, especially in sandy areas ('Plaggen soils'). In *Man–Made Soils* (edited by W Groenman-van Waateringe and M Robinson). British Archaeological Reports, Oxford, IS 410, 5–20.

Wetterstrom, W (forthcoming) Climate, diet and population at a prehistoric pueblo in New Mexico. In *Presenting Environmental Archaeology* (edited by N Balaam and J Rackham). Institute of Archaeology, London.

White, J M, Mathewes, R W and Mathews, W H (1985) Late Pleistocene chronology and environment of the 'ice free corridor' of North-West Alberta. *Quaternary Research*, **24**, 173–186.

Whitford, P B (1983) Man and the equilibrium between deciduous forest and grassland. In *Man's Impact on Vegetation* (edited by W Holzner, M J A Werger and I Ikusima). Junk, The Hague, 163–172.

Whitten, D G A and Brooks, J R V (1977) *A Dictionary of Geology*. Penguin, London.

Whittle, A W R (1989) Two later Bronze Age occupations and an Iron Age channel on the Gwent foreshore. *Bulletin of the Board of Celtic Studies*, **36**, 200–223.

Wigley, T M L and Barnett, T P (1990) Detection of the Greenhouse Effect in the observations. In *Climate Change. The IPCC Scientific Assessment* (edited by J T Houghton, G J Jenkins and J J Ephraums). Cambridge University Press, Cambridge, 239–255.

Wigley, T M L and Jones, P D (1989) Influences of precipitation changes and direct CO_2 effects on streamflow. *Nature*, **34**, 149–152.

Wigley, T M L and Kelly, P M (1990) Holocene climatic changes, ^{14}C wiggles and variation in solar irradiance. *Philosophical Transaction of the Royal Society, London*, **A330**, 547–560.

Wigley, T M L and Raper, S C B (1987) Thermal expansion of sea water associated with global warming. *Nature*, **330**, 127–131.

Wigley, T M L, Ingram, M J and Farmer, G (1981) *Climate and History*. Cambridge University Press.

Wigley, T M L, Jones, P D and Kelly, P M (1986) Empirical climate studies. In *SCOPE 29, The Greenhouse Effect, Climatic Change and Ecosystems* (edited by B Bolin, B R Döös, J Jäger and R A Warrick). John Wiley, Chichester and New York, 271–322.

Wilke, P J (1978) *Late Prehistoric Human Ecology at Lake Cahuilla, Coachella Valley, California*. Contributions to the University of California Archaeological Research Facility, 38.

Wilkes, G (1989) Maize: domestication, racial evolution and spread. In *Foraging and Farming* (edited by D R Harris and G C Hillman). Unwin Hyman, London, 440–455.

Wilkins, D A (1984) The Flandrian woods of Lewis (Scotland). *Journal of Ecology*, **72**, 251–258.

Williams, C T (1985) *Mesolithic Exploitation Patterns in the Central Pennines*. British Archaeological Reports, Oxford, BS 139.

Williams, D F, Thunell, R C, Tappa, E, Rio, D and Raffi, I (1988) Chronology of the Pleistocene oxygen isotope record: 0–1.88 m.y. BP. *Palaeogeography, Palaeoclimatology, Palaeoecology*, **64**, 221–240.

Williams, M (1990) Clearing of the forests. In *The Making of the American Landscape* (edited by M P Conzen). Unwin Hyman, Boston, 146–168.

Williams, R B G (1975) The British climate during the last glaciation: an interpretation based on periglacial phenomena. In *Ice Ages: Ancient and Modern* (edited by A E Wright and F Moseley). Seel House Press, Liverpool, 95–120.

Wilson, A J (1972) Theoretical geography: some speculations. *Transactions of the Institute of British Geographers*, **57**, 31–44.

Wilson, A T, Hendy, C H and Reynolds, C P (1979) Short-term climatic change and New Zealand temperatures during the last millennium. *Nature*, **279**, 315–317.

Wilson, C (1985) The Little Ice Age on eastern Hudson/James Bay: the summer weather and climate at Great Whale, Fort George and Eastmain, 1814–1821, as derived from Hudson's Bay Company records. *Syllogeus*, **55**, 147–190.

Wiltshire, P E J and Moore, P D (1983) Palaeovegetation and Palaeohydrology in upland Britain. In *Background to Palaeohydrology* (edited by K J Gregory). John Wiley, Chichester, 433–452.

Woillard, G (1978) Grande Pile peat bog: a continuous pollen record for the last 140,000 years. *Quaternary Research*, **9**, 1–21.

Woillard, G (1979) Abrupt end of the last interglacial SS in north-east France. *Nature*, **281**, 558–562.

Woillard, G and Mook, W G (1982) Carbon-14 dates at

Grande Pile: correlation of land and sea chronologies. *Science*, **215**, 159–161.

Wolff, E and Peel, D (1985) The record of global pollution in polar snow and ice. *Nature*, **313**, 535–540.

Wollin, G, Kukla, G J, Ericson, D B, Ryan, W B F and Wollin, J (1973) Magnetic intensity and climatic changes 1925–1970. *Nature*, **242**, 34–36.

Woodman, P C (1985) Prehistoric settlement and environment. In *The Quaternary History of Ireland* (edited by K J Edwards and W P Warren). Academic Press, London, 251–278.

Worsley. P (1980) Problems in radiocarbon dating the Chelford Interstadial of England. In *Timescales in Geomorphology* (edited by R A Cullingford, D A Davidson and J Lewin). John Wiley, Chichester and New York, 289–304.

Worsley, P (1987) Permafrost stratigraphy in Britain – a first approximation. In *Periglacial Processes and Landforms in Britain and Ireland* (edited by J Boardman). Cambridge University Press, 89–99.

Wright, H E Jr (1977) Environmental change and the origin of agriculture in the Old and New Worlds. In *Origins of Agriculture* (edited by C A Reed). Mouton, The Hague, 282–320.

Wright, H E Jr (1987) Synthesis: the land south of the ice sheets. In *The Geology of North America. Volume K–3. North America and Adjacent Oceans During the Last Deglaciation* (edited by W F Ruddiman and H E Wright Jr). Geological Society of America, Boulder, Colorado, 479–488.

Wright, H E Jr (1989) The amphi-Atlantic distribution of the Younger Dryas palaeoclimatic oscillation. *Quaternary Science Reviews*, **8**, 295–306.

Yapp, C J and Epstein, S (1977) Climatic implications of D/H ratios of meteoric water over North America (9500–22,000 BP) as inferred from ancient wood cellulose C–H hydrology. *Earth and Planetary Science Letters*, **34**, 33–350.

Yen, D E (1989) The domestication of environment. In *Foraging and Farming* (edited by D R Harris and G C Hillman). Unwin Hyman, London.

Yesner, D R (1988) Island biogeography and prehistoric human adaptation on the southern coast of Maine (USA). In *The Archaeology of Prehistoric Coastlines* (edited by G Bailey and J Parkington). Cambridge University Press, 53–63.

Zagwijn, W (1974) Vegetation, climate and radiocarbon datings in the Late Pleistocene of the Netherlands II. Middle Weichselian. *Mede Rijks. Geol. Dienst. New Series*, **25**, 101–110.

Zagwijn, W H (1984) The formation of the Younger Dunes on the west coast of the Netherlands (AD 1000–1600). In *Geological Changes in the Western Netherlands During the Period 1000–1300 AD* (edited by H J A Berendensen and W H Zagwijn). *Geologie en Mijnbouw*, **3**, 259–268.

Zagwijn, W (1985) An outline of the Quaternary stratigraphy of the Netherlands. *Geologie en Mijnbouw*, **64**, 17–24.

Zeist, W van (1974) Palaeobotanical studies of settlement sites in the coastal area of the Netherlands. *Palaeohistoria*, **16**, 223–271.

Zeist, W van and Bottema, S (1982) Vegetational history of the eastern Mediterranean and the Near East during the last 20,000 years. In *Palaeoclimates, Palaeoenvironments and Human Communities in the Eastern Mediterranean Region in Later Prehistory Part ii* (edited by J L Bintliff and W van Zeist). British Archaeological Reports, Oxford, IS 133 ii, *277–321*.

Zeist, W van and Casparie, W A (1984) *Plants and Ancient Man*. Balkema, Rotterdam.

Zohary, M (1983) Man and vegetation in the Middle East. In *Man's Impact on Vegetation* (edited by W Holzner, M J A Werger and I Ikusima). Junk, The Hague, 287–295.

Zohary, D (1989) Domestication of the Southwest Asian Neolithic crop assemblage of cereals, pulses and flax: the evidence from the living plants. In *Foraging and Farming* (edited by D R Harris and G C Hillman). Unwin Hyman, London, 358–373.

Zoltai, S C and Vitt, D H (1990) Holocene climatic change and distribution of peatlands in western interior Canada. *Quaternary Research*, **33**, 231–240.

Zvelebil, M (1986) Mesolithic prelude and Neolithic revolution. In *Hunters in Transition* (edited by M Zvelebil). Cambridge University Press, 5–16.

Zvelebil, M and Rowley–Conwy, P (1986) Foragers and farmers in Atlantic Europe. In *Hunters in Transition* (edited by M Zvelebil). Cambridge University Press, 67–93.

Index